S0-CAS-519

Fundamentals of Molecular Similarity

MATHEMATICAL AND COMPUTATIONAL CHEMISTRY

Series Editor: PAUL G. MEZEY
University of Saskatchewan
Saskatoon, Saskatchewan

FUNDAMENTALS OF MOLECULAR SIMILARITY
Edited by Ramon Carbó-Dorca, Xavier Gironés, and Paul G. Mezey

MANY-ELECTRON DENSITIES AND REDUCED DENSITY MATRICES
Edited by Jerzy Cioslowski

A Continuation Order Plan is available for this series. A continuation order will bring delivery of each new volume immediately upon publication. Volumes are billed only upon actual shipment. For further information please contact the publisher.

Fundamentals of Molecular Similarity

Edited by

Ramon Carbó-Dorca
Xavier Gironés
University of Girona
Girona, Catalonia, Spain

and

Paul G. Mezey
University of Saskatchewan
Saskatoon, Saskatchewan, Canada

Kluwer Academic / Plenum Publishers
New York • Boston • Dordrecht • London • Moscow

ISBN 0-306-46425-X

233 Spring Street, New York, New York 10013

http://www.wkap.nl

10 9 8 7 6 5 4 3 2 1

A C.I.P. record for this book is available from the Library of Congress

Printed in the United States of America ac✓

Contributors

ALLAN, N.L. SCHOOL OF CHEMISTRY, UNIVERSITY OF BRISTOL, CANTOCKS CLOSE, BRISTOL BS8 1TS, UNITED KINGDOM

AMAT, L. INSTITUT DE QUÍMICA COMPUTACIONAL, UNIVERSITAT DE GIRONA, GIRONA 17071, CATALONIA, SPAIN

ANGLADA, J.M. INSTITUT D'INVESTIGACIONS QUÍMIQUES I AMBIENTALS. C.I.D. - C.S.I.C., JORDI GIRONA SALGADO 18 - 26, E-08034 BARCELONA, CATALUNYA, SPAIN

ARENAS, A. DEPARTAMENT D'ENGINYERIA QUÍMICA, ESCOLA TÈCNICA SUPERIOR D'ENGINYERIA QUÍMICA, DEPARTAMENT D'ENGINYERIA INFORMÀTICA I MATEMÁTICA (ETSE), UNIVERSITAT ROVIRA I VIRGILI, TARRAGONA, CATALUNYA, SPAIN

BARRIL, X. DEPARTAMENT DE BIOQUÍMICA I BIOLOGIA MOLECULAR, FACULTAT DE QUÍMICA, UNIVERSITAT DE BARCELONA, MARTÍ I FRANQUÈS 1, BARCELONA 08028, SPAIN

BESALÚ, E. INSTITUT DE QUÍMICA COMPUTACIONAL, UNIVERSITAT DE GIRONA, GIRONA 17071, CATALONIA, SPAIN

BOFILL, J.M. DEPARTAMENT DE QUÍMICA ORGÀNICA I CENTRE ESPECIAL DE RECERCA EN QUÍMICA TEÒRICA, UNIVERSITAT DE BARCELONA, MARTÍ I FRANQUÈS 1, E-08028 BARCELONA, CATALUNYA, SPAIN

CARBÓ-DORCA, R. INSTITUT DE QUÍMICA COMPUTACIONAL, UNIVERSITAT DE GIRONA, GIRONA 17071, CATALONIA, SPAIN

CIOSLOWSKI, J. DEPARTMENT OF CHEMISTRY AND CSIT, FLORIDA STATE UNIVERSITY, TALLAHASSEE, FL 32306-3006, USA

COOPER, D.L. DEPARTMENT OF CHEMISTRY, UNIVERSITY OF LIVERPOOL, P.O. BOX 147, LIVERPOOL L69 7ZD, UNITED KINGDOM

CREHUET, R. INSTITUT D'INVESTIGACIONS QUÍMIQUES I AMBIENTALS. C.I.D. - C.S.I.C., JORDI GIRONA SALGADO 18 - 26, E-08034 BARCELONA, CATALUNYA, SPAIN

DE CÀCERES, M. GRUP DE RECERCA EN INFORMÀTICA MÈDICA, INSTITUT MUNICIPAL D'INVESTIGACIÓ MÈDICA, UNIVERSITAT POMPEU FABRA,C/ DR. AIGUADER 80, E-08003 BARCELONA, SPAIN

ESPINOSA, G. DEPARTAMENT D'ENGINYERIA QUÍMICA, ESCOLA TÈCNICA SUPERIOR D'ENGINYERIA QUÍMICA, DEPARTAMENT D'ENGINYERIA INFORMÀTICA I MATEMÁTICA (ETSE), UNIVERSITAT ROVIRA I VIRGILI, TARRAGONA, CATALUNYA, SPAIN

GÁLVEZ, J. UNIDAD DE INVESTIGACIÓN DE DISEÑO DE FÁRMACOS Y CONECTIVIDAD MOLECULAR SEC. DEPT. QUÍMICA FÍSICA. FACULTAD DE FARMACIA, UNIVERSITAT DE VALENCIA

GARCÍA-DOMENECH, R. UNIDAD DE INVESTIGACIÓN DE DISEÑO DE FÁRMACOS Y CONECTIVIDAD MOLECULAR SEC. DEPT. QUÍMICA FÍSICA. FACULTAD DE FARMACIA, UNIVERSITAT DE VALENCIA

GELPÍ, J.L. DEPARTAMENT DE BIOQUÍMICA I BIOLOGIA MOLECULAR, FACULTAT DE QUÍMICA, UNIVERSITAT DE BARCELONA, MARTÍ I FRANQUÈS 1, BARCELONA 08028, SPAIN

GERBER, P.R. PHARMACEUTICAL RESEARCH AND DEVELOPMENT, F. HOFFMANN-LA ROCHE, BASEL, SWITZERLAND

GIRALT, F. DEPARTAMENT D'ENGINYERIA QUÍMICA, ESCOLA TÈCNICA SUPERIOR D'ENGINYERIA QUÍMICA, DEPARTAMENT D'ENGINYERIA INFORMÀTICA I MATEMÁTICA (ETSE), UNIVERSITAT ROVIRA I VIRGILI, TARRAGONA, CATALUNYA, SPAIN

GIRONÉS, X. INSTITUT DE QUÍMICA COMPUTACIONAL, UNIVERSITAT DE GIRONA, GIRONA 17071, CATALONIA (SPAIN)

HALL, L.H. DEPARTMENT OF CHEMISTRY, EASTERN NAZARENE COLLEGE, QUINCY, MASSACHUSETTS

JULIÁN-ORTIZ, J.V. UNIDAD DE INVESTIGACIÓN DE DISEÑO DE FÁRMACOS Y CONECTIVIDAD MOLECULAR SEC. DEPT. QUÍMICA FÍSICA. FACULTAD DE FARMACIA, UNIVERSITAT DE VALENCIA

KARADAKOV, P.B. DEPARTMENT OF CHEMISTRY, UNIVERSITY OF SURREY, GUILDFORD, SURREY GU2 5XH, UNITED KINGDOM

KARWOWSKI, J. INSTYTUT FIZYKI, UNIVERSYTET MIKOŁAJA KOPERNIKA, GRUDZIĄDZKA 5, 87-100 TORUŃ, POLAND

KIER, L.B. DEPARTMENT OF MEDICINAL CHEMISTRY, VIRGINIA COMMONWEALTH UNIVERSITY, RICHMOND, VIRGINIA, 23298

LUQUE, F.J. DEPARTAMENT DE BIOQUÍMICA I BIOLOGIA MOLECULAR, FACULTAT DE QUÍMICA, UNIVERSITAT DE BARCELONA, MARTÍ I FRANQUÈS 1, BARCELONA 08028, SPAIN

MEZEY, P.G. INSTITUTE FOR ADVANCED STUDY, COLLEGIUM BUDAPEST,1014 BUDAPEST,I. SZENTHÁROMSÁG U. 2, HUNGARY AND MATHEMATICAL CHEMISTRY RESEARCH UNIT, DEPARTMENT OF CHEMISTRY AND DEPARTMENT OF MATHEMATICS AND STATISTICS. UNIVERSITY OF SASKATCHEWAN, 110 SCIENCE PLACE, SASKATOON, SK, CANADA, S7N 5C9

MUÑOZ, J. DEPARTAMENT DE BIOQUÍMICA I BIOLOGIA MOLECULAR, FACULTAT DE QUÍMICA, UNIVERSITAT DE BARCELONA, MARTÍ I FRANQUÈS 1, BARCELONA 08028, SPAIN

OROZCO, M. DEPARTAMENT DE BIOQUÍMICA I BIOLOGIA MOLECULAR, FACULTAT DE QUÍMICA, UNIVERSITAT DE BARCELONA, MARTÍ I FRANQUÈS 1, BARCELONA 08028, SPAIN

RICHARDS, W.G. PHYSICAL AND THEORETICAL CHEMISTRY LABORATORY, THE UNIVERSITY OF OXFORD, OXFORD, UNITED KINGDOM

ROBERT, D. INSTITUT DE QUÍMICA COMPUTACIONAL, UNIVERSITAT DE GIRONA, GIRONA 17071, CATALONIA, SPAIN

ROBINSON, D.D. PHYSICAL AND THEORETICAL CHEMISTRY LABORATORY, THE UNIVERSITY OF OXFORD, OXFORD, UNITED KINGDOM

SANZ, F. GRUP DE RECERCA EN INFORMÀTICA MÈDICA, INSTITUT MUNICIPAL D'INVESTIGACIÓ MÈDICA, UNIVERSITAT POMPEU FABRA,C/ DR. AIGUADER 80, E-08003 BARCELONA, SPAIN

SELLO, G. DIPARTIMENTO DI CHIMICA ORGANICA E INDUSTRIALE, UNIVERSITA' DEGLI STUDI DI MILANO

VILLÀ, J. GRUP DE RECERCA EN INFORMÀTICA MÈDICA, INSTITUT MUNICIPAL D'INVESTIGACIÓ MÈDICA, UNIVERSITAT POMPEU FABRA,C/ DR. AIGUADER 80, E-08003 BARCELONA, SPAIN; DEPARTMENT OF CHEMISTRY, UNIVERSITY OF SOUTHERN CALIFORNIA, LOS ANGELES, CA, 90089, USA.

WILLETT, P. KREBS INSTITUTE FOR BIOMOLECULAR RESEARCH AND DEPARTMENT OF INFORMATION STUDIES, UNIVERSITY OF SHEFFIELD, WESTERN BANK, SHEFFIELD S10 2TN, UK.

WINN, P.J. PHYSICAL AND THEORETICAL CHEMISTRY LABORATORY, THE UNIVERSITY OF OXFORD, OXFORD, UK

Foreword

This volume edited by Ramon Carbó-Dorca, Xavier Gironés and Paul G. Mezey deals with Molecular Similarity, a very specialised topic, which provides, however, an answer, even if limited, on a very modern, broad and relevant trend in the evolution of the scientific method. Let me start a bit far from this specific topic in order to put it into perspective.

Today society expects science to be useful toward the solution of its (real or imaginary) survival needs, blurring in this process the traditional separation between "pure" and "applied" sciences. Thus, "theory" and "applications", "models" and "experimental reality", "verification" and "prediction" are interwoven components in the flow of our scientific-technical-industrial-commercial -social life.

In the last half of the twentieth century "computers and computations" appeared as new tools of ever increasing strength, originally envisaged mainly to solve complex numerical or statistical problems and to simulate a limited range of natural events, thus connecting pure and formal scientific theories to laboratory data, with the dual ability to verify theoretical hypotheses and predict phenomena.

In general, theory attempts to rationalise events in our ordinary x,y,z, t, space-time frame. The motion of objects did capture man's interest and fascination since the beginning of civilisation, and its rationalisation required notions of geometry, and related algebra but mainly a variety of hypotheses concerning the origin and nature of the causes responsible for the transit from A to B for the object. Soon it appeared that different kinds of objects are in motion due to different but not unrelated causes, and science was born.

Thus, "grosso modo", recalling Pythagoras, Democritus, Galilei and Kepler, and later Newton, we arrive at the theory of gases, at the systematisation of the atomic table of the elements, and eventually to modern science.

Despite the fact that the space-time frame encompasses an incredibly vast number of events, today, we assume that we can account for all of them with relatively few simple laws. These describe the motions of objects of different kinds, where the difference is related mainly to the complexity of the objects, and the way we choose to describe them (re-normalisation). Thus, neglecting the theoretical description of sub-atomic particles, for which the theory is in continuous evolution, we model (a) systems with both charged particles - nuclei and electrons - and uncharged ones - photons and the like - using the Schroedinger or Dirac equations, (b) systems composed of an ensemble of atoms or molecules using statistical dynamics and (c) systems made of ensembles of control volumes (filled with atoms and molecules of given mass, density and velocity) using the fluid dynamics equations. There is no predetermined upper value on the number of nuclei and electrons, when we use quantum mechanics; at a given complexity level we simply decide to shift from a quantum mechanical to the statistical description, and here too, when we reach a given level of complexity we simply shift to fluid dynamics equations. Note that these three types of objects (a) nuclei, electrons and photons, (b) ensembles of molecules and atoms and (c) solids, liquids, and gases represent three different ways to describe the same "reality". Note, in addition, that complexity is a rather relative criterion, when related to computability; indeed, today complexity is, in part, yielding ground to computations, due to the enormous advances of computer technology.

Therefore, one might think that we have models and equations to solve "any" problem. Most unfortunately, there are still areas which could be labelled *hic sunt leones*, where our equations seems not to be applicable. Paradoxically and ironically these hostile areas are the most vital ones to we "humans": indeed we lack a scientific model to explain the interaction between living objects, in general, and human objects, in particular. To deal with these interactions, since the early days of civilisation, we made use of a different kind of knowledge and reasoning, namely we have developed ethics and aesthetics. It will be the task of the next millennium to re-describe the above interactions, substituting and integrating present ethical and aesthetical paradigms with scientific "laws", likely less constrained than the present ones, possibly due to the expected advances of relatively new disciplines, like artificial intelligence, fuzzy algebra and logic, information retrieval and formation of data banks, advances in the understanding of brain mechanisms, etc. In this very broad context "drug design" and rationalisations of relations between "chemical structures and chemical or even biochemical activity" becomes an obvious privileged focal point in modern research, and an important test of new directions in modern science.

A vague and faint indication of these future trends is already at hand. Studies on interactions of the many organs in the living bodies as well as the understanding of the many biological tissues and of the enormous variety of cells are becoming more and more detached from empiricism and purely phenomenological descriptions and are moving towards the "scientific method". Among some important moments of this progress we recall last century's evolutionary theories, and more important, the contemporary discovery of "molecular disease", the industrial syntheses of new drugs, the beginning and then the enormous advances of molecular genetics: likely applications of the latter will compel society to advance a critical attitude to re-examine today's ethical and aesthetical paradigms thus favouring concepts derived from the "scientific method"; in turn it will influence novel horizons and methodologies in science.

To relate molecular structure to physicochemical or biochemical activity one generally starts with a quantum chemical description of a molecule or of a molecular complex. But here one encounters a first difficulty, since modern quantum chemists have *de facto* forgotten that molecules are made of atoms, rather than of nuclei and electrons. Indeed, in the first attempt to apply quantum theory to molecules, the central idea was that molecules are made up of atoms in an appropriate valence state. But soon it was realised that the Valence Bond method was too difficult even with the use of the early computers and also it failed to explain excited states. Thus quantum chemists turned to the Hartree-Fock method, even if conceived for atomic structures; thus molecules are modelled as a fixed frame of nuclei on which we glue electrons expressed as one-particle functions, the elements of a determinant. The Hartree-Fock, HF, method too has errors, the best known being the electron correlation error, which accounts for about one third of the molecular binding energy. The interested reader is referred to the Volume 5 of "Lecture Notes in Chemistry", "A general SCF Theory" by R. Carbó and J. M. Riera, Springer-Verlag N.Y., 1978, for details.

The two main avenues of quantum theory were explored some time ago to solve the correlation energy problem: either variational techniques using more than one determinant or perturbation expansions, using large number of terms. Both techniques were clearly formulated and tested before quantum chemists became interested in their use. The applications of these methods were dependent on the availability of large computational facilities at the time of the early computers (1950-1970) to solve very simple chemical systems, or on generous access to computational facilities of present computers to solve medium-small chemical systems, but eventually will use future computers in a moderate way to solve medium-large molecules: the driving force of these methods is the enormous progress of computer technology, witnessed in the last fifty years (of course efficient computer codes and improved numerical algorithms are also important).

But many chemists need and request a reliable and computationally cheap solution for large systems "here and now!"; thus the acceptance in the last half of a century of a variety of pre-Hartree-Fock semi-empirical approximations, quickly neglected as soon the computer power increased enough as to make other, superior quality, approximations computationally feasible.

There are various levels of realism ("decency"?) in proposing approximations. The HF approximation is a realistic approximation and therefore has gained a special place in quantum theory. Indeed, it considers all the particles of the system (even if it replaces the sub-nuclear particles with a continuous nucleus, generally, but not necessarily, assumed the size of a point) and its solutions are correct quantum mechanical eigenfunctions. Further, it yields wave functions with total energy about 99.X percent of the correct value, molecular binding energies about 65 to 75 percent relative to laboratory data, and reasonable other-than-energy expectation values. More important to the applied quantum-chemist, HF solutions are available using commercial codes, which include push-button geometry optimisation, frequency analysis, graphics, etc. Even more relevant to the applied quantum-chemist, the HF solution today requires relatively small amounts of computer time (from minutes to few hours) on a modern work station (one or more 600 MHz processors, starting at 250 MB of RAM, and few Gigabytes of disk space). To the quantum chemist, the HF solutions provide a well-defined zero-order approximation, with variationally optimal one-electron functions ‘the orbitals’ nicely orthogonal to one another (this helps algebraic manipulations, but deters from more general and unconstrained one-electron functions). Finally, the correlation error can be easily expressed in terms of functionals of the HF density, as we know from Wigner since 1935.

From all this it was natural to propose the addition of a density functional operator to the Hartree-Fock effective hamiltonian (for a review, see Clementi in IBM J. Res. and Dev.9, 2, 1965)). One can add a new flavour and introduce several specific density functional operators: a density functional operator to correct the "atoms in the molecule" model and different density functional operators for covalent and for ionic bonds, recognised by appropriate descriptors, thus connecting the HF- description to concepts originated by the valence bond theory. This is done in the HF-CC approximation, which yields binding energies in agreement with experimental data to within one Kcal/mol, and requires computational time equal to that for standard HF solutions. In addition in the HF-CC method the total molecular energy is partitioned into bond energies which are in agreement with available laboratory data. We call this type of approaches "dfa", an abbreviation for "density functional approximations".

Following the dfa proposal and numerical tests, an alternative route was explored: one can approximate atoms and molecules as semi-classical objects, thus with no quantum mechanical degeneracy; one can then (1) adopt the Hartree-Fock formalism and equations, (2) replaces the HF wave function notation by its density equivalent (assumed unique) (3) replace the exchange HF operator with the Slater exchange operator, a well known function of the density. At this point one has a HF method which yields a variartionally optimal electronic density (recall Poisson's equation relating energy and electronic density). Further, at this point it is also natural to take over dfa methodology to get the correlation energy; since both the Slater exchange operator and the correlation operator are functions of the density and contain empirical parameters, one can be tempted to construct a unique exchange-correlation operator, even if the HF exchange energy does not need empirical factors and is in general by far larger than the correlation energy. All this is done in contemporary Density Functional Theory, DFT. The DFT energies are as good as the corresponding exchange-correlation operator allows, and can yield binding energies within about one or two Kcal/mole relative to laboratory data; the computer time needed by DFT computations is larger than that for dfa computations, since the exchange-correlation operator contains gradients of the density, but shorter than that of ab-initio post-HF- methods. Incidentally, statistical mechanics has also made great progress in the last half of a century: indeed, Monte Carlo and Molecular Dynamics simulations have become routine computational tasks.

In conclusion, chemists today can reliably and routinely obtain energies and electronic densities for molecules with up to about two dozen atoms; in a near future computer technology will extend this limit towards notably larger molecules. We stress that many important chemicals, including drugs, are constituted of molecules with up to two dozen atoms. Further, larger molecules can be decomposed into fragments which often are up to two dozen atoms in size. (Of course one must have the knowledge to recognise appropriate sub-structures within the primary structure.)

The fourteen Chapters of "The Fundamentals of Molecular Similarity" represent a successful attempt to establish a new discipline with its own language, hypotheses and algorithms aimed at comparing sets of molecules of different chemical structures, yielding similar response to a given measurement.

In Chapter 12, Carbó-Dorca, Amat, Besalú, Gironés and Robert present a volume within the volume (over 100 pages) with both the theory and sample applications of Quantum Molecular Similarity. The aim is to build a rigorously logical frame based on mathematical methods and quantum theory in order to associate descriptors to molecules, with only the molecular geometry and its electronic density as starting points.

Stated differently, the chapter's aim is to define, compute, and use quantum mechanical descriptors fully "unbiased" in order to generate "unbiased" relationships between Quantum Structure and Properties, either physicochemical, or chemical, or pharmaceutical or biochemical or even medicinal chemical. This approach was very futuristic when it started, since quantum chemistry was at a very early stage. The authors lucidly present definitions, rules, and algorithms with which a set of "real objects" can be uniquely related to a set of "mathematical objects" each in its appropriate space. The preoccupation to maintain throughout the chapter a high level of logical rigor is apparent and this makes the reading at times a bit more difficult than needed; however, this is likely a small price to pay to ensure an unbiased and logical foundation to a field fertile for different (even contradictory) and most often intuitive paradigms as well as "ad hoc" proposals, offered to correlate selected sets of molecules to "chemical activity".

We would like to see this chapter reprinted into an extended stand-alone volume, where the present appendices are expanded to full chapters, with the added effort to prove that the various SAR proposals of todays literature can be unified under the same set of master rules.

Of course master rules must evolve and adapt. Seeds in this direction are evident, for example in Chapters 6, 7, and 8. G. Sello (Chapter 6) calls attention "to the other side of the coin"; if similarity measures are important, the same must be for dissimilarity measures, being these complementary to a full characterisation of the molecule. Cioslowski and Karwoski (Chapter 7) remind us that emphasis on the Schroedinger equation is insufficient for molecules containing heavy atoms; we recall that cancer drugs, Pt containing molecules, have grossly incorrect computed geometry, if the relativistic correction formalism is not used. Mezey in Chapter 8 provocatively proposes that knowledge of a small fragment of a molecule contains the full information of the entire molecule (but in part neglects to critically and pragmatically discuss the "accuracy" requirements needed to reach this goal, since the smaller the fragment, the larger the "accuracy requirement" to maintain constant the information quality). Mention should also be given to Chapter 9 by Bofill, Angada, Besalú, and Crehuet, where the discussion on an essential aspect of chemical activity, namely chemical reactivity, brings necessarily in the forefront the "time" parameter, somewhat disregarded by a large sector of today's quantum-chemistry. Later we shall return to this point.

The reading of Chapter 11 returns to a question which has been vexing the mind of un-inhibited quantum chemists for many years. The dominance of HF related methods is partly the result of an historical accident related to the timing for the availability of computers. The Valence Bond methodology is based on the Atoms in Molecules hypothesis, which in turn is at the base of any SAR; it appears once more now when the VB method could be re-evaluated.

In order to compare physicochemical, chemical, biochemical, etc. activity, the availability of reliable and quantitatively accurate database with friendly access is essential. In the computer era databases are the cornerstones of SAR; this involves the organisation, information retrieval and data base use discussed in Chapter 3 by Kier and Hall. The organisation of databases requires a vision on how to characterise an atom in a molecular environment, vision which must precede and not follow the development of a quantum mechanical description of a molecule. Indirectly, we restate some uneasiness in the tout-court acceptance of today HF dominated quantum chemistry as the most appropriate avenue to deal with chemistry.

Chapter 4 by Willet deals with additional aspects related to data search; it advocates an eclectic approach by using a variety of similarity measurements methods, "data fusion" methodology; a very pragmatic evaluation of results. In time this pragmatic approach likely will have to come to terms with the "intransigent" rationalism of Chapter 12 and vice-versa.

Specific and very important sub-fields of SAR are addresses in Chapter 5 by Gerber 'topological pharmacophore description' and in Chapter 10 by Muñoz, Barril, Luque, Gelpi and Orozco 'partitioning of free energy of solvents into fragment distributions, and in Chapter 14 by Sanz, De Cáceres and Villà' distributions of molecular interactions.

In Chapter 13, Winn, Robinson, and Richards deal squarely with QSAR's impact on medicinal-chemistry, a very complex area which calls for an eminently pragmatic approach, with emphasis on a reliable technology to perform measurements for data analyses, robust algorithms with low, and therefore affordable computational cost. In some respect, we could classify medicinal-chemistry as the forefront and physicalchemistry as the old-guard of this "new scientific approach", namely, "structure-activity-relationship" studies. Indeed, one could note that traditional statistical mechanics simulations are certainly capable to describe most physicochemical properties of ensembles of molecules, and to offer reliable predictions. Similarly, traditional fluid- dynamics is certainly able to describe or simulate mechanical properties of matter. But presently neither statistical mechanics nor fluid dynamics seems to be capable to answer problems in medicinal-chemistry considered with SAR.

Therefore, one wonders whether one should consider SAR as today's "trail" which tomorrow will be covered by the "highways" of traditional physical sciences or as a pointer to a novel way to understand complexity, as the emerging top of an iceberg, which will compete and complement today's science. One could suggest that there seems to exist a trend: yesterday physical science did "represent" reality, today physical science "is only a temporary model" to mimic reality; perhaps tomorrow we will consider traditional physical sciences as a temporary model to mimic only an "idealised reality", like we look today at the laws of perfect gases.

There are, however, open questions on QSR development; perhaps we have over-stressed quantum chemistry (actually, only some aspects of it). Can we expect to establish general relationships between molecular structure and chemical activity still ignoring time, temperature, entropy, namely some of the basic dimensions of the space in which activity manifest itself?

With prudence we observe that man created science to his own advantage, slowly over the last many centuries; today there are new and drastically different needs, and man will develop sciences or if needed propose drastically new paradigms, again to his own advantage. Certainly SAR appears to us as a magnificent landscape and we are confident it is not a "mirage".

Enrico Clementi
May 26, 2000

Preface

In recent years the fundamental concepts and applied methodologies of molecular similarity analysis have experienced a revolutionary development. Motivated by the increased degree of understanding of elementary molecular properties on the levels ranging from fundamental quantum chemistry to the complex interactions of biomolecules, and aided by the spectacular progress in computer technology and access to computer power, the area has opened up to many new ideas and new approaches.

The current status of the field can be characterized by a peculiar duality. On the one hand, rigorous quantum similarity approaches and electron density shape analysis methods, among others, provide new inroads to a better theoretical understanding of molecular similarity, where scientific rigor and a nearly axiomatic approach are the goals. On the other hand, the applications of molecular similarity approaches are often analogous to traditional experimental science, where phenomenological descriptions, even without much understanding of the causes, have proved to be extremely useful. Such is the case in the generation of correlations between some molecular electron density properties and a complicated biochemical effect, where nearly nothing is known about the actual chemical mechanism. Nevertheless, such correlations have been proven exceptionally useful in quantitative shape – activity relations (QShAR) and in the prediction of the pharmacological or toxicological effects of molecules in the related context of quantum QSAR (QQSAR).

This volume, written by some of the experts in the various subfields of molecular similarity, provides a collection of the most recent ideas, advances, and methodologies. It is the hope of the Editors that by presenting these topics within a single volume, the readers will find a balanced overview of the status of the field. We also hope that the book will serve as a tool for selecting and assessing the best approach for various new types of problems of molecular similarity that may arise and it will provide a set of easy references for further studies and applications.

Ramon Carbó-Dorca, Xavier Gironés and Paul G. Mezey

Acknowledgments

The editors wish to warmly acknowledge the generous financial support from the following institutions:

- Fundació María Francisca de Roviralta (www.roviralta.org)
- Fundació Catalana per a la Recerca (www.fcr.es)
- Centre Europeu per a la Paral·lelització de Barcelona (CEPBA) (www.cepba.upc.es)
- Generalitat de Catalunya (CIRIT) (www.gencat.es)
- Ministerio de Educación y Ciencia (CICYT) (www.mec.es)
- Diputació de Girona (www.ddgi.es)
- Ajuntament de Girona (www.ajuntament.gi)

which sponsored in 1999 the celebration of the IV Girona Seminar on Molecular Similarity and, in this manner, made possible the publication of the present volume.

Chapter 1

Prediction of boiling points of organic compounds from molecular descriptors by using backpropagation neural network

G. Espinosa, A. Arenas, and Francesc Giralt
Departament d'Enginyeria Química, Escola Tècnica Superior d'Enginyeria Química (ETSEQ), Departament d'Enginyeria Informàtica i Matemática (ETSE), Universitat Rovira i Virgili, Tarragona, Catalunya, Spain

1. INTRODUCTION

The design and optimisation of industrial process require the knowledge of thermophysical properties. Available data banks can provide this information. However in specific cases, such as those related to drug activity or enviromental impact assessment, data are scarce and difficult or expensive to obtain experimentally. To overcome this lack of ready information, several thermodynamic models and correlations have been developed for a wide range of conditions. Among these models, the methods based on quantitative structure property relationships (QSPR) are promising. The basic concept of QSPR is to relate the structure of a compound with the property of interest. The compound's structure is expressed in terms of molecular descriptors that characterise a given molecular feature. Molecular descriptors, such as the connectivity indices and the corresponding valence connectivity indices, that encode features such as size, branching, unsaturation, heteroatom content and cyclicity [1,2] are useful. For example, the first order connectivity index was used in 1982 to correlated the solubility of hydrocarbons in water [3]. The connectivity indices are based

Fundamentals of Molecular Similarity, Edited by Carbó-Dorca *et al.*
Kluwer Academic/Plenum Publishers, New York 2001

on local molecular properties and are bond-additive quantities so that in bonds of different kinds make different contribution to the overall molecular descriptors. The key step is to build the structure property relationship. This involves two major activities: 1 The representation of compounds using molecular descriptors and multivariate statistical methods or artificial neural networks [4,5]. And 2 the mapping of the descriptors to built a relationship with the properties of interest. Among the physical properties correlated by QSPR are boiling points, [1,6,7], melting points, [7], solubilities, [3], partition coefficients, [8]. The success of regression analysis in QSPR model building depends upon the degree of linearity between the physical property of interest and the descriptors selected. As the number of descriptors increases the capability of regression analysis decreases due to the redundancy of information incorporated by the different descriptors. Some techniques, such as principal component analysis and partial least square regression, have been used to minimise this problem. Nevertheless, these techniques require the a priori assumption of the form of the model. To solve this issue, multilineal regression analyses (MLR) is commonly used as an alternative. Recently, artificial neural networks have become an option to build QSPR models. The purpose of the current study is to apply QSPR and neural networks to better correlate the boiling points of organic compounds.

2. BOILING POINT

The boiling point of organic compound is useful for identifying substances and for estimating other physical properties [9]. There are different methods to predict boiling points. For example, group contribution methods are widely used for this purpose [10]. These contribution methods are limited to the class of compounds for which the groups have been established. The QSPR approach employs descriptors derived solely from molecular structure [4,11]. One of the pioneering works to predict the boiling points of paraffins was carried out by Wiener [2]. Other topological indices, such as the connectivity indices [1,4] and the Randic indices [2], have been successfully applied to correlate the boiling points of alkanes and alcohols.

In the present study, 1116 organic compounds were considered. The boiling points were taken from the Design for Physical Property Data (DIPPR) database. Two subsets that contained the same data as those used by Kier and Hall [1] and Hall and Story [12] were chosen to validate the results. The complete set is structurally heterogeneous, includes saturated and unsaturated hydrocarbons, aromatic, and halogenated compounds, with groups cyano, amino, ester, ether, carbonyl, hydroxyl, and carboxyl. The structures and connectivity indices of these compounds were obtained using the Molecular Modelling Pro software. Four molecular connectivity indices and four valence molecular connectivity indices for each compound were considered.

3. ARTIFICIAL NEURAL NETWORKS

The next step is to establish the relationship between molecular descriptors and the boiling points by using artificial neural networks. A standard neural architecture consists of many simple interconnected processors (units). The weight of each connection or synapse stores the information learned from examples [13]. The successful application of neural networks depends on three factors. First the design is critical because the network will overfit data if too few hidden units are used. Second, the size of the training set must be correct to avoid over or under training. Finally, it is important to select an appropriate training set, because it has to represent the entire dataset.

Two supervised neural algorithms, where input patterns are associated with known output patterns, were used. The first one was backpropagation architecture [14,15]. Its implementation involves a forward pass through the layers of units (nodes) to estimate the error, followed by a backward pass that modifies the weights (synapses or connections) to decrease the error. Networks with one input and one output layer, and with one or two hidden layers with different nodes in each layer, were examined. Six or eight nodes in the input layer were considered so that input vectors with six or eight connectivity indices could be presented to the network. The output node was the boiling point. To train the network, the weights of the synapses between the nodes of each layer and those of the next layer were optimised with a steepest descent method by propagating the error back throughout the layers.

One problem with backpropagation is to find the appropriate topology. To overcome this constrain an auto constructive algorithm was also implemented. The cascade correlation method was selected because it is one of the most relevant constructive algorithms [16,17]. The hidden units are added to this network one at a time without changing the connection weights after they have been added. It supports a variety of learning algorithms, but a backpropagation scheme was used for consistency. An initial, minimal network with only input and output units was trained. Training continued until a given criteria, such as the maximum number of epochs or a patience indicator, were met. If the network did not fulfilled the error criteria during the initial phase, a new hidden unit was added to maximise the correlation. This hidden unit should account for missing features.

4. RESULTS AND DISCUSSION

Three sets of compounds were used to evaluate the present model and to compare its performance with previous proposals reported in the literature. The first set of compounds include 242 alcohols, with up to ten carbon atoms, and all the alkane isomers with five to ten carbon atoms. The range of boiling points is 282.65K-504.15K. A number of 42 compounds were selected for the testing phase. The results obtained with several backpropagation architectures indicated that expansion of the input space yielded better results than a contraction in all cases. This shows that extra dimensionality can represent better additional features of the training set, and that these extra features make a favourable contribution to the performance of the network. The best configuration for this set of alkane isomers and alcohols is a 8-12-1, i.e., the combination of eight connectivity indices as input (four molecular connectivity indices plus four valence connectivity indices), one hidden layer with twelve units, and the boiling point as output.

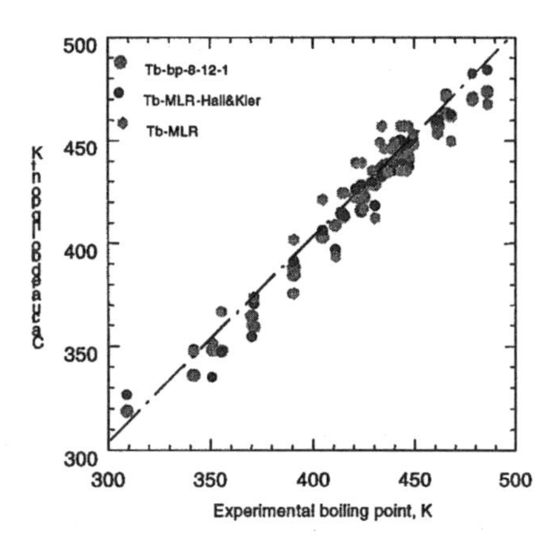

Figure 1. Boiling points for the 42 alcohols and alkanes used for testing with an 8-12-1 backpropagation architecture (bp). Comparison with Hall and Kier, [1], and with MLR with connectivity indices.

The absolute mean error between the predicted and experimental boiling points was 4.3K. Only two compounds (heptane and nonanol) yielded residuals greater than 10K The standard deviation of the predictions was

3.3K and the average relative error 2.9%. For the same set of connectivity indices, a multilineal regression analysis yielded predictions with a mean absolute error of 10.3K, a standard deviation of 5.5K and a relative mean error of 7.7%. It is not possible to make a direct comparison with the previous work of Hall and Kier [1] for a similar set of compounds, because they used different molecular descriptors. Still it should be noted that those authors reported a mean absolute error of 5.9K. This corresponds to a 4.1% relative error which is higher than for the model presented here. The predicted and measured boiling points are plotted in Fig. 1. The data in Fig.1 were correlated with a coefficient r^2=0.983, which validates the model established by the network. These results show that the non linear relationship between molecular structure and boiling point is well extracted by the neural network, increasing the extrapolation capabilities of this model to other different but similar sets of compounds.

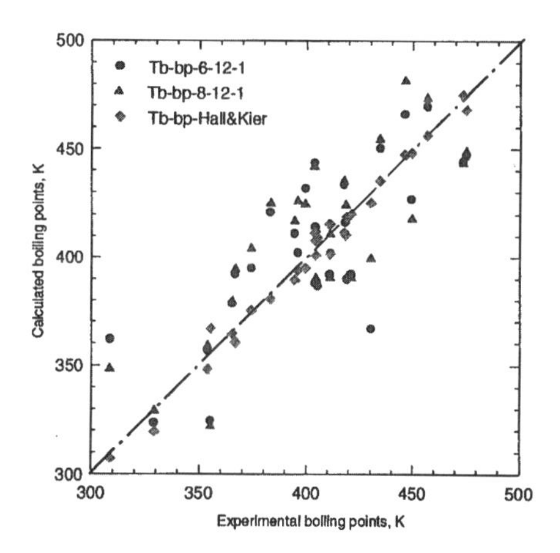

Figure 2. Boiling points for the 30 organic compounds used for testing with an 8-12-1 and 6-12-1 backpropagation architectures (bp). Comparison with the neural architecture 19-5-1 of Hall and Story, [12] (electrotopological indices)

The second set is formed by 220 heterogeneous organic compounds with three to nineteen carbon atoms, including saturated and unsaturated hydrocarbons, and the groups ester, ether, carbonyl, hydroxyl, and carboxyl. Their boiling point range was 225.51K-608K. Training was carried out with 30 randomly selected compounds. In this case the expansion of the input space also yielded better results than contraction. The best configuration was

8-12-1. The standard deviation between predictions and measurements was 11.3K, with an absolute mean error of 21.8K and an average relative error of 5.38%. Hall and Story [12] reported a mean absolute error of 4.57K, for a19-5-1 architecture (nineteen electrotopological indices as input) for the same group of compounds, which correspond to a 1.12% relative error. This better performance is due to the type of indices used by these authors, which allows the complete characterisation of all functional groups involved. Fig. 2 depicts the boiling points predicted versus the experimental data, for the set of heterogeneous organic compounds.

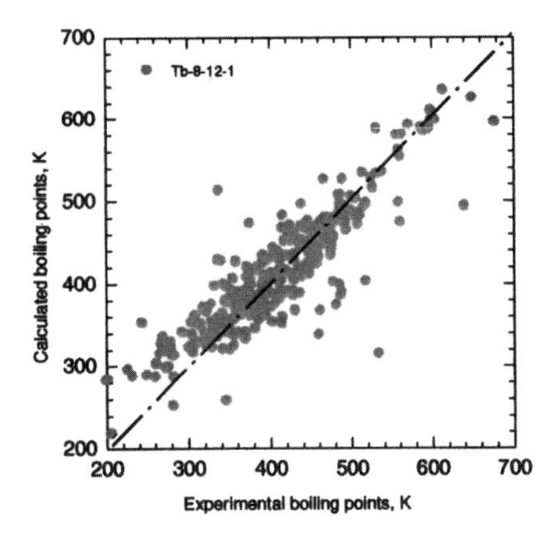

Figure 3. Boiling points for the 416 organic compounds used for testing with an 8-12-1 backpropagation architecture (bp).

The data in Fig. 2 were correlated with a coefficient r2=0.8. This low correlation coefficient indicates that increasing the diversity of compounds without increasing the information about the functional groups involved decreases the capability of neural of networks based on processing units. It should be noted that a neural network requires a minimum number of data for training and that the training set has to represent the majority of characteristics of the whole set of compounds considered. Also, connectivity indices alone do not provide enough information to characterise a heterogeneous set. To solve the first issue, the two previous sets were unified and incremented to 1116 heterogeneous organic compounds, with boiling points spanning the range 111.7K-711.5K. About 60% of the compounds were used for training and the rest were used to evaluate the model. The best configuration was again 8-12-1. The mean absolute error was 28 K and the relative mean error 7%. In all trials carried out to select the best architecture the compounds that could be consider as outliers (compounds with high residuals) were the polyfluorine compounds, substituted aromatics, and pyridines. The results are summarised in Fig. 3.

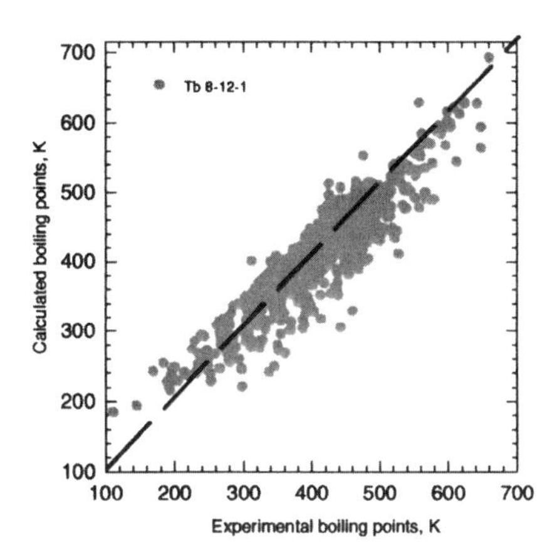

Figure 4. Boiling points for the 509 organic compounds used for testing with an 6-12-1 backpropagation architecture (bp) and the minimum training set.

The performance of the architectures tested did not improve by randomly selecting different compounds for training the network, within 60% of the total. To overcome this deficiency, the size and the content of training set was optimised using a specific algorithm reported in the literature for image classification problems [18]. This means the minimum number of patterns with the maximum of information was finally determined. The best configurations for this set were 8-12-1 and 6-12-1, eight or six inputs respectively, twelve units in the hidden layer and the boiling point of the output layer. The standard deviations of the predicting set were respectively 12.8K and 11.3K, absolute mean errors were 19.7K and 11.6K and the average relative errors were 4.62% and 4.33%. The predictions obtained with the 8-12-1 and 6-12-1 architectures and the best training set are shown in Figs. 4 and 5. The comparison of these results with Figure 2 show the importance of the definition of the training set to build a good model.

To overcome the second issue raised above about the insufficient molecular information of topological indices, Espinosa et al., [21] considered the dipole moment and the kappa index as additional descriptors to model de boiling points of a homogeneous set of aliphatic hydrocarbons. The inclusion of these two indices in the input vector does not improve the correlation of the boiling points of the current heterogeneous sets with backpropagation algorithm. Thus, the use of cognitive systems such as FuzzyARTMAP, [22] should be considered, together with additional three-dimensional molecular information. Finally, it is worth noting that contrary to previous reports, cascade correlations does not improve the performance of backpropagation, [16,17]. For the compounds in Fig 6, the cascade correlation algorithm yields a mean absolute error of 33K, compared to the 12.8K and 11.3K obtained with the 6-12-1 (Fig. 4) and 8-12-1 (Fig. 5) backpropagation architectures, respectively.

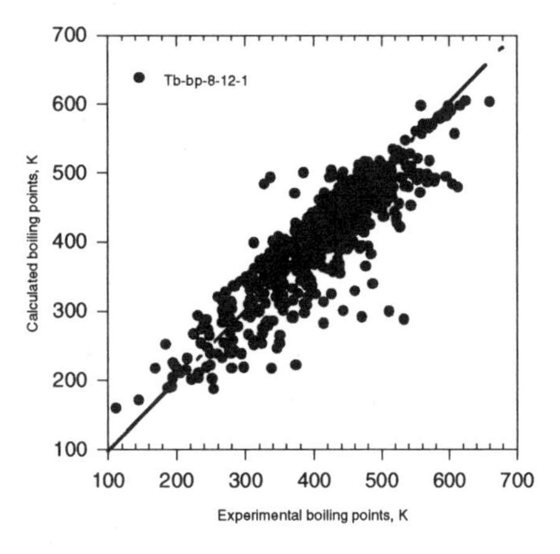

Figure 5. Boiling points for the 536 organic compounds used for testing with an 8-12-1 backpropagation architecture (bp) and the minimum training set.

5. CONCLUSION

The present model, which combines neural networks with QSPR, performs better than previous correlations for similar input information. The best backpropagation configurations to predict the boiling points of 1116 organic compounds were 8-12-1 and 6-12-1. This implies using eight or six connectivity indices as input nodes, twelve middle nodes, and a single node for boiling point. The cascade correlation constructive algorithm didn't yield better results than backpropagation. The determination of the minimum training set reduces the absolute mean error and improves predictions.

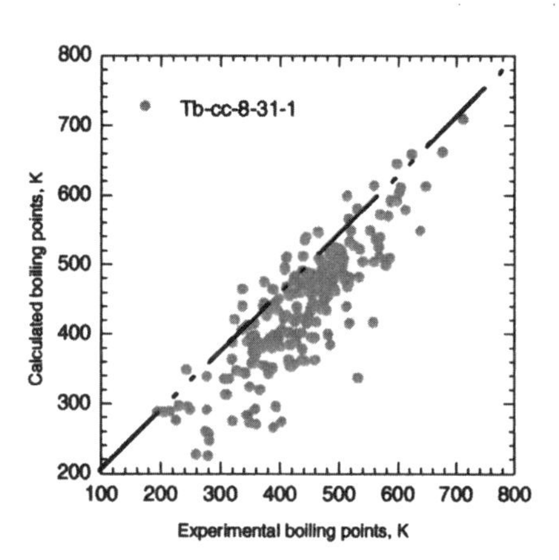

Figure 6. Boiling points for the 416 organic compounds used for testing with an 8-33-1 cascade correlation architectures (cc).

REFERENCES

1. L. Hall and L. Kier, *J. Chem. Inf. Compt. Sci.* **35**, pp 1039(1995).
2. M. Randic and N. Trinajstic, *J. Mol Struct.* **284**, 209 (1993).
3. M. Medir and F. Giralt, *AICHE Journal* **28**, 341 (1982).
4. Katritzky, M. Karelson and V. Lobanov, *Pure Appl. Chem.* **69**, 245 (1997).
5. P. Jurs, 214 th ACS National Meeting, 1997.
6. Katritzky, Lan Mu, and V. Lobanov, *J. Phys. Chem.* **100**, 10400 (1996).
7. Katritzky,Lan Mu, and M. Karelson, *J. Chem. Inf. Compt. Sci.* **38**, 293 (1998).
8. Patil,. *J. Hazard. Mater.* **19**, 35 (1994)
9. R. Reid, J. Prausnitz and B.Poling, *The Properties of Gases and liquids*, 4th ed., McGraw-Hill, New York, 1987.
10. Joback and R. Reid, *Chem. Eng. Commun.* **57**, 233 (1987).
11. Bünz, B. Braun, and R. Janowsky, *Ind. Eng. Chem. Res.* **37**, 3043 (1998).
12. Hall and C. Story, *J. Chem. Inf. Compt. Sci.* **36**, 1004 (1996).
13. Hertz and K. Palmer, *Introduction to the Theory of Neural Computation*, Addison Wesley, The Advanced Book Program, pp. 115 (1991).
14. D. Rumelhart, G. Hinton, and R.Williams, *Nature* **323**, 533 (1986).

15. S. Falhman, *An Empirical Study of Learning Speed in Backpropagation Networks*, Technical Reports CMU-CS-88-162 (1988).
16. S. Fahlman, and C. Lebiere, *The Cascade Correlation Learning Architecture, Advances in Neural Information Processing System* II, pp. 524 (1990).
17. Squieres and J. Savlik, *Experimental Analysis of Aspects of the Cascade Correlation Learning Architectures*, Machine Learning Research Group Working, paper 91-1 (1991).
18. F.Tamburini and R. Davoli, *An Algorithm Method to Build Good Training Sets for Neural Networks Classifiers*, Technical Report UBLCS-94-18, (1994).
19. G. Carpenter and S. Grossberg, *Computer Vision, Graphics, and Image Processing*. **37**, 54 (1987).
20. G. Carpenter and S. Grossberg, *Computer*, **21**, 77 (1988).
21. Espinosa G., Yaffe D., Cohen, Y., Arenas, A., and Giralt F., *J. Chem. Inf. Compt. Sci.* **40**, 859 (2000).
22. Giralt, F., Arenas, A., Ferre-Gine, J., Rallo, R. and Kopp, G., *Physics of Fluids*, **12**, 1826 (2000).

Chapter 2

Some Relationships between Molecular Energy-Topology and Symmetry

J. Gálvez, R. García-Domenech and J. V. de Julián-Ortiz
Unidad de Investigación de Diseño de Fármacos y Conectividad Molecular Sec. Dept. Química Física. Facultad de Farmacia, Universitat de Valencia

1. INTRODUCTION

Molecular Topology (M.T.) has demonstrated its efficiency in the prediction of many experimental parameters. The application of the topological indices to the prediction of pharmacological properties, and, above all, to its inverse problem, the drug design, is particularly interesting [1,2]. Several attempts to explain the reasons of such efficiency have been carried out [3,4]. So far, we are not close to a definitive answer.

The introduction of new topological indices has focused the efforts of the scientists working on M.T. [5-7]. In the present work, we find analogies between physical models and graph invariants and, eventually, introduce new topological indices which are able to encode information regarding a given physicochemical property. These descriptors could provide a link between molecular topology and its physical interpretation.

2. VIBRATIONAL AND ELECTRONIC ENERGIES

In graph theory, molecular structures are represented as hydrogen-depleted graphs whose vertices and axes act as atoms and covalent bonds, respectively. Chemical graphs representing a carbon skeleton of an alkane are called carbon trees. The number of carbon atoms bonded to a given carbon i is δ_i, degree of the vertex i. This constitutes a topological analog of

Fundamentals of Molecular Similarity, Edited by Carbó-Dorca *et al.*
Kluwer Academic/Plenum Publishers, New York 2001

the chemical valence. We will introduce a relationship between topological valences and the probability to find a given electron in a given σ C-C bond.

2.1 Topological Valence and Probability

Let $K_{1,n}$ be the star carbon trees. Consider one electron belonging to the central atom before the formation of the molecule and that is now occupying any σ C-C bonding level. This electron has a probability of $1/\delta_i$ to be in a given σ C-C bond.

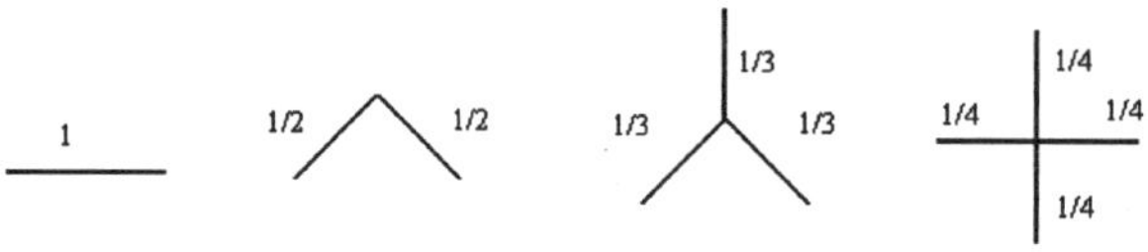

1 1/2 1/2 1/3 1/3 1/3 1/4 1/4 1/4 1/4

$K_{1,1}$ $K_{1,2}$ $K_{1,3}$ $K_{1,4}$

For a general carbon graph, the probability P_i to find a fixed electron near the atom *i* is inverse to its respective δ value. Similarly, the probability to find two "fixed" electrons in C-C bonds, forming a given σ C_i-C_j bond is:

$$P_{ij} = P_i \cdot P_j = \frac{1}{\delta_i \delta_j}$$

As this probability is proportional to a squared wavefunction that could describe the bond, then taking the real part, it results:

$$\Psi_{ij} \propto (\delta_i \delta_j)^{-1/2}$$

This expression probably constitutes a quantum mechanical link with the Randic´ index. However, Altenburg [8] and Randic´ [9] studied other exponents for the terms different from -1/2, obtaining better correlations in several cases.

For chemical multigraphs, representing unsaturated hydrocarbons, $\delta_i = 4 - h_i$, where h_i is the number of hydrogen atoms attached to the carbon *i*.

2.2 Electronic Energies of Conjugated Hydrocarbons

Let *i* and *j* be two bonded atoms in a conjugated hydrocarbon. Following the variational method to calculate the bond energy,

$$E_{ij} = \frac{<\psi_{ij}|H|\psi_{ij}>}{<\psi_{ij}^2>}$$

Where E_{ij} represents the bond energy, ψ_{ij} is the molecular orbital wave function, and H is the Hamiltonian operator.

According to the preceding paragraph, the probability to find an electron close to the atoms *i* or *j* is inverse to the product of the corresponding δ values. This assumption is reasonable since the greater the number of bonded hydrogens, the higher the probability is to find an electron near to the atom. Thus, if we start from the following trial wave function:

$$\Psi_{ij} = \frac{1}{(\delta_i)^{1/2}}\Phi_i + \frac{1}{(\delta_j)^{1/2}}\Phi_j$$

Φ_i and Φ_j are orthonormalized:

$$<\Phi_i \mid \Phi_i> = <\Phi_j \mid \Phi_j> = 1$$
$$<\Phi_i \mid \Phi_j> = <\Phi_j \mid \Phi_i> = S = 0$$

With all these assumptions, finally, the bond energy results:

$$E_{ij} = 2\alpha + \frac{4\delta_i\delta_j}{\delta_i + \delta_j}\frac{H_{ij}}{(\delta_i\delta_j)^{1/2}}$$

where α is the standard Coulomb integral, assuming the same for all the atoms, and H_{ij} is the resonance integral for each i-j bond.

If we define the average resonance integral, β as:

$$\beta = \frac{\delta_i\delta_j H_{ij}}{\delta_i + \delta_j}$$

β represents a weighted value of the resonance integrals for single and double bonds, and should be equivalent to the Hückel β value. Consequently,

$$E_{ij} = 2\alpha + \frac{4\beta}{(\delta_i\delta_j)^{1/2}}$$

This is the general equation which is able to evaluate the C-C bond energy in any alternant hydrocarbon, including σ and π bonds.

Thus, the global π energy may be expressed as a function of Randic´ index, $^1\chi$:

$$E_\pi = N_\pi \alpha + 4\beta\,^1\chi$$

where N_π is the number of π electrons.

On the other hand, the global π energy considering the double bonds as localized is:

$$E_\pi{}^* = N_\pi \alpha + N_\pi \beta$$

The resonance energy is defined as the difference between these values.

$$E_{res} = \beta(4\,^1\chi - N_\pi)$$

Applying this equation to a set of conjugated hydrocarbons, we obtain the following regression equation:

$$\begin{aligned} &E_{res}(kcal/mol) = 16.89(4\,^1\chi - N_\pi) - 2.70 \\ &\mathrm{N} = 8 \quad \mathrm{R} = 0.9899 \quad \mathrm{SE} = 4.51 \quad \mathrm{F} = 289 \end{aligned} \tag{1}$$

Table 1 gives a comparison between resonance energy values obtained through Eq. 1 and those calculated by the HMO formalism [10]. It should be noted the similarity in the predictions reached, and that the β value calculated is very close to the assumed in the HMO calculations.

For comparative purposes, we include the predictions reached through an equation obtained using topological charge indexes (TCI) [11]:

$$\begin{aligned} &E_{res}(kcal/mol) = -11.51G_1 + 6.59G_2 + 0.99 \\ &\mathrm{N} = 8 \quad \mathrm{R} = 0.9984 \quad \mathrm{SE} = 1.98 \quad \mathrm{F} = 759 \end{aligned} \tag{2}$$

The HMO formalism as well as Eq. 1, fail in the prediction of the resonance energies in nonalternant hydrocarbons. Table 2 shows the predictions for Eqs. 1 and 2 in a cross-validation group of such compounds.

In Eq. 2, the use of TCI makes possible not only a better prediction of resonance energies in the training group (Table 1), but also in the cross-validated prediction of nonalternant hydrocarbons not considered in the obtention of Eq. 2. The Randic´ index reasonably reproduces the values for π and resonance energies.

Table 1. Comparison between "experimental" resonance energy values [10], those obtained by Hückel-MO calculations, and from Equations 1 and 2. All values are expressed in kcal/mol.

Compound	E_{res}(exp.)	Hückel´s E_{res} (β =16.5)	$E_{res}{}^{a}$ (calc)	$E_{res}{}^{b}$ (calc)
Benzene	36	33	31	36
Naphthalene	61	61	58	60
Anthracene	83	87	86	83
Phenantrene	91	90	86	91
Styrene	38	40	38	35
Stilbene	74	80	81	76
Biphenyl	71	72	70	72
Butadiene	3.5	7.8	7.4	5.5

[a] *Calculated from Eq. 1*
[b] *Calculated from Eq. 2*

2.3 Vibrational Energies

The frequencies of vibration of double bonds C=C can be predicted as a function of the number of carbons attached to each sp^2 carbon.

The vibrations in a diatomic molecule can be approximated by a harmonic oscillator. The vibrational frequency is given by:

$$\nu = \left[k/(4\pi^2\mu)\right]^{1/2}$$

where k is the strength constant, and μ is the reduced mass,

$$\mu = \frac{m_1 m_2}{m_1 + m_2}$$

m_1 and m_2 are the masses of the two atoms.

In our model, we consider a similar approximation for the bond C=C, and substitute μ by the quotient $\delta_1\delta_2/(\delta_1+\delta_2)$, reduced vertex degree, where $\delta_1$and δ_2 are the respective vertex valence degrees corresponding to the two carbon atoms considered. This assumption is defensible because the more ramification there is, the more mass there is. It is also logical to think that this influence will be lower with the increase in the distance of the substituted groups to the double bond, and in a first approximation, we can ignore the other contributions. The constant k loses its primitive sense in this new context, and we introduce an additional freedom degree to render a straight line, function of the inverse root squared reduced vertex degree:

$$\nu = b_0 + b_1[(\delta_1 + \delta_2)/\delta_1\delta_2]^{1/2} = b_0 + b_1[(1/\delta_1) + (1/\delta_2)]^{1/2}$$

Table 2. Comparison between "experimental" resonance energy values and those obtained by $^1\chi$ and topological charge indices, for a set of nonalternant hydrocarbons in a cross-validation test

Compound	E_{res}(exp.)	$E_{res}{}^{a}$(calc)	$E_{res}{}^{b}$(calc)
Fluorene	76	144	76
1,3,5- triphenylbenzene	149	134	142
9,10-dihydroanthracene	72	207	69
9,10-diphenylanthracene	152	154	180
9,9´-bianthryl	167	172	192
Diphenylmethane	67	132	63
Toluene	35	94	31
o-xylene	35	151	32
m-xylene	35	150	25
p-xylene	35	150	25
1,2,3-trimethylbenzene	34	209	34
1,3,5-trimethylbenzene	34	207	19
Ethylene	0.0	-1.1	1.0

[a] *Calculated from Eq. 1*
[b] *Calculated from Eq. 2*

Applying this equation to the set of six possible general alkenes, we obtain the following regression equation:

$$\nu(cm^{-1}) = 1780.6 - 147.2[(1/\delta_1) + (1/\delta_2)]^{1/2} \qquad (3)$$
$$N = 6 \quad R = 0.9953 \quad SE = 1.22$$

Table 3 displays the prediction of vibrational frequencies for the different substitution patterns in alkenes using Eq. 3.

Good predictions of the “observed” values are obtained through this approach. The more ramification there is, the more frequency there is. This feature explains the negative sign for the variable term.

Further refinement of this method perhaps can make it useful in the theoretical calculation of double bonds vibration frequencies for definite structures.

Table 3. Vibrational frequencies in cm^{-1} for different substituted C=C.

Substitution	Observed value[a]	Calculated value[b]
CH_2= CH_2	1634	1633.4
CH_2= CHR	1645	1646.2
CH_2= CRR'	1652	1653.1
CHR= CHR'[c]	1662	1660.4
CHR= CR'R''	1670	1668.2
CRR'= CR''R'''	1675	1676.5

[a] *Averaged experimental value [18]*
[b] *Calculated from Eq. 3*
[c] *Mean value of cis and trans isomers*

3. ELECTRONIC INTERFERENCES IN AROMATIC MOLECULES

The concept of topological distance between vertices is normally used in Graph Theory. This is the number of edges that separates any pair of vertices within a graph. But a chemical graph also can be assimilated to a finite one-dimensional Euclidean space. Using this representation, benzene can be intended as an Euclidean 1-torus. In such space, let us suppose that a wave originated in the vertex A is moving through two different paths p_1 and p_2 to a neighbour vertex B.

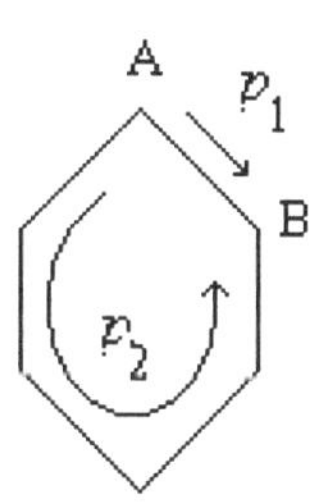

Paths p_1 and p_2 have Euclidean distances s_1 and s_2 respectively, in the considered space. In the case of a constructive interference in B:

$$s_1 - s_2 = k\lambda$$

being k an integer number.

Considering the distance d as a mean value for the length of all the edges of the graph, we have that:

$$s_1 - s_2 = d(p_1 - p_2)$$

Then it follows that:

$$p_1 - p_2 = \frac{\lambda}{d} k$$

Here, p_1 and p_2 are the topological distances of the paths that the perturbation follows from A to B.

In a quantical context, we can suppose that the considered wave can be the one associated to an electron, then, according to the Broglie's hypothesis, $\lambda = h/mv$, where h is the Planck constant, m the electron mass, and v its velocity. It leads to:

$$p_1 - p_2 = \frac{h}{m \cdot v \cdot d} k = \Delta p_{12}$$

From this formula, we can calculate the kinetic energies required by the electron to reach a constructive interference:

$$E_{12} = \frac{m \cdot v^2}{2} = \frac{h^2}{2m \cdot d^2} \frac{k^2}{(\Delta p_{12})^2}$$

The kinetic energy of the electron results to be proportional to $(k/\Delta p_{12})^2$. If we take a minimal value for E_{12} (then k=1), and we add the contributions from the difference of path lengths for which one of the paths is always of length one, we obtain the following expression:

$$E = \sum_{i=1}^{N} \sum_{j=1}^{N} E_{ij} = \frac{h^2}{2m \cdot d^2} \sum_{i=1}^{N} \sum_{j=1}^{N} \frac{\delta(1, D_{ij})}{(\Delta p_{ij})^2}$$

E is the total energy required for a constructive interference of all the π electrons; N, the number of vertices; δ, the Kronecker symbol; and D_{ij} the topological distance i-j.

We introduce the difference of path lengths indices DP as a parameter useful in QSAR studies that is proportional to the previous result. By analogy, we generalize the definition to any graph representing an aromatic system, and to any length of paths. This extrapolation pursues to simplify the computation of the new indices. A DP index of order k is defined as:

$$DP_k = \sum_{i=1}^{N}\sum_{j=1}^{N}\frac{\delta(k, D_{ij})}{(\Delta p_{ij})^2}$$

Also, the DP of valence indices, DP^v, are defined as:

$$DP_k^{\,v} = \sum_{i=1}^{N}\sum_{j=1}^{N}\frac{\delta(k, D_{ij})}{(\Delta p_{ij})^2}(1+\Delta(EN_{ij}))$$

where $\Delta(EN_{ij})$ are the differences of electronegativities between the i and j vertices placed at a topological distance k, considering pondered values of Pauling electronegativity taking the chlorine value as 2.

The DP_K indices have been successfully used for predicting the resonance energies of aromatic hydrocarbons (12). The electron mobility clearly determines other molecular properties, such as polarizability, and many biological and pharmacological characteristics are concerned with it. To test the ability of the DP indices to modelize this kind of properties, we tried to discriminate the antibacterial activity by Linear Discriminant Analysis (LDA) [12]. A set of 355 structurally heterogeneous compounds with either antibacterial or nonantibacterial activity have been analyzed. Each group was separated in two sets, one for training and the other for testing.

The discriminant functions chosen was:

$$F_1 = 0.089DP_4^V - 2.674 \qquad (4)$$
$$N = 355 \quad F = 136 \quad U = 0.649$$

A better fit was obtained with the addition of connectivity indices:

$$F_2 = -0.934\left({}^0\chi^V - {}^0\chi\right) + 5.993DP_4^V - 3.635 \qquad (5)$$
$$N = 355 \quad F = 98 \quad U = 0.56$$

The selection of the descriptors was made according to the Snedecor F, and the classification criteria was the shortest Mahalanobis distance (distance of each case to the mean of all cases used in the regression equation). Table 4 summarizes the discrimination achieved. By using only one DP index, Eq. 4, an overall accuracy of about 80% is obtained for both training and test sets. However, if a second index is included, ${}^{0}\chi^{v}$ -${}^{0}\chi$, the overall accuracy increases up to over 85% (Eq. 5). A given compound is classified as antibacterial if $F_2 > 0$.

Table 4. Classification results from the linear discriminant analysis study of antibacterial activity using Eqs. 4 and 5.

Group	Percent of Success	Number of Cases Classified as: Active	Incative
		$F_1= 0.089\ DP_4{}^{v} - 2.674$	
Active	71.6	116	46
Inactive	79.1	19	72
Test active	79.4	54	14
Test inactive	80.0	14	28
	$F_2 = -0.934\ ({}^{0}\chi^{v}-{}^{0}\chi) + 5.993\ DP_4{}^{v} - 3.635$		
Active	82.0	132	29
Inactive	90.1	9	82
Test active	88.2	60	8
Test Inactive	91.4	3	32

In order to check Eq. 5 for detection of antibacterial activity, we made a virtual screening on a large home-made database including about 15 000 commercial compounds, from both the Merck Index [13] and the Aldrich catalog [14]. It resulted in the identification of some new potentially active compounds, which were tested against different strains from both gram-positive and negative germs. Table 5 shows the results of the antibacterial activity test for the best six of them: mordant brown 24, neohesperidine dihydrochalcone, silymarine, hesperetine, morine and niflumic acid, showing significant activity against two species of bacteria. It must be emphasized that no previous report on antibacterial activity was found in the literature for these compounds and most of these compounds can be considered new *leads*, since no structural similarity to any other known antibacterial was found.

Table 5. Results for antibacterial activity susceptibility in a set of compounds selected using Eq. 5.

Compound	Prob.[a]	*E. coli*	*S. aureus*	*P.aeruginosa*	*P. mirabilis*
Mordant Brown 24	0.747	++	++	++	++
Neohesperidine	0.989	++	++	++	++
Silymarine	0.965	-	++	++	++
Hesperetine	0.623	++	++	++	++
Morine	0.778	+	++	++	++
Niflumic acid	0.526	+	++	++	+

[a] Calculated probability of activity.

4. VAPOR PRESSURE

We obtained a curious correlation between the log P and $1/{}^{1}\chi^{v}$, for a set of plaguicides found in the bibiliography [15]:

$$\log P(\text{torr}) = -0.84 - 22.43/{}^{1}\chi^{V} \tag{6}$$

$$N = 13 \quad R = 0.9786 \quad SE = 0.52$$

Predictions can be seen in Table 6. If we compare this expression with the Clausius-Clapeyron equation:

$$\log P = C - \Delta H_{V}/2.303RT$$

Table 6. Comparison between experimental vapor pressures (-log P_{exp}) and those obtained from Eq. 6 (-log P_{calc}) for a set of organophosphorus plaguicides.

Compound	- log $P_{exp.}$	- log $P_{calc.}$
Bromophos	3.886	3.841
Fenitrothion	5.222	4.926
Fenthion	4.523	4.457
Parathion-methyl	5.013	5.261
Bromophos-ethyl	4.337	3.433
Parathion	4.223	4.431
Chlorpiriphos	4.728	3.969
Diazinon	3.854	3.884
Pirimiphos-ethyl	3.538	3.617
Etrimphos	4.310	4.348
Chlorpiriphos-methyl	4.375	4.583
Pirimiphos-methyl	4.000	4.089
Triazophos	3.974	3.983

Applying the Trouton rule:

$$\log P = C - 21T_{b}/2.303RT$$

from Eq. 6, it is possible to obtain, for a fixed temperature:

$$T_{b} = A + \left(B/{}^{1}\chi^{V} \right)$$

where A and B are constants.

This latter expression can be explained by the decreasing of the dipole moment as ${}^1\chi^v$ increases, which implies the decreasing in the boiling point and increasing of the vapor pressure.

This result is important because it shows inverse of connectivity indices can be useful descriptors and contribute to our understanding and prediction of some experimental phenomena.

5. THERMODYNAMIC PROPERTIES

5.1 Molecular Accessibility and Symmetry Index

The concept of chemical accessibility was introduced by Kier [16] as a measure of the interaction of a given molecule, group, or atom with its neighbors, and therefore as a measure of its chemical reactivity. By applying this concept to a thermodynamic context, we will define a new *accessibility index* derived from the Kier-Hall connectivity indices ${}^0\chi^v$ and ${}^1\chi^v$ [17]. It will be useful to predict free energy changes in isomerization of hydrocarbons. On the other hand, part of the free energy change in these reactions is due to the symmetry relationships between both isomers. This latter contribution may be also easily evaluated though a simple formalism.

Let us consider the case of hydrocarbons. The higher the substitution in a carbon atom is, the lower its possibilities for interaction with other atoms. For example, the central carbon of neopentane should be less accessible than the four others placed around it. A given molecule will be more accessible as it had a low number of highly substituted carbon atoms. According to this idea, which would be the molecule containing carbon atoms with the lowest accessibility? It is clear that it should be the *molecule* of diamond. We can use it to define a very simple index which quantifies the accessibility using ${}^0\chi^v$ and ${}^1\chi^v$. Considering in a diamond structure a subgraph of topological length one, we can assign it a null accessibility, so A = 0. For this subgraph, the condition is satisfied by using the following definition:

$$A = 4\,{}^1\chi^V - {}^0\chi^V$$

Symmetry is important for determining free energy values. A symmetry index σ is defined as the maximum number of symmetry axes of the graph in the plane. Symmetry can be determined by deforming the graph as necessary without altering the topology of the graph. This is a simple algorithm and its applicability is completely general.

Table 7 shows the values for A for a set of carbon trees. The values of σ are also included, as well as the values for accessibility per atom A/N, a pondered value of accessibility.

Table 7. Values of the accessibility index A and the symmetry index σ for a set of hydrocarbons.

Carbon tree	$^1\chi^v$	$^0\chi^v$	A	A/N	σ
Diamond 1-subgraph	0.250	1.000	0.00	0.00	1
Ethine	0.333	1.155	0.18	0.09	1
Ethene	0.500	1.414	0.59	0.295	1
Ethane	1.000	2.000	2.00	1.00	1
Propane	1.414	2.707	2.95	0.98	1
n-Butane	1.914	3.414	4.24	1.06	1
Isobutane	1.732	3.577	3.35	0.84	3
n-Pentane	2.414	4.121	5.53	1.11	1
Isopentane	2.270	4.284	4.79	0.96	1
Neopentane	2.000	4.500	3.50	0.70	4
n-Hexane	2.914	4.828	6.83	1.14	1
2-methylpentane	2.770	4.992	6.09	1.02	1
3-methylpentane	2.808	4.992	6.24	1.04	1
2,2-dimethylbutane	2.561	5.207	5.04	0.84	1
2,3-dimethylbutane	2.643	5.155	5.42	0.90	2

The index A is good to evaluate accessibility. For example, in the series of pentanes the value for n-pentane is the highest, and the lower one corresponds to neopentane.

The parameter A/N is somehow a shape factor. Its value tends to $2-2^{-1/2} = 1.29...$ for N-alkanes for $N \rightarrow \infty$. Only *long* molecules show values higher than 1.

5.2 Prediction of stability from accessibility and symmetry

There is a established criteria on spontaneity and equilibrium in chemical reactions: The change of free energy or Gibbs potential, ΔG. A process is spontaneous, in standard conditions and assuming an activity value equal to 1 for all the components, if $\Delta G^0{}_{298} < 0$. Where $\Delta G^0{}_{298}$ is the change of standard free energy, i.e., the difference between the sum of the formation free energies values for the products and the sum of these energies for the reactants.

Similarly, we define the change in accessibility as:

$$\Delta A = A_P - A_R$$

where the subindex P stands for products, and the R for reactants.

We also define an accessibility corrected by the symmetry factor as:

$$A_S = A + 0.08\sigma \tag{7}$$

The weight of 0.08 is introduced in Eq. 7 for a better correlation with the experimental free energy values. Furthermore,

$$\Delta A_S = A_{S_P} - A_S$$

Proposed equations for ΔG^0 predictions are the following:

$$\Delta G^0_{298}(\text{kJ/mole}) = 10\Delta A \tag{8}$$

$$\Delta G^0_{298}(\text{kJ/mole}) = 10\Delta A_S \tag{9}$$

In Table 8 we can see experimental ΔG^0, ΔA, ΔAs and predictions of ΔG^0 for different imaginary processes that reveal the relative stability of some hydrocarbon isomer pairs. It is notable that in this table, even the worst predicted values, have the right sign.

The consideration of the symmetry does not allow better predictions for ΔG^0. However, when $\Delta A=0$, the consideration of σ allows us to predict the right sign for predicted ΔG^0, as can be seen in the last two examples in Table 8.

In this relationship, the term ${}^1\chi^v$ is the *enthalpic* term. The relationship between ${}^1\chi^v$ and enthalpy for diverse reactions has been published [19]. However ${}^0\chi^v$ and σ seem to act as *entropic* terms. The higher ${}^0\chi^v$, the higher the ramification and the formation entropy are. Similarly, σ is clearly an entropic term: The number of possible *complexions* or ways to arrange the atoms within the molecule depends on σ.

From these results one can draw the following conclusions:

It is possible to predict the sense of isomerization processes, and the relative stability within a pair of isomers, by computing the changes in the *accessibility index* A. Thus a general law could be expressed as:

The isomerization between positional isomers occur in the sense of the decreasing of accessibility.

The interest of this assertion is that the free energy change is expressed as a function of a simple topological concept as *accessibility* is. This reinforces the hypothesis that the most important components of molecular energies depend on more essential topological and symmetrical magnitudes.

Table 8. Values of ΔG^0_{298} (kJ/mol) experimental and ΔG^0_{298} calculated from Eqs. 8 and 9 together ΔA and ΔAs for different processes.

Process	ΔG^0_{298} (Exp.)[a]	ΔA	ΔAs	ΔG^0_{298} (Eq. 8)	ΔG^0_{298} (Eq. 9)
n-butane → isobutane	-4.13	-0.565	-0.405	-5.65	-4.05
n-pentane → isopentane	-4.24	-0.413	-0.413	-4.13	-4.13
isopentane → neopentane	-4.75[b]	-0.86	-0.620	-8.60	-6.20
n-hexane → 2methylpentane	-6.09	-0.558	-0.558	-5.58	-5.58
2-methylpentane → 3-methylpentane	2.13	0.152	0.152	1.52	1.52
3-methylpentane → 2,2-dimethylbutane	-8.04	-1.02	-1.02	-10.2	-10.2
2,2-dimethylbutane → 2,3-dimethylbutane	3.24	0.380	0.460	3.80	4.60
Methylcyclopropane → cyclobutane	-	0.588	0.908	5.88	9.08
Methylcyclopentane → cyclohexane	-	0.587	0.987	5.87	9.87
1-butene → 2-butene	-6.89[c]	-0.306	-0.266	-3.06	-2.66
propene → cyclopropane	41.67	2.221	2.461	22.21	24.61
1-hexene → cyclohexane	25.1	2.069	2.549	20.69	25.49
Propylamine → isopropylamine	-7.70	-0.675	-0.675	-6.75	-6.75
o-methylaniline → *m*-methylaniline	-2.20	-0.024	-0.024	-0.24	-0.24
m-methylaniline → *p*-methylaniline	2.20	0.000	0.080	0.00	0.80
cis-2-butene → *trans*-2-butene	-2.90	0.000	-0.080	0.00	-0.800

[a] *Experimental data obtained from ref. [18].*
[b] *Calculated from enthalpy and entropy, assuming Trouton rule, ΔS_{vap} = 21 cal/°K mol.*
[c] *Mean value of cis and trans isomers*

Interestingly, other processes unrelated to isomeric transformations also follow this rule, as for example:

$$2\,\text{Ethane} \rightarrow \text{Butane} + \text{H}_2 \quad \Delta G^0_{298} = 19.6\,\text{kJ/mole}, \Delta A = 0.24$$

5.3 Wiener number and symmetry

Considering the number of possible orderings for the vertices in a molecular graph, we have established a relationship between molecular symmetry and Wiener number [20].

We assume the contribution of the hydrogen atoms as embedded in the information given by every graph vertex. The *formation* of the chemical graph of ethane from two initially isolated vertices A and B, allows only one ordination: A-B that is identical to B-A.

For the chemical carbon tree corresponding to propane, there are three possibilities: A-B-C , A-C-B and B-A-C. We call Ω the number of such possible ordinations. Table 9 illustrates the different values of Ω, for several chemical graphs.

Table 9. Comparison of true W, and W calculated from σ, Ω and f_v; and values for σ_{calc}.

Compound	σ	Ω	f_v	W	$W_{calc.}$ (Eq. 10)	σ_{calc} (Eq.11)
Ethane	1	1	1	1	1	1.00
Propane	1	3	3	4	4.3	1.26
Cyclopropane	3	1	3	3	3	3.00
Butane	1	12	6	11	11.2	1.06
Isobutane	3	4	6	9	8.5	2.37
Methylcyclopropane	1	12	6	8	11.2	3.80
Cyclobutane	4	3	6	8	7.9	3.80
Pentane	1	60	9	20	20.4	1.11
Isopentane	1	60	9	20	20.4	1.11
Cyclopentane	5	12	9	15	14.8	4.66
Neopentane	4	15	9	16	15.5	3.38
Spiropentane	2	30	9	14	17.8	6.59
Methylcyclopentane	1	360	12	28	32.0	2.23
Cyclohexane	6	60	12	27	23.7	2.77

The symmetry index σ, defined as the maximum number of symmetry axes of the graph in the plane, plays a role in this domain. It is possible to demonstrate that for a graph with N vertices, with a given σ value, Ω is:

$$\Omega = N!/2\sigma$$

We postulate that: $\Omega \cong (W/f_v)^N$, where: W, Wiener number [20], f_v, vibrational degrees of freedom, and N, number of vertices of the corresponding chemical graph.

From the two expressions for the number of different possible ways for arranging N vertices into a graph Ω, we obtain an expression for W as a function of size and symmetry of the graph:

$$W \cong \left(N!/2\sigma\right)^{1/N} f_v \qquad (10)$$

f_V = 3N-6, vibrational degrees of freedom; for ethane, f_V = 3N-5, by analogy with the vibrational degrees of freedom for linear molecules.

From Eq.10 we can recalculate the symmetry index introducing the true value for W:

$$\sigma_{calc} = \left({f_V}/{W} \right)^N \left({N!}/{2} \right) \quad (11)$$

Table 9 displays the different predictions obtained for W and σ_{calc} within this framework.

Wiener number is reasonably predicted by this approach, at least for the hydrocarbons considered. The values for σ_{calc} are well predicted except for cycles with low symmetry. Probably the driving factor for the reproducibility of W is "cyclicity" more than symmetry.

6. CHIRALITY

A common assumption in chemistry is that chiral behavior is associated with 3-D geometry. Chirality is not a geometrical feature. The description of a chiral molecule through geometrical parameters such as bond lengths and bond angles is necessarily partial. On the other hand, describing the molecule by means of a orthogonal 3-D coordinate system will be possible only if there is definite a previous criterion for the algebraic sense of the axes. In addition, the concept of chirality is not restricted to the third dimensional space [21], as can be seen in the following examples presented in Figure 1.

At any rate, the chiral information must be supplied externally to the geometric one in order to characterize completely the static molecular structure. With this assumption in mind, it is conceivable that a pregeometrical molecular paradigm, such as a topological model, could incorporate chiral information [22-24].

Every vertex in a graph can be associated to a number called local vertex invariant (LOVI) [25]. This number can characterize the kind of atom, and thereby chiral information can be introduced [24]. This permits us to calculate new descriptors giving a weight to the corresponding entry in the main diagonal of the topological matrix.By operating in this way, it is possible to obtain a set of modified connectivity indices called chiral topological indices (CTIs) that provides for the proper consideration of the asymmetric carbon isomerism [26]. For these indices we use the notation: ${}^{m}\chi_t{}^{*}$, ${}^{m}\chi_t{}^{v*}$. They are able to differentiate the pharmacological activity between pairs of enantiomers.

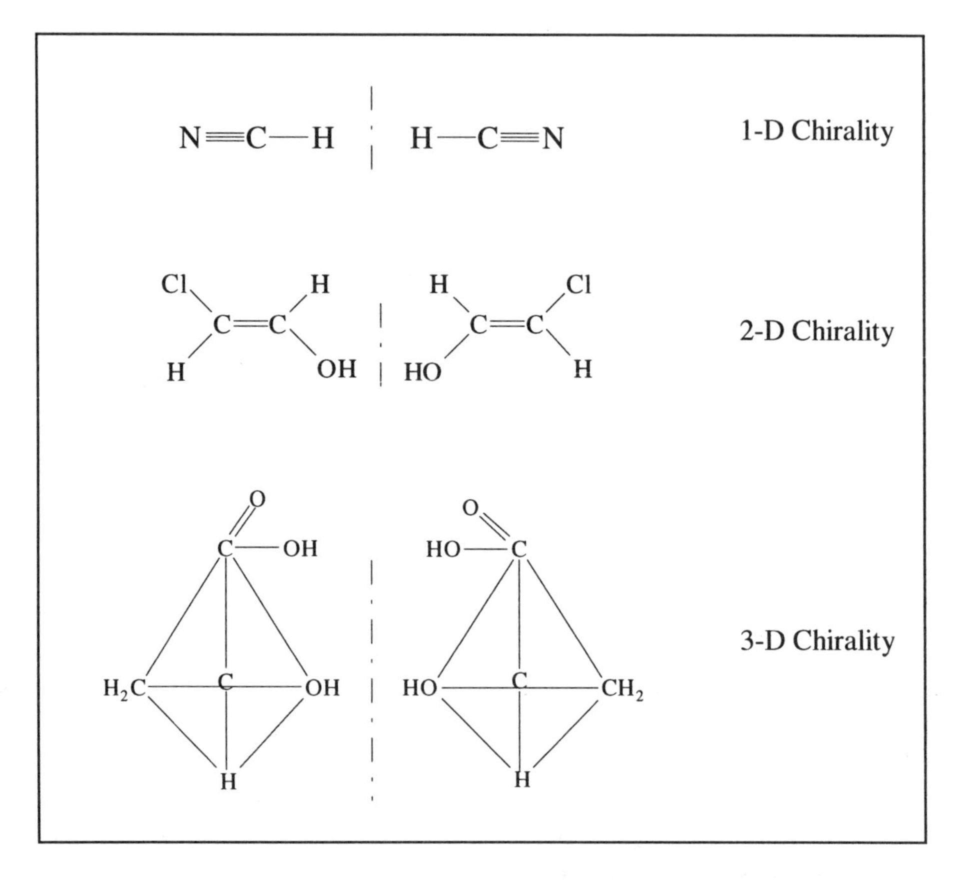

Figure 1. Graphical representation of the concept of chirality applied to different dimensional spaces

The model developed distinguishes which the active enantiomer is, in a group of chiral analgesics.

It was obtained using the following procedure. Through linear discriminant analysis (LDA) we obtained a classification function using Randic´-Kier-Hall type connectivity indices [27], with a group of 80 analgesics. In the equation obtained we substituted every connectivity index by its corresponding CTI. In this case, the modification that generates the CTIs consisted of introducing a weight of +1 for an R carbon and a value of -1 for an S carbon into the corresponding entries.

The empirical relation obtained was:

$$D = 0.41\,{}^{0}\chi^{*} - 1.60\,{}^{2}\chi^{*} + 0.57\,{}^{2}\chi^{V*} - 0.44\,{}^{3}\chi^{*}_{P}$$
$$+ 1.47\,{}^{3}\chi^{*}_{C} + 1.09\,{}^{4}\chi^{*}_{P} - 0.69\,{}^{4}\chi^{*}_{PC} - 0.24$$
$$N = 80 \quad F = 9.45 \quad U = 0.818 \qquad (12)$$

The results in the prediction of activity for a group of chiral analgesics are shown in Table 10. Reliable predictions can be obtained using this procedure. Globally, it is more than 83%, but considering the set of inactive enantiomers, the success reaches 100%. False positives are not obtained.

Table 10. Values of the classification function according to the discriminant function D, of a set of enantiomeric pairs of chiral analgesics.

Compound	D	Classification (Eq.12)	Classification (Experimental)
R-Benoxaprofen	-0.22	Inactive	Inactive
S-Benoxaprofen	-0.10	Inactive	Active
R-Fenoprofen	-0.11	Inactive	Inactive
S-Fenoprofen	0.01	Active	Active
R-Flurbiprofen	-0.04	Inactive	Inactive
S-Flurbiprofen	0.065	Active	Active
R-Ibuprofen	-0.01	Inactive	Inactive
S-Ibuprofen	0.09	Active	Active
R-Ketoprofen	-0.02	Inactive	Inactive
S-Ketoprofen	0.09	Active	Active
R-Naproxene	-0.4	Inactive	Inactive
S-Naproxene	-0.06	Inactive	Active

7. CONCLUSIONS

This work demonstrates that solid links between physicochemical properties and topological indices may be outlined. Relationships with quantum mechanical parameters, such as wavefunctions, classical thermodynamic magnitudes, as for instance free energy, enthalpy and vapor pressure, as well as number of complexions of N atoms arranged within a molecule, may be established. Other phenomena such as the electronic interferences may be also expressed as a function of graph theoretical descriptors.

All these features permit us to explore the essential nature of molecular topology, allowing us to speculate whether topological description is necessarily derived from the physicochemical nature of molecules or vice-versa. There is a lot of work to do in this field. Every relationship even if it's a simple one, reveals that we are only in the beginning of a greater understanding of the chemical structure, and consequently a greater knowledge of the world in which we are immersed.

ACKNOWLEDGMENTS

This work was funded by the Spanish Ministry of Education (CICYT, SAF96-0158-C02-02) and Generalitat Valenciana (GV99-91-1-12). We are indebted to Jennifer Chai Chang for careful reading of this manuscript and linguistic advice.

REFERENCES

1. Gálvez, J., García-Domenech, R., Julián-Ortiz, J. V. de, Soler, R. Topological Approach to Drug Design. *J. Chem. Inf. Comput. Sci.,* **1995,** *35,* 272-284.
2. Milne, G. W. A. Mathematics as a Basis for Chemistry. *J. Chem. Inf. Comput. Sci.,* **1997,** *37,* 639-644.
3. Stankevich, I. V., Skovortsova, M. I., Zefirov, N. S. On a quantum chemical interpretation of molecular connectivity indices for conjugated hydrocarbons. *J. Mol. Struc. (TEOCHEM),* **1995,** *342,* 173-179.
4. Gálvez, J. On a topological interpretation of electronic and vibrational molecular energies. *J. Mol. Struc. (TEOCHEM),* **1998,** *429,* 255-264.
5. Balaban, A. T., Motoc, I., Bonchev, D., Mekenyan, O. Topological indices for structure-activity correlations. *Top. Curr. Chem.,* **1983,** *114,* 21-25.
6. Balaban, A. T. Application of Graph Theory in Chemistry. *J. Chem. Inf. Comput. Sci.,* **1985,** *25,* 334-343.
7. Hansen, P. J., Jurs, P. C. Chemical Application of Graph Theory. Part I. Topological Indexes. *J. Chem. Educ.,* **1988,** *65,* 574-580.
8. Altenburg, K. Eine Bemerkung zu dem Randic´schen "Molekularen Bindungs-Index (Molecular Connectivity Index) *Z. Phys. Chem. Leipzig,* **1980,** *261,* 389-393.
9. Randic´, M., Hansen, P. J., Jurs, P. C. Search for Useful Graph Theoretical Invariants of Molecular Structure. *J. Chem. Inf. Comput. Sci.,* **1988,** *18,* 60-68.
10. Klages, F. Über eine Verbesserung der additiven Berechnung von Verbrennungswärmen und der Berechnung der Mesomerie-Energie aus Verbrennungswärmen *Chem. Ber.,* **1949,** *82,* 358-375.
11. Gálvez, J., García, R., Salabert, M.T. and Soler, R. Charge Indices. New Topological Descriptors. *J. Chem. Inf. Comput. Sci.,* **1994,** 34,N°3, 520-525.
12. Gálvez J., García-Domenech, R., Gregorio Alapont. Indices of Differences of Path Lengths. Novel Topological Descriptors Derived from Electronic Interferences in Graphs. *J. Comput. Aid. Mol. Des.* (In press).
13. *The Merck Index.* 12th ed., S. Budavari, Ed., Merck, Rahway, NJ, 1996.
14. *The Aldrich Structure Index.* Aldrich Chemical. Milwaukee, WI, 1992.
15. Worthing, C.R. *The Pesticide Manual: A World Compendium,* Croydok., UK 1979.
16. Kier . L. B. Personal communication, (Nov 17, 1998).
17. Kier, L. B., Hall, L. H. General Definition of Valence Delta-Values for Molecular Connectivity. *J. Pharm. Sci.,* **1983,** 72, 1170-1173.

18. Lide, D.R. *CRC Handbook of Chemistry and Physics*, 73rd Edition. CRC Press, 1992-1993, p. 9-154.
19. Kier, L. B., Hall, L. H. *Molecular Connectivity in Chemistry and Drug Research.* Academic Press. NY. 1976. Chapter 4.
20. Wiener, H. Structural Determination of Paraffin Boiling Points. *J. Am. Chem. Soc.*, **1947,** *69,* 17-20.
21. Randic´ M., Razinger M. Molecular Shapes and Chirality *J. Chem. Inf. Comput. Sci.*, **1996**, *36*, 429-441.
22. Wipke, W. T., Dyott T. M. Stereochemically Unique Naming Algorithm *J. Am. Chem. Soc.*, **1974,** *96,* 4834-4842.
23. Akutsu, T. A New Method of Computer Reperesentation of Stereochemistry. Transforming a Stereochemical Structure into a Graph. *J. Chem. Inf. Comput. Sci.*, **1991,** *35,* 864-870.
24. Schultz H. P., Schultz, E. B., Schultz T. P. Topological Organic Chemistry. 9. Graph Theory and Molecular Topological Indices of Stereoisomeric Organic Compounds. *J. Chem. Inf. Comput. Sci.*, **1995,** *35,* 864-870.
25. Balaban, A. T. Using Real Numbers as Vertex Invariants for Third-Generation Topological Indexes. *J. Chem. Inf. Comput. Sci.,* **1992,** *32,* 23-28.
26. Julian-Ortiz, J.V. de, Gregorio Alapont, C. de, Ríos-Santamarina, I., García-Domenech, R. and Gálvez, J. Prediction of Properties of Chiral Compounds by Molecular Topology *J. Mol. Graphics Mod.,* **1998**, 16, 14-18.
27. Julián-Ortiz, J.V de., Gálvez, J., Muñoz-Collado, C., García-Domenech, R. and Gimeno-Cardona, C. Virtual Combinatorial Syntheses and Computational Screening of New Potential Anti-Herpes Compounds. *J. Med. Chem.* **1999**, 42, N°17, 3308-3314.

Chapter 3

Database Organization and Similarity Searching with E-State Indices

Lemond B. Kier
Department of Medicinal Chemistry, Virginia Commonwealth University, Richmond, Virginia, 23298

Llowell H. Hall
Department of Chemistry, Eastern Nazarene College, Quincy, Massachusetts

1. ATOM INFORMATION FIELDS

Any approach to the quantitation of atoms or fragments within a molecule must be built upon the context and their relationships operating within this complex system. We can view each atom present in a molecule as existing in a field within a molecule in which all other atoms share and participate. The methyl group in the toluene molecule is different from the methyl group in acetic acid by virtue of its different molecular environment and connectivity, in spite of its intrinsic state as a methyl group. Quantifying the methyl group requires both an identity as a methyl group and it's relationship to all other atoms present in the molecule in which it resides. The influence of all other atoms present in the toluene molecule makes the methyl group unique relative to other methyl groups in all other possible molecules which include a methyl group. There are also intrinsic characteristics of the methyl group which transcend its molecular context. These inner characteristics must be identified in order to correctly describe

Fundamentals of Molecular Similarity, Edited by Carbó-Dorca *et al.*
Kluwer Academic/Plenum Publishers, New York 2001

this molecular fragment in a quantitative way. We consider each of these characteristics in turn.

2. THE INTRINSIC STATE OF AN ATOM

In toluene we recognize that the methyl groups possess attributes that are identified as *intrinsic*. The common attributes of methyl groups that comprise its intrinsic state are those we believe have an important influence on chemical, physical and biological properties. Three attributes are immediately apparent: elemental content, electonic organization, and topological state. The elemental content, in the case of the methyl group, is a composite of carbon and hydrogen atoms. A single atom in a molecule is described as a particular element. The electronic organization is the hybrid state or, more simply, the valence state of the atom or group in a molecule. This includes the counts of sigma, pi, and lone pair electrons comprising the valence electrons of the atom. In the case of a group such as methyl, we must encode the number of hydrogens to distinguish it from the other hydrides of carbon. The degree of adjacency or, more generally, the topological state is important for defining the position of the atom relative to the topology of a molecule. For example, when we compare carbon atoms in the three isomers of pentane, it is apparent that the spatial domain or accessibility of some carbons are different. The methyl groups in pentane are mantle fragments residing on the periphery of the molecule hence they are easily accessible to interactions with neighboring molecules. In contrast, the methylene group in pentane is located within the structure of the molecule with somewhat diminished accessibility to an intermolecular contact. Finally, the branched atom of isopentane is buried within the molecule. Its accessibility to any intermolecular contact is much smaller than that of the methyl or methylene group.

3. THE GRAPH REPRESENTATION OF A MOLECULE

A common way to the describe a molecule is to use a chemical graph to represent it. The bonding scheme of atoms is depicted as a network using lines to represent bonds and vertices to represent atoms. It is common to omit hydrogen atoms from carbon vertices because these are implied by valence rules. Hydrogen atoms associated with heteroatoms are retained in the scheme for clarity. Multiple bonds are explicitly represented. Aromatic rings include pi bonds using canonical representation or a circle denoting the pi orbital annulus.

4. THE DELTA VALUES

We begin our description of a molecule by writing the conventional chemical structure of a molecule such as ethyl acetate (Figure 1), and then deleting hydrogens associated with carbon atoms to form the chemical graph. We now calculate for each atom two values previously used in molecular connectivity descriptions [1]. The first is the simple delta (δ) value, which is the count of adjacent atoms other than hydrogen. This is equivalent to describing just the sigma bond skeleton structure. A second set of delta values is calculated for each atom based on the total number of valence electrons on the atom minus those bonding hydrogen. These are called the valence delta values (δ^v). These two sets of delta values, shown in Figure 1, are the basic ingredients for the definition of the intrinsic state of atoms in molecules.

5. THE ELECTRONIC INFORMATION IN THE DELTA VALUES

The pair of delta values just discussed provide a rich resource of information that describes atoms in molecules in our quest for a structural description. It is summarized as follows:

The simple delta value, δ, encodes:

a) the count of adjacent atoms excluding hydrogen
b) the count of sigma electrons on an atom excluding hydrogen
c) the count of bonds joining an atom other than hydrogen
d) the topological environment of the atom in the molecule

The valence delta value (δ^v) encodes:

a) the count of valence electrons on an atom other than to hydrogen
b) the count of sigma, pi, and lone pair electrons excluding bonds to hydrogen

Table 1 summarizes the δ and δ^v values for carbon, nitrogen, and oxygen atoms in a molecule. It is evident that $\delta^v - \delta$ = pi + lone pair electrons on an atom in a molecule. This information provides a quantitative measure of the potential of the atom for intermolecular interaction and reaction. These values comprise the Kier/Hall electronegativity [2], which has a high correlation with the Mulliken/Jaffe electronegativity of atoms in their valence states [3,4], shown in Table 1. The electronegativity of an atom in a molecule is of major importance within the context of the general information field described earlier. This simple statement of structure has a significant role in encoding the intrinsic state of the atom.

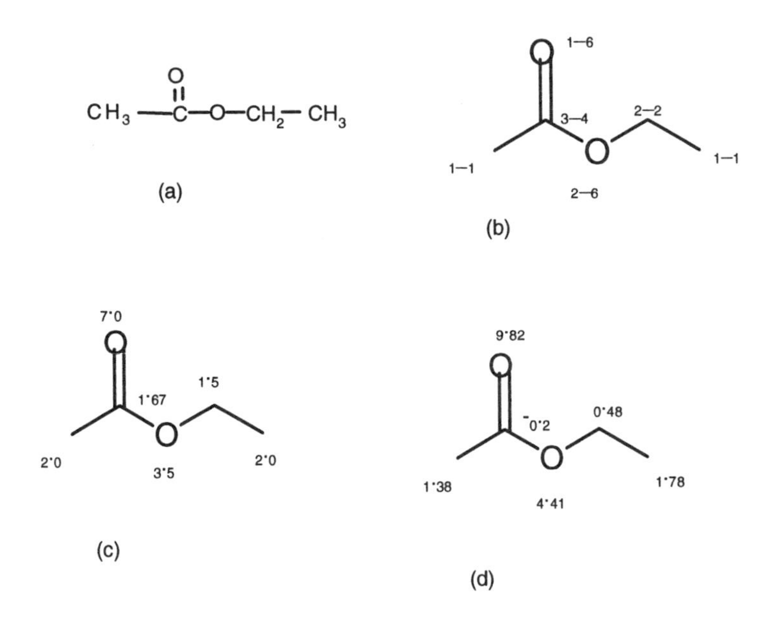

Figure 1. 1a. The molecule of ethyl acetate, 1b. ethyl acetate presented as a chemical graph with δ (first number) and δ^v (second number) values. 1c. ethyl acetate with calculated intrinsic state values, I, and 1d. ethyl acetate with calculated E-State values

Table 1. Delta, Kier-Hall Electronegativity and Intrinsic State Values of Atoms.

Atom	δ	δ^v	X_{KH} [(a)]	X_M (eV)	I
=O	1	6	1.25	17.07	7.00
-O-	2	6	1.00	15.25	3.50
≡N	1	5	1.00	15.68	6.00
=N-	2	5	0.75	12.87	3.00
>N-	3	5	0.50	11.54	2.00
≡C-	2	4	0.50	10.39	2.50
=C<	3	4	0.25	8.79	1.67
>C<	4	4	0.00	7.98	1.25
-Cl	1	7	0.67	11.54	4.11
=S	1	6	0.55	10.88	3.67
-S-	2	6	0.44	10.14	1.83

[(a)] $X_{KH} = (\delta^v - \delta) / n^2$

n = principal quantum number

X_{KH} = Kier-Hall electronegativities

X_M = Mulliken-Jaffe electronegativities expressed in electron volts.

6. THE INTRINSIC STATE ALGORITHM

The possibility of encoding a close approximation of the valence state electronegativity with such a simple index is of great value. Table 1 lists the values of δ^v and δ for several covalently bound atoms in their valence states. The derivation of an intrinsic state index labeled I, begins with the use of the δ^v - δ term. Of equal importance in defining an intrinsic state is the adjacency or topology of the atom in the molecule. Accordingly the intrinsic state encodes two attributes: 1) the availability of the atom or group for intermolecular interaction, and 2) the manifold of bonds over which adjacent atoms may influence, and be influenced by its state.

The adjacency, encoded by the simple delta value, δ, must therefore be a companion descriptor with the electronegativity in defining the intrinsic state. One possibility is to use the reciprocal of the adjacency, $1/\delta$, as an index of accessibility. The larger this value, the greater the accessibility of an atom or group. The product of the two terms produces an initial description of the intrinsic state value, I.

$$I = \left(\delta^{V} - \delta\right)/\delta \qquad (1)$$

The values for several hybrid atoms and groups are shown in Table 1. The intrinsic state, I, may be viewed as the ratio of the pi and lone pair electron count to the count of avenues of intramolecular interaction, the number of sigma bonds in the skeleton for this atom. That is, due to the intermolecular interaction associated with the atom, pi and lone pair electron density may be influential across the bonding network which is the set of sigma bonds in the molecular skeleton.

The value found for I in Equation 1 for the carbon sp^3 atom is zero since $\delta^v = \delta$ in every case. If we scale the δ^v - δ term by one, the zero values are eliminated and there is a discrimination among the various hydrides of carbon, arising from the different values of δ. This modification leads to Equation 2:

$$I = \left(\delta^{V} - \delta + 1\right)/\delta \qquad (2)$$

This expression achieves the objective of encoding electronic structure, topology, and a fortuitous effect, an approximation of the valence state electronegativity. An alternative form of this expression reveals the pi and lone pair count explicitly: (pi + n + 1)/(σ - H). A simplification of this expression can be made by adding 1 to the entire term to produce the intrinsic state for an atom or group in a molecule.

$$I = (\delta^{v} + 1)/\delta \quad (3)$$

Table 1 shows the intrinsic states of second row atoms and groups. To calculate the I values for higher quantum level atoms, the valence delta value is calculated as $(2/N)^2\delta^v$ where N is the quantum level.

7. FIELD INFLUENCES ON THE INTRINSIC STATE

We have derived the intrinsic state of an atom or group, but this expression does not reflect its position or influence within the field of other atoms in a molecule. This influence may take the form of a perturbation of the intrinsic state using some characteristic of every other atom in the molecule. A reasonable choice of attributes producing a perturbation on an atom is the intrinsic states of all other atoms in the molecule. This is equivalent to using electronegativities of other atoms to modify the state of each atom within the field of the molecular structure. The vehicle for this influence is the network of bonds linking each atom with all others in the molecule. This network is synonymous with the chemical graph model of the molecule over which electronegativity influence manifests itself.

A second consideration is the influence of two atoms in a molecule on the intrinsic state of the other. Since the chemical graph is the model of the presence and connectivity of atoms within the molecule, the count of bonds or atoms in paths separating two atoms was chosen for the unit of distance between any two atoms in a molecule. More precisely, the count of atoms in the minimum path length, r_{ij}, separating two atoms, i and j, is the distance selected to encode the influence between two atoms. Note that this count is equal to the usual graph distance plus 1. From this model we have chosen the difference between the intrinsic states of atom i and atom j, $(I_i - I_j)$, as the perturbation on each other. This effect is assumed to diminish by some power, m, of the distance, hence, the perturbation of I_i, called ΔI_i, is expressed as:

$$\Delta I_i = (I_i - I_j)/r_{ij}^m \quad (4)$$

The choice of the value of m results in a variable influence of distant and close atoms in the graph. Most studies to date have employed a value of m = 2. The total perturbation of atom i is a consequence of the influence of all other atoms in the molecule. Accordingly, the total perturbation of the intrinsic state of atom i, ΔI_i, should be a sum of these individual perturbations. The term, $\Sigma\Delta I_{ij}$, is a sum of all perturbation terms expressed by Equation 4. The state of atom i in a molecule is the intrinsic state, I_i, plus the sum of all perturbations included in ΔI_{ij}. This bonded state of atom i is called the electrotopological state, S_i, and is expressed as:

$$S_i = I_i + \sum \Delta I_{ij} \qquad (5)$$

For brevity, the S_i term is called the E-State for atom i. Figure 1e shows the E-State values calculated for ethyl acetate. The E-State index was introduced in a series of articles [4-8] and is fully described with many examples in a recent book [9].

7.1 Atom-Type E-State Indices

An extension of the E-State method is the use of an atom-type index, making it possible to study molecules of noncongeneric structure. Each atom is classified according to its valence state, the number of bonded hydrogens, and aromaticity [10]. For an atom-type index, the E-State values are summed for all atoms of the same type in the molecule. The symbol for an atom-type index is $S^T(X)$ where X denotes the atom or hydride group. As examples we have $S^T(Cl)$ for a chloro, $S^T(-OH)$ for a hydroxy, and $S^T(...CH...)$ for an aromatic CH. The program Molconn-Z recognizes 80 atom-types [11].

Atom-type indices encode three distinct types of chemical structure information: 1) electron accessibility for the atom-type; 2) presence/absence of the atom-type; and 3) count of atom-type present in the molecule. The atom-type E-State indices are used for heterogeneous data sets for structure-activity and for similarity analyses.

8. ORGANIZATION OF DATABASES

The atom type E-State values for an atom or groups in a molecule may be thought of as numerical components of a space or basis vectors in a manifold containing all possible atoms or groups. Each dimension is a parameter calculated for a particular atom or group. As an example, a set of molecules made up of alkanes, alcohols, and glycols are defined by the atom-types, -CH_3, -CH_2-, and –OH, and shown in Table 2. The realm of this set of molecules can be defined by the three atom-type indices corresponding to these groups. This is shown in Figure 2.

Within database subsets of molecules there are patterns of structure variation that are of interest in compound design. These patterns characterize the relative similarity or diversity within the subset. The atom-type indices make it possible to organize the subset in some manner which facilitates the design of other modifications, the selection of diverse structures for testing, cluster analysis based on structure and to conduct structure-activity analyses. To illustrate this organization and how the E-state indices can accomplish this, we look at the notorious polychlorobiphenyls, (the PCBs) (I).

Table 2. Sum Atom Type E-State Values for a Set of Alcohols and Glycols

Molecule	ST(-CH_3) [a]	ST(-CH_2-) [b]	ST(-OH) [c]
CH_3-OH	1.00	0.00	7.00
CH_3-CH_2-OH	1.68	-0.25	7.57
CH_3-CH_2-CH_2-OH	1.93	1.19	7.88
CH_3-CH_2-CH_2-CH_2-OH	2.05	2.38	8.07
HO-CH_2-CH_2-OH	0.00	-0.25	15.25
HO-CH_2-CH_2-CH_2-OH	0.00	0.69	15.81
CH_3-CH_2-CH_3	4.25	1.25	0.00
CH_3-CH_2-CH_2-CH_3	4.36	2.64	0.00

[a] Symbol for the sum of E-State values for -CH_3 groups in the molecule.
[b] Symbol for the sum of E-State values for -CH_2- groups in the molecule.
[c] Symbol for the sum of E-State values for -OH groups in the molecule.

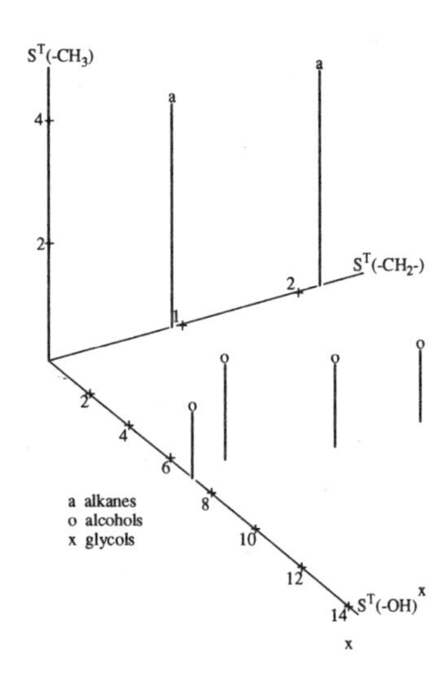

Figure 2. Three-dimensional realm of a set of alkanes, alkanols and glycols using atom-type indices for CH_3CH_2 and OH

8.1 The Polychlorobiphenyls

Three atom types are present in this series, each designated by it's atom type E-State code. These atom types are the chlorine atom (S^T-Cl), the aromatic -CH- group , and the substituted aromatic carbon atom -C< (S^T--C--). Using the sum of E-states for each atom type, a parameter space is created. To illustrate the structure organization possible, the two parameters, (S^T-Cl) and (S^T--CH--) are used to create a two-dimensional space. A view of this space over a relatively wide range reveals many of the possible polychlorobiphenyls from the unsubstituted through the tetrasubstituted derivatives (Figure 3). The major parameter governing the position in this space is the count of the number of chlorine atoms. To extract more useful information, we examine subsets with common numbers of chlorine atoms by viewing a restricted range of parameter values.

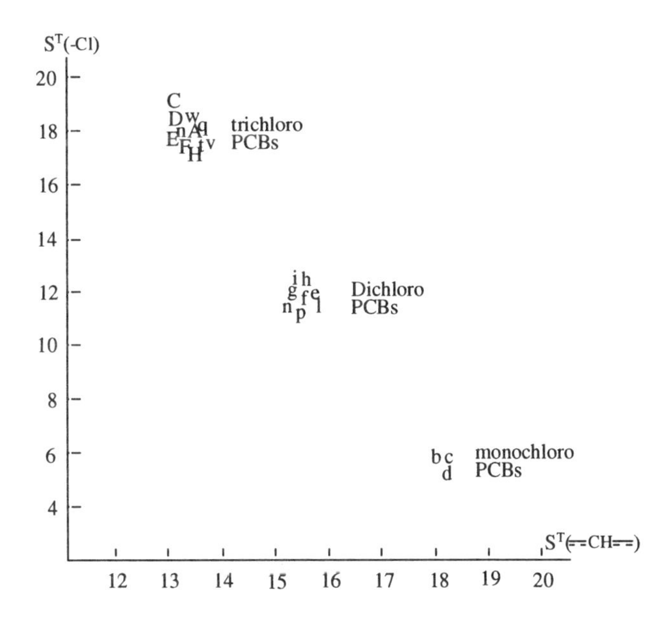

Figure 3. Structurally meaningful distribution of monochloro-, dichloro-, trichloro PCBs in atom type E-State space, based on two indices: Chlorine [S^T(-Cl)] and aromatic CH, [S^T(--CH--)].

In Figure 4 are shown the dichlorobiphenyls in the sum of atom type E-State space for Cl and aromatic -CH- coded as SsCl, SaaCH space. This figure presents an opportunity to assess the structure information as organized by these two parameters. The disubstituted biphenyls are arrayed in a pattern that can be interpreted in structural terms. The structures with their letter codes are shown in Table 3. The upper box contains molecules h (2,2') and i (2,6), which are substituted on the rings at the 2 or 2' positions in (I). The lower box contains molecules with chlorine atoms far removed from the juncture of the two rings. In between these two subsets lie biphenyls substituted at intermediate positions. At the top of the middle box, one chlorine is in the ortho position while the other is in a meta position. At the bottom of the box, one chlorine is ortho and the other is in the para position. This set of PCBs is clearly organized in a meaningful way in terms of molecular structure, and it represents chemically relevant information.

The organization of these molecules in this parameter space is more evident in the case of trichlorobiphenyls shown in Figure 5. As in Figure 4, the topmost substitution pattern contains chlorines only in the ortho position. An examination of this figure shows a cluster of derivatives with all three chlorine atoms on the same ring. These are labeled t, s, r, and q corresponding to the derivatives 2,3,6; 2,4,6; 2,3,5 and; 2,3,4 Within this cluster the molecules with chlorine atoms closer to the ring juncture 2,6,2' in (I), are higher in the grid, ranking in this order. As can be seen here, the pattern of organization for trichloro compounds is similar to that found for the dichloro compounds.

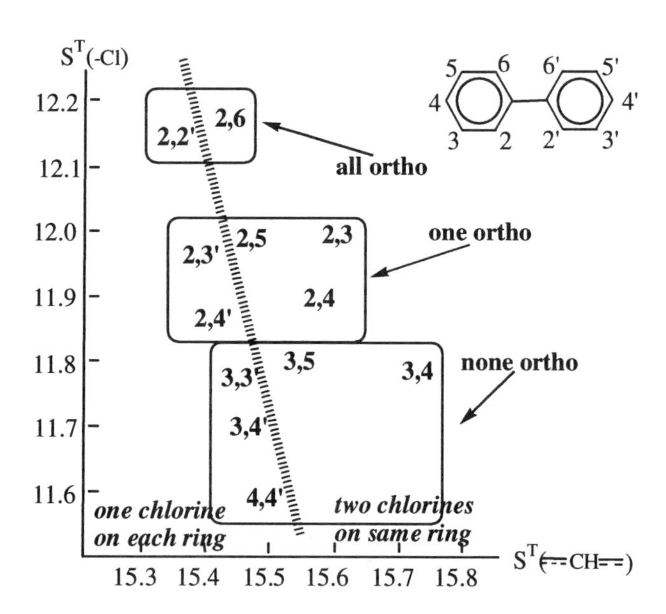

Figure 4. Structurally meaningful distribution of dichloro-PCBs in E-State space as a projection onto two dimensions of E-State space, based on two atom type indices: Chlorine [S^T(-Cl)] and aromatic CH [S^T(--CH--)]. See text for structure analysis

A second pattern in the grid of trisubstituted PCBs is found after an examination of the location of compounds substituted in the positions near the ring junctures. This includes the 2,3,6 and the 2,4,6 just discussed and also the 2,3,2' and 2,4,2' derivatives. At the other extreme, if substituent locations are far from the ring junctures as in the 3,4,4'; 3,5,4'; and the 3,4,3' derivatives, these molecules are located the lower region of the grid. A central cluster of molecules in the grid are the derivatives of both rings including a 2,3 substitution pattern. Below these are the mixed ring derivatives with 3,3' substituents always present.

Table 3. E-State Indices for Polychlorobiphenyls (PCBs)

Obs	Symbol	PCB	Cl atom count	S^T (-Cl)	S^T (--CH--)	S^T (--C--)
1	a	biphenyl	0	0.000	20.781	0.0000
2	b	2-Cl	1	6.059	18.005	0.7997
3	c	3-Cl	1	5.895	18.078	0.7790
4	d	4-Cl	1	5.799	18.114	0.7775
5	e	2,3-diCl	2	12.028	15.604	1.2022
6	f	2,4-diCl	2	11.900	15.527	1.3457
7	g	2,5-diCl	2	11.980	15.429	1.4249
8	h	2,6-diCl	2	12.159	15.418	1.3680
9	i	2,2'-diCl	2	12.144	15.355	1.4456
10	j	2,3'-diCl	2	11.972	15.392	1.4696
11	k	2,4'-diCl	2	11.871	15.405	1.4959
12	l	3,4-diCl	2	11.768	15.713	1.1800
13	m	3,5-diCl	2	11.831	15.565	1.3266
14	n	3,3'-diCl	2	11.803	15.442	1.4768
15	o	3,4'-diCl	2	11.704	15.463	1.4937
16	p	4,4'-diCl	2	11.607	15.488	1.5050
17	q	2,3,4-triCl	3	17.942	13.429	1.3717
18	r	2,3,5-triCl	3	17.991	13.217	1.5959
19	s	2,3,6-triCl	3	18.155	13.144	1.6166
20	t	2,4,5-triCl	3	17.895	13.254	1.5943
21	u	2,4,6-triCl	3	18.042	13.130	1.6825
22	v	3,4,5-triCl	3	17.778	13.502	1.3510
23	w	2,3,2'-triCl	3	18.131	13.045	1.7389
24	x	2,3,3'-triCl	3	17.954	13.059	1.7908
25	y	2,3,4'-triCl	3	17.851	13.057	1.8357
26	z	2,4,2'-triCl	3	17.999	12.945	1.9103
27	A	2,4,3'-triCl	3	17.823	12.967	1.9528
28	B	2,4,4'-triCl	3	17.721	12.969	1.9920
29	C	2,6,2'-triCl	3	18.271	12.895	1.8599
30	D	2,6,3'-triCl	3	18.091	12.896	1.9287
31	E	2,6,4'-triCl	3	17.986	12.886	1.9829
32	F	3,4,2'-triCl	3	17.859	13.095	1.7893
33	G	3,4,3'-triCl	3	17.687	13.130	1.8150
34	H	3,4,4'-triCl	3	17.586	13.140	1.8447
35	I	3,5,2'-triCl	3	17.927	12.969	1.9080
36	J	3,5,3'-triCl	3	17.753	12.996	1.9431
37	K	3,5,4'-triCl	3	17.651	13.002	1.9785

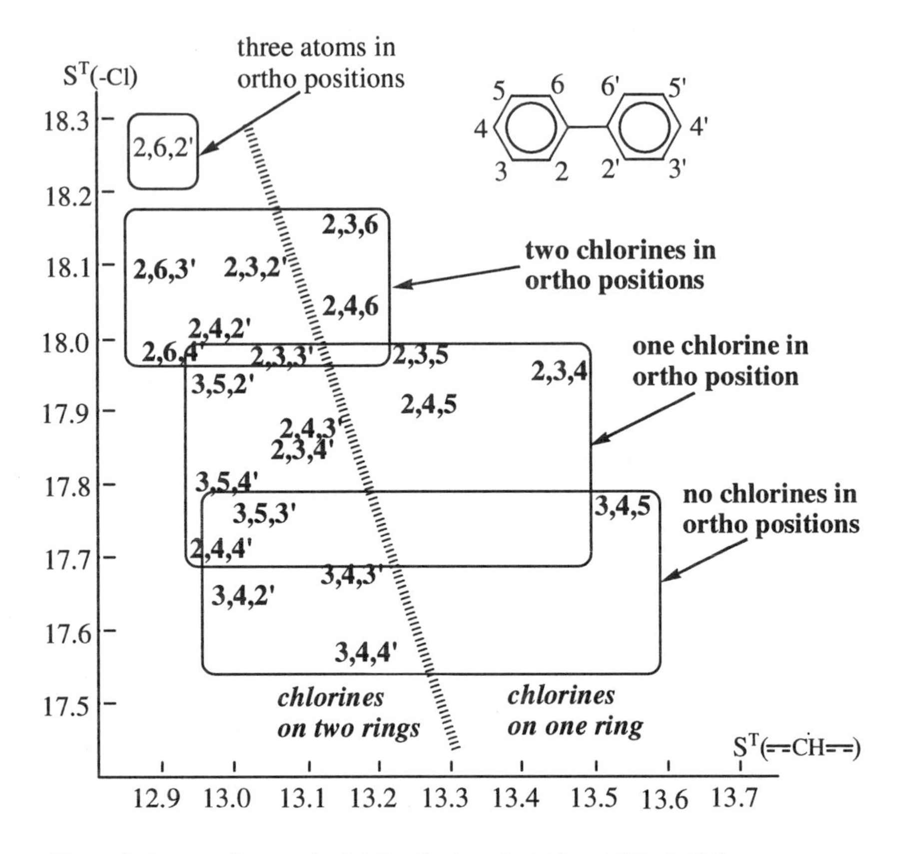

Figure 5. Structurally meaningful distribution of trichloro-PCBs in E-State space as a projection onto two dimensions of E-State space, based on two atom type indices: Chlorine [S^T(-Cl)] and aromatic CH [S^T(--CH--)]. See text for structure analysis

8.2 Ester Group Derivatives

Structural variations on the ester group often influence the rates of hydrolysis with enzymes or chemical reagents. The variations may be present on either the alcohol or the acid moiety, may be near or distant from the ester fragment, and may be of strong or weak electronic and steric influence. This matrix of variables presents an opportunity and a challenge to organize a virtual database of derivatives for the purpose of rational choices in compound design.

A few variations of the ester fragment are calculated and presented in the structure realm, using the atom-type E-States for the carbonyl oxygen and the ester oxygen atom-types (Figure 6). The influence of F, OH, NH_2, and CH_3 groups on the acid or the alcohol moiety is presented in the figure. It is apparent that a systematic organization of this database is possible from these indices. Along the direction A-B, the electronegativity of X' increases with a corresponding decrease in the E-State values of both oxygens, but to a much greater extent. The influence of the functional groups on the two oxygen atoms is portrayed in such a way that a selection of appropriate variation can be made with a desired goal in mind. Rational compound design is greatly facilitated by this organizational ability.

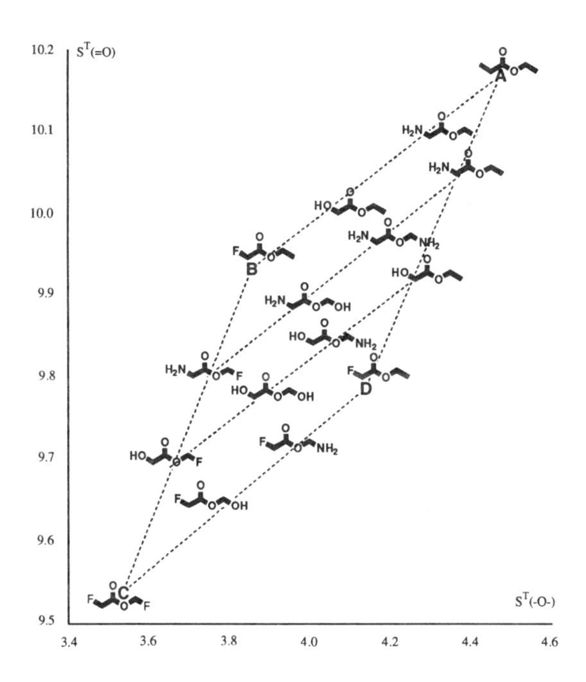

Figure 6. Esters of the form X-CH_2-C(=O)-O-CH_2-Y in atom type E-State space using the S^T(=O) and S^T(-O-) atom type indices.

9. SIMILARITY SEARCHING USING THE ATOM-TYPE E-STATES

A commonly held view is that molecules that are structurally similar, by some criteria, may have properties of comparable magnitude. In the context of biological properties, similarity presages comparable activities as ligands, substrates, inhibitors or other agents engaged in intermolecular encounters in living systems. In the area of drug design, this is the guiding principal in structure-activity modeling. A mathematical relationship is sought between a measured activity and a molecular property or structure representation. The purpose is to predict another structure, representing a candidate molecule, and to shed light on the structure-activity relationship. The rationale is that similarity may portend comparable behavior. The central element in this dialectic is similarity. This concept and associated methodology is explored in this section.

The E-state concept is a prime candidate for the definition of similarity among molecules, molecular fragments, and atoms-in-molecules [12]. One reason for this conjecture is that the E-state analysis produces numerical values which encode the extent of some attributes. The structure information may be represented at the atom level, the atom-type level and the bond type level, providing a broad basis for encoding molecular structure. We show here two examples of molecular similarity searches using two drug molecules as references and the atom-type E-State indices for the criteria of similarity. The database used is a modification of the Pomona MedChem database, which contain 21,000 molecular structures. The atom type E-State indices were added to the database and Euclidean distances computed between a target drug molecular structure and each molecule in the database. The structures found closest are listed in the two tables.

9.1 Sulfisoxazole

The antimicrobial agent sulfisoxazole is a prototype for this class of drugs where there is a continuing need to find new drugs that conquer resistant organisms. Using this molecule as a reference, we can search a database described by atom-type indices identifying structures that are similar on the basis of low numerical Euclidean distances. Table 4 shows a few of the closest molecules found in a database of about 25,000 drug-like molecules. There is a strong showing of molecules, modified by bioisosteric replacement in the connecting chain and in the heteroaromatic ring.

Table 4. Database Similarity search with reference compound Sulfisoxazole, using all nine atoms types in the structure

Molecule structure	Distance (arbitrary units)
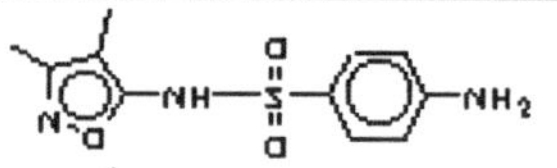	reference
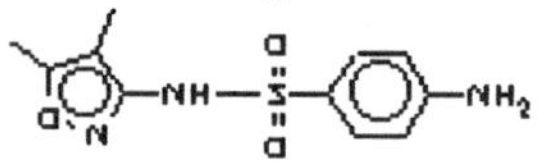	0.03
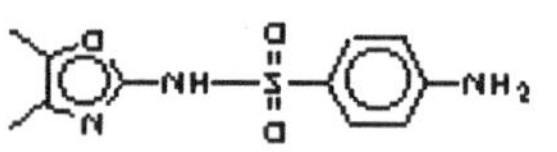	0.33

Molecule structure	Distance (arbitrary units)
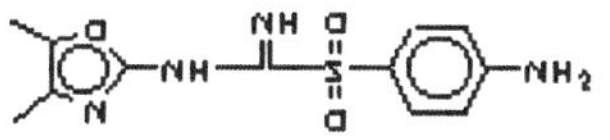	0.88
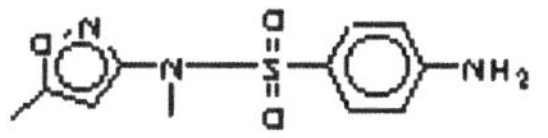	1.26
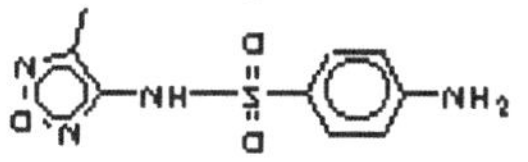	1.39
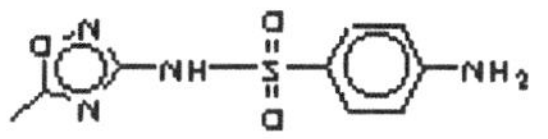	1.49
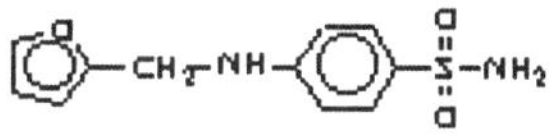	1.59
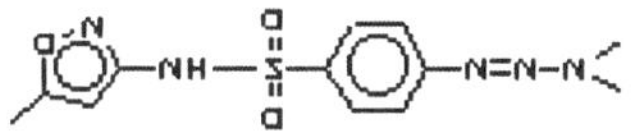	2.04
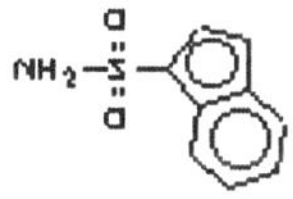	2.10

9.2 Chlorzoxazone

Chlorzoxazone is a central nervous system depressant with sedative and skeletal muscle relaxing properties. The need for better and safer drugs is never more evident than in this class of palliatives. Using the same database as in the previous example, compounds similar to chlorzoxazone are identified using atom-type E-state indices and Euclidean distances for the similarity metric. In table 5, the results of the search are revealed. Significantly different compounds based upon chemical intuition are identificd. These present opportunities to break away from simple modifications and to identify with some rationality, different structures that should be evaluated.

Table 5. Database Similarity search with reference compound Chlorzoxazone, using all six atoms types in the structure

Molecule structure	Distance (arbitrary units)
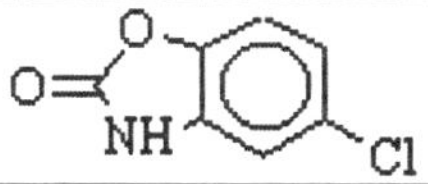	reference

Molecule structure	Distance (arbitrary units)
O, O=, NH, Br, Cl	0.52
O, N, O, N–N, N, NH	1.05
O, O=, NH	1.27
N, N, NH, O, NO_2	1.47
N, N, NH, O, NO_2	1.49

10. CONCLUSION

The E-State index is shown to contain information reflecting intermolecular accessibility of atoms and groups in a molecule. This information is encoded into a numerical value reflecting the electronegativity and the topology of structural fragments. The index for an atom is sensitive to the electronegativities of all other atoms in the molecule. This perturbing effect is carried through the network of atoms, described by a chemical graph.

The E-State indices may be used as individual atom-type descriptors thus making it possible to conduct structure-activity analyses with diverse structures. The E-State atom-type indices may also be used to organize a very diverse database of molecules into a coherent mosaic of molecules with a strong potential for navigating and searching in a logical way. This is demonstrated by the organizing of the PCBs into a database wherein any compound may be found by inspection. This capability forms the basis of the description of similarity by any criteria that is chosen. This is demonstrated by the search for similar compounds in a database, relative to a

context of small sets of compounds [13-19]. Over the last few years, work in Sheffield has studied the use of such measures for similarity searching in large chemical databases, with studies being conducted of: the effectiveness of various standardisation methods and similarity coefficients for field-based similarity searching [20]; the automatic alignment of pairs of MEP fields (two fields must be aligned before similarity can be calculated) by means of graph theory [21] and by means of a genetic algorithm (hereafter a GA) [22]; and the extent to which these two approaches to the generation of molecular alignments can handle conformationally flexible molecules [23]. Here, we discuss the extension of this work to encompass other types of molecular field and the use of the resulting field-based similarity measures for identifying *bioisosteres*, *i.e.*, structurally different molecules that exhibit the same biological effect [24].

Given a molecular property P that can be calculated at any point around a molecule, a field may be created around the molecule by integrating P with respect to volume. The similarity between a pair of molecules may then be calculated based on the overlap of the corresponding fields of the two molecules, using a similarity coefficient such as the Carbo index [13], a form of the long-established cosine coefficient. The Carbo index is defined to be

$$R_{AB} = \frac{\int P_A P_B dv}{(\int P_A^2 dv)^{1/2} (\int P_B^2 dv)^{1/2}}$$

where P_A and P_B are the properties of the two molecules that are being compared and where the integrations are over the 3D space surrounding these two molecules. If P denotes the electrostatic potential, then the potential, P_r, at a point r for a molecule of n atoms is calculated from the point charges q_i on each atom i in the molecule, so that

$$P_r = \sum_{i=1}^{n} \frac{q_i}{(r - R_i)}$$

where R_i denotes the position of the i-th atom and where r-R_i denotes the Euclidean distance between the two points. The resulting P_r values can then be inserted back into the Carbo index and the integral calculated using grid-based methods, but this is very time-consuming unless a coarse grid spacing is used (in which case it is unlikely that the calculated similarities will give an accurate representation of the structural resemblances that are present in a dataset). Instead, we have used the approach described by Good *et al.* [16]. Here, the $1/r$ term in the calculation of the potential is replaced by a

Gaussian approximation that is inserted into a version of the Carbo equation that can then be evaluated far more quickly than in the conventional approach to the calculation of field-based similarities. The Gaussian approach hence provides an elegant means of efficiently calculating the similarity between two molecules' MEPs, and comparable procedures can be employed to calculate steric and hydrophobic similarities. The electron density, ρ_r, at a point *r* around a molecule can be calculated from the sum of the contributions from each of the atoms in the molecule, *i.e.*,

$$\rho_r = \sum_{i=1}^{n} E_i(r - R_i)$$

where $E_i(d)$ is the electron density contribution of atom *i* at distance *d* from the nucleus, and where *r* and R_i are defined as previously. Good and Richards [17] have shown that it is possible to use Gaussian approximations to fit the curve of electron density against distance from the atomic nuclei, and suggested that the resulting expressions can be used for the calculation of a shape-similarity version of the Carbo index. A measure of lipophilicity potential may be gained by assigning partial lipophilicity constants to atoms, and designing a distance function that relates lipophilicity to distance from atomic centres, in much the same way as the MEP is calculated. Specifically, Gaillard *et. al.* [25] define the molecular lipophilic potential (MLP) at a point *r* around a molecule by

$$MLP_r = \sum_{i=1}^{n} f_i e^{-(r-R_i)/2}$$

where f_i is the atomic lipophilicity constant. This function was fitted to a Gaussian function, analogous to those used for the MEP and shape similarities, with the f_i values being calculated using a scheme described by Croizet *et al.* [26]. In this way, it is possible to generate electrostatic, steric and hydrophobic field descriptors that can then be used to quantify the degree of similarity between pairs of 3D structures, given an appropriate alignment of the molecular fields.

2.2 Use of a genetic algorithm

The alignments are generated using a GA that aligns two molecules' fields so as to maximise the value of the Carbo index [22]. In brief, each chromosome in this GA encodes the rotations and translations that are to be applied to a database structure to align it with the target structure. If no account is taken of conformational flexibility then just the rigid-body rotations are encoded; alternatively, if the molecules are allowed to flex, then the chromosome additionally encodes the torsional rotations [23] (although the experiments reported here consider only rigid molecules). The fitness function is the value of the Gaussian similarity coefficient resulting from that particular encoded alignment. The algorithm has been used previously to align molecules on the basis of their MEPs but it is simple to modify it to encompass the Gaussian versions of the shape and MLP similarity measures described above, thus allowing the calculation of all three types of similarity by the same basic procedure. Alignments may be made based on a single field-type, or on any combination of the three types of field; here, we report the results when all three of them were combined. During the execution of the GA, each alignment of a target structure and a database structure was used to calculate each of the three individual types of field-based similarity (MEP, shape and MLP) and then the fitness for the chromosome encoding that alignment was the mean of the three resulting Gaussian similarity values. No types of weighting or standardisation are applied to the individual similarity measures, so that all three types of field are assumed to contribute equally to the overall score of an individual database structure.

2.3 Experimental details and results

The experiments used the 1995 edition of the *World Drugs Index* (WDI), which contains the 2D structures and activity classes for about 41,000 drugs [27]. The GA is quite slow in operation, taking about 4 CPU days to search the entire file, and the experiments hence used a randomly-selected 1-in-10 subset of the full database. This was searched using the 10 target structures employed in a recent study of property-based similarity measures [28]. These target structures are listed in Figure 1, together with their associated WDI activity classes. For the purposes of these experiments, drugs that lie within the same activity class or classes as the target are considered *actives*. For each of the target structures, all of the actives were added to the 1-in-10 subset and then a similarity search was carried out to retrieve the nearest neighbours for the target structure.

Target	Structure	WDI Classification
Apomorphine	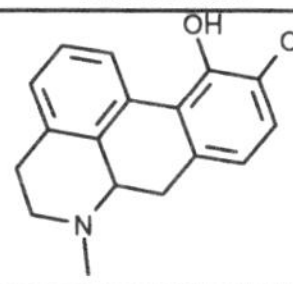 OH OH N	DOPAMINERGICS

Captopril	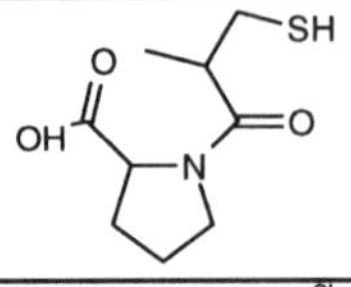	ANGIOTENSIN ANTAGONISTS
Cycliramine	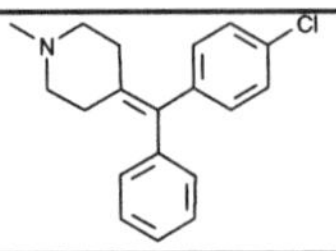	ANTIHISTAMINE-H1
Diazepam	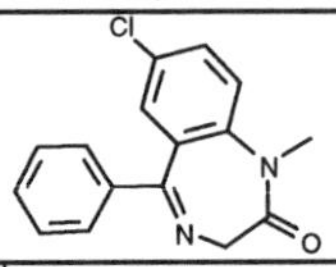	TRANQUILIZER or BENZODIAZEPINE-AGONIST
Diethylstilbesterol	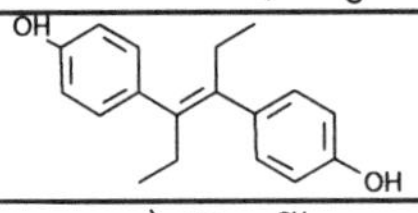	ESTROGENS
Fenoterol	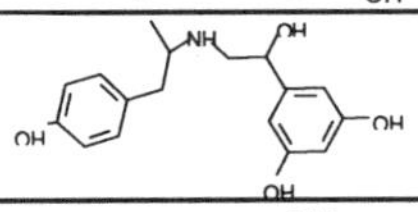	SYMPATHO-MIMETICS-BETA
Gaboxadol	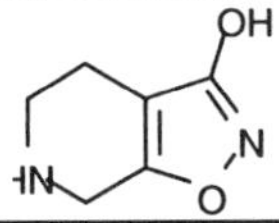	GABAMINERGICS
Morphine	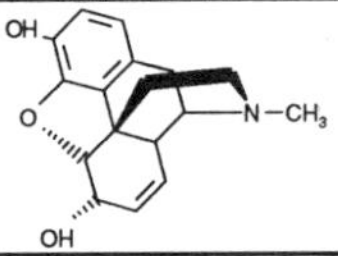	NARCOTIC or OPIOID
RS86	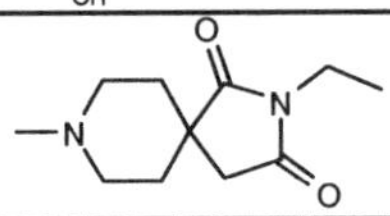	PARASYMPATHO-MIMETICS
Serotonin	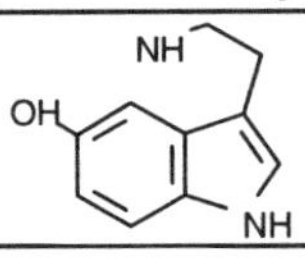	SEROTONINERGICS

Figure 1. The ten target structures for the similarity searches, and their WDI activity classes.

Three types of search were carried out for each target structure: a field-based similarity search (hereafter referred to as FBSS), which, as noted previously, involved all three types of field; a conventional 2D, fragment-based similarity search that was implemented using the search routines in the UNITY chemical information management system produced by Tripos Inc. [29]; and a 3D similarity search that was implemented using *atom mapping*, a distance-based measure of geometric similarity that maps the most similar atoms in two molecules to each other, based on the 3D geometric environment and the hydrogen-bonding type of the atom [30]. In each case, the search performance was measured by the number of actives present in the top-ranked 300 nearest neighbours, and by the number of *unique actives*, *i.e.*, actives that were not retrieved by any other of the similarity measures.

The left-hand portion of Table 1 shows the number of actives in the top-300 hits returned by each of the methods, where it will be seen that the 2D similarity measure is the most effective at retrieving active molecules. This finding might appear counter-intuitive, given the known importance of 3D information in general (and 3D field information in particular) in determining biological activities [12], but account must also be taken of the types of molecule in the database used here. Many of the molecules with a given activity class in the WDI are topologically very similar, including many close analogues and so-called "me-too" drugs. These molecules would be very easy to retrieve using a similarity measure that explicitly encodes topological information; however, they might well not represent the full range of bioactive structural types, whereas such diverse sets of molecules might be retrieved by measures that do not focus on the specific patterns of atoms and bonds in molecules. As Briem and Kuntz note [31], such measures permit the identification of molecules that are complementary to a particular biological target whereas 2D measures are well suited to finding molecules with a common substructural moiety.

An inspection of the various search outputs certainly suggests that the 2D measure results in less diverse sets of nearest neighbours than do the other measures, and we have sought to quantify this finding by means of a diversity index based on the fragment bit-strings used for the 2D similarity search. The index used here is the mean pairwise dissimilarity when averaged over all of the molecules in a dataset [32]. This index was calculated for the sets of active structures retrieved in each of the searches and the results are listed in the right-hand portion of Table 1, where it will be seen that the FBSS measure results in more diverse sets of compounds (*i.e.*, larger values of the diversity index) than do the established topological and geometric measures; similar results are obtained if all of the nearest neighbours are considered, rather than just the active nearest neighbours as here. This indicates that whilst the FBSS measure generally finds fewer actives, those that it does find are better able to suggest novel structural classes that are additional to the close analogues usually retrieved by similarity searches. This finding provides some quantitative support for the suggestion that FBSS provides an attractive way of identifying bioisosteres in chemical databases; current work in Sheffield using the BIOSTER database [33] provides further evidence to support this conclusion [34]. Indeed, the various types of search are complementary in nature, as evidenced by the numbers of unique actives that are found by each of the measures (listed in brackets in the left-hand portion of Table 1).

Table 1. Analysis of the active molecules retrieved in the top-300 rank positions by the 2D, 3D and FBSS similarity measures, using ten different target molecules. The left-hand part of the table contains the number of actives (with the unique actives in brackets) retrieved, and the fourth column lists the total number of molecules with the same activity classification as the target. The right-hand part of the table contains the diversity of the retrieved active molecules.

Target	Numbers of (Unique) Active Molecules				Calculated Diversities		
	2D	3D	FBSS	Total	2D	3D	FBSS
Apomorphine	60 (18)	31 (2)	49 (15)	117	0.32	0.32	0.56
Captopril	68 (33)	28 (1)	26 (0)	86	0.41	0.33	0.70
Cycliramine	94 (17)	141 (28)	120 (27)	261	0.59	0.61	0.60
Diazepam	46 (8)	45 (4)	37 (6)	113	0.51	0.51	0.55
Diethylst'ol	100 (34)	91 (29)	100 (2)	189	0.38	0.31	0.35
Fenoterol	58 (19)	31 (0)	33 (0)	65	0.42	0.45	0.44
Gaboxadol	7 (2)	16 (4)	14 (2)	34	0.49	0.53	0.52
Morphine	46 (7)	43 (3)	30 (5)	128	0.31	0.33	0.34
RS86	26 (20)	18 (4)	14 (3)	93	0.62	0.72	0.70
Serotonin	21 (4)	19 (2)	22 (3)	38	0.33	0.42	0.52
Mean Actives	52.6 (16.2)	46.3 (7.7)	44.5 (6.3)	112.4	0.44	0.46	0.53

3. COMBINATION OF SIMILARITY MEASURES USING DATA FUSION

3.1 Background

Many different types of similarity measure have been described in the literature [1-5] but the great majority of published studies have considered the use of only a single type of similarity measure: in many cases, indeed, a description of a new type of similarity measure forms the principal focus of the publication. Even where this is not the case, multiple measures have typically been employed only as the input to a comparative study that seeks to identify the "best" measure, using some quantitative performance criterion (see, e.g., [8]). Such comparisons, of which there are many in the literature, are limited in that they assume, normally implicitly, that there is some specific type of structural feature, weighting scheme or whatever that is uniquely well suited to describing the type(s) of biological activity that are being sought for in a similarity search. The assumption cannot be expected to be generally valid, given the multi-faceted nature of biological activities, and we have thus undertaken a study of the use of *data fusion* [35] methods for combining multiple similarity measures.

3.2 Combination of rankings

Data fusion involves combining the inputs from several different sensors, with the expectation that using the combined information will enable more effective decisions to be made than if just a single sensor was to be employed. The methods are widely used, in applications such as establishing the friend-or-foe nature of an incoming missile or aeroplane, surveillance operations by law enforcement agencies, real-time control of continuous manufacturing processes, the provision of all-weather visibility for aircraft pilots, and multi-imaging systems for the analysis of medical images. Our interest in data fusion methods arose from work on their application to information retrieval, specifically to the combination of the rankings produced by different retrieval mechanisms when applied to databases of textual documents. An early study is that by Belkin *et al.* [36], in which data fusion was used to combine the results of a series of searches of bibliographic databases, conducted in response to a single query, but employing different indexing and searching strategies. A query was processed with several different retrieval strategies, each of which was used to produce a ranking of a set of documents in order of decreasing similarity with the query. The ranks for each of the documents were then combined using one of several different fusion rules (including the MIN, MAX and SUM rules discussed below); the output of the fusion rule was taken as the document's new similarity score and the fused lists were then re-ranked in descending order of similarity. The combination of document rankings is now a well-established technique in several areas of information retrieval

The work on chemical data fusion reported here is based directly on the study by Belkin *et al.* [36], and involves the simple procedure shown in Figure 2, where a user-defined target structure is searched against a database using several different similarity measures. The fusion rules that we use here are summarised in Figure 3. It will be seen that the MIN and MAX rules represent the assignment of extreme ranks to database structures and it is thus hardly surprising that both can be highly sensitive to the presence of a single "poor" retrieval system amongst those that are being combined. The SUM rule is expected to be more stable against the presence of a single poor or noisy input ranking; here, each database structure is assigned the sum of all the rank positions at which it occurs in the input lists. This report considers just these three rules but there are clearly many others that could be considered, *e.g.*, the median, the product, the harmonic mean, *etc.* of the individual rankings. Whatever the fusion rule employed, the combined scores output by the rule are then used to re-order the database structures to give the final ranked output. In many cases, especially with SUM, the application of the fusion rule may result in the assignment of the same score to two or more items. When this happens, it is necessary to specify a further sort key to allow the resolution of the tied structures, *e.g.*, alphabetical ordering of the canonicalised connection tables describing the tied database structures or the allocation of weights to individual rankings (perhaps based on past performance in similarity searches) so that a high position in one ranking would differ in importance from that same position in another ranking.

1. Execute a similarity search of a chemical database for some particular target structure using two, or more, different measures of inter-molecular structural similarity.

2. Note the rank position, r_i, of each database structure in the ranking resulting from use of the i-th similarity measure.

3. Combine the various rankings using a fusion rule to give a new combined score for each database structure

4. Rank the resulting combined scores, and then use this ranking to calculate a quantitative measure of the effectiveness of the search for the chosen target structure.

Figure 2. Combination of similarity rankings using data fusion

Name	Fusion Rule
MIN	minimum $(r_1, r_2, \ldots r_i \ldots r_n)$
MAX	maximum $(r_1, r_2, \ldots r_i \ldots r_n)$
SUM	$\sum_{i=1}^{n} r_i$

Figure 3. Fusion rules for combining n ranked lists, where r_i denotes the rank position of a specific database structure in the i-th $(1 \leq i \leq n)$ ranked list.

Our initial studies of data fusion were carried out as part of a comparison of similarity measures based on the EVA descriptor, which characterises a molecule by its fundamental vibrational fingerprint, and on conventional 2D fingerprints [37]. Simulated property prediction experiments using logP data suggested that data fusion could be used to improve the performance of similarity searching in chemical databases: the remainder of this section describes further experiments that have been undertaken to ascertain the accuracy of this conclusion.

3.3 Experimental details and results

The first set of experiments used the 2D, 3D and FBSS searches of the WDI dataset described previously. Specifically, we noted the number of top-50 nearest neighbours that had the same activity class as one of the target structures, these again being the set of 10 structures used by Kearsley *et al.* [28]. Table 2 lists the numbers of actives identified in the original and fused searches for each of the 10 target structures. It will be seen that while the fused results are not always as good as the *best* individual result, they provide a generally high, and thus robust, level of effectiveness whereas the best original measure varies from target to target. This is particularly clear if one inspects the mean activities and ranks at the bottom of the table, which demonstrate the effectiveness of the SUM rule, in particular, for the combination of rankings.

Table 2. The number of actives found in the top-50 rank positions for searches in the WDI database for the original similarity methods (columns 3D, 2D and FBSS) and after data fusion (columns MAX, MIN and SUM). The shading indicates a fused result at least as good as the best original similarity measure for that target structure.

Target	3D	2D	FBSS	MAX	MIN	SUM
Apomorphine	15	23	14	24	16	26
Captopril	23	34	12	26	27	31
Cycliramine	43	31	36	43	42	45
Diazepam	27	27	15	23	23	22
Diethylst'ol	44	33	34	42	38	42
Fenoterol	19	33	17	28	29	31
Gaboxadol	6	2	6	5	6	5
Morphine	20	28	16	19	24	16
RS86	0	8	5	10	6	14
Serotonin	13	19	13	13	20	15
Mean Actives	21.0	23.8	16.8	23.3	23.1	24.7
Mean Rank	3.60	3.05	5.15	3.40	3.05	2.75

The utility of the SUM rule was confirmed with other datatsets [10], and the second set of experiments reported here hence considered just SUM in a study of data fusion when a larger number of original similarity measures is available. The dataset is that described by Kahn in a discussion of descriptors for the analysis of combinatorial libraries [38]: it contains 75 compounds each belonging to one of 14 well-defined activity classes (angiotensin-converting enzyme inhibitors, acetylcholine receptor inhibitors, antagonists of 2-aminoproprionic acid, aldose reductase inhibitors, angiotensin-II receptor antagonists, beta adrenergic blockers of the type-3 receptor, cyclo oxygenase 2 receptor antagonists, dopamine 3 receptor (ant)agonists, endothelin receptor (ant)agonists, histamine 2 antagonists, neurokinase-1 receptor antagonists, HIV-1 protease inhibitors, non-nucleoside HIV reverse transcriptase inhibitors, and steroid aromatase inhibitors).

Six similarity measures were used to generate rankings: the Molecular Simulations Inc. [39] Jurs descriptors; FBSS (as discussed in the previous section); two types of ChemX 3D flexible fingerprints [40]; and two types of Daylight 2D fingerprints [41]. The Jurs descriptors are part of the Cerius^2 software package, and describe shape and electronic charge by mapping the atomic partial charges onto the solvent accessible areas of the individual atoms within a molecule. In what follows, the inclusion of the Jurs rankings in a fusion combination is indicated by "J". The FBSS similarity measure has been described previously: its inclusion in a fusion combination is denoted by "F". The ChemX 3D flexible fingerprint keys record the presence or absence of potential pharmacophoric patterns (consisting of three pharmacophore centres and the associated inter-atomic distances) in any of the low-energy conformations identified by a rule-based conformational analysis of a molecule. Two sets of similarity scores were generated from these fingerprints: the Tanimoto coefficient scores and the Tversky similarity scores [8, 42], the inclusion of these in a fusion combination being denoted by "3" or by "T", respectively. The Daylight fingerprints were based on unfolded fingerprints considering pathlengths of up to 7, the inclusion of these in a fusion combination being denoted by "2" (for a standard fingerprint where a bit is either set or not set) or by "N" (for a fingerprint where a count is kept of how many times each bit is set), respectively. Thus 23F, for example, represents the fusion of the standard Daylight, Tanimoto ChemX and FBSS rankings. Each of the 75 molecules was used in turn as the target structure for a similarity search, with the similarity scores being calculated using either the binary or non-binary versions of the Tanimoto coefficient (as appropriate) or the Tversky similarity in the case of the "T" searches.

The SUM rule was used to generate all possible combinations of rankings from the six similarity measures. Table 3 details the mean numbers of actives (*i.e.*, molecules with the same activity as the target structure) found in the top-10 nearest neighbours when averaged over all 75 target structures.

The values of c at the top of the table denote the number of similarity measures that were fused (so that, *e.g.*, c=1 represents the original measures and c=2 represents the fusion of a pair of the original measures) and a shaded element indicates a fused combination that is at least as good as the best original individual measure (which was the ChemX keys with the Tanimoto coefficient, i.e., the "3" searches). It will be seen that very many of the fused combinations in Table 3 are shaded, thus providing further support for fusing individual similarity rankings. The table also shows that the fraction of the combinations that are shaded increases in line with c, so that all combinations with $c \geq 4$ perform at least as well as the best of the individual similarity measures. However, it is not the case that, *e.g.*, the c=5 combinations are invariably superior to the c=4 combinations, and the best result overall was obtained with 23FJT (rather than with 23FJNT, the combination involving all of the individual measures). Thus, while simply fusing as many individual measures as are available in a similarity investigation would appear to perform well, superior results may be obtained (for this dataset at least) from fusing a subset of the individual measures.

Table 3. Mean number of actives found in the ten nearest neighbours when combining various numbers, c, of different similarity measures for searches of the Kahn dataset. The shading indicates a fused result at least as good as the best original similarity measure.

c=1		c=2		c=3		c=4		c=5		c=6	
2	0.80	23	1.10	23F	1.28	23FJ	1.52	23FJN	1.45	23FJNT	1.43
3	1.12	2F	1.04	23J	1.39	23FN	1.23	23FJT	1.69		
F	0.89	2J	1.01	23N	1.04	23FT	1.43	23FNT	1.36		
J	1.08	2N	0.68	23T	1.24	23JN	1.31	23JNT	1.43		
N	0.63	2T	0.95	2FJ	1.35	23JT	1.45	2FJNT	1.43		
T	0.69	3F	1.09	2FN	1.08	23NT	1.25	3FJNT	1.51		
		3J	1.25	2FT	1.28	2FJN	1.28				
		3N	1.00	2JN	1.03	2FJT	1.53				
		3T	1.32	2JT	1.10	2FNT	1.28				
		FJ	1.20	2NT	0.95	2JNT	1.17				
		FN	0.91	3FJ	1.40	3FJN	1.35				
		FT	1.11	3FN	1.19	3FJT	1.55				
		JN	0.89	3FT	1.33	3FNT	1.41				
		JT	0.93	3JN	1.25	3JNT	1.36				
		NT	0.85	3JT	1.45	FJNT	1.32				
				3NT	1.20						
				FJN	1.11						
				FJT	1.21						
				FNT	1.11						
				JNT	1.12						

4. CONCLUSIONS

In this paper we have considered two ways of improving the similarity measures that are used for searching chemical structure databases. The first part considers the use of molecular field information for the calculation of inter-molecular structural similarities. Searches of the WDI database demonstrate that field-based searches encompassing steric, electrostatic and hydrophobic information can return sets of compounds that are noticeably different and more diverse than those resulting from a conventional, fragment-based measure of structural similarity (although the latter approach retrieves a much larger total number of actives). Field-based similarity searching may thus perhaps be best viewed as an "ideas generator", providing novel structural types to complement the analogues that result from the established, and computationally far less expensive, fragment-based methods. The second part of the paper discusses the use of data fusion methods to combine the rankings resulting from similarity searches of chemical datasets. The experiments reported here demonstrate that the use of a fusion rule such as SUM will generally result in a level of performance (however this is quantified) that is at least as good (when averaged over a number of searches) as the best individual measure: since the latter often varies from one target structure to another in an unpredictable manner, the use of a fusion rule will generally provide a more consistent level of searching performance than if just a single similarity measure is available.

ACKNOWLEDGEMENTS

I thank the following: the Biotechnology and Biological Sciences Research Council, the Engineering and Physical Sciences Research Council and GlaxoWellcome Research and Development Limited for funding; Tripos Inc. for software support; Derwent Information for providing the WDI database; and John Bradshaw, Claire Ginn, David Wild and Matt Wright for their contributions to the work reported here. The Krebs Institute for Biomolecular Research is a designated Biomolecular Sciences Centre of the Biotechnology and Biological Sciences Research Council.

REFERENCES

1. Downs, G.M. and Willett, P., *Rev. Comput. Chem.*, 7 (1995) 1.
2. Willett, P., Barnard, J.M. and Downs, G.M., *J.Chem.Inf.Comput. Sci.*, 38 (1998) 983.

3. Johnson, M.A. and Maggiora, G.M. (editors), *Concepts and Applications of Molecular Similarity*, John Wiley, New York (1990).
4. Special issue devoted to molecular similarity, *J. Chem. Inf. Comput. Sci.*, 32 (1992) 577.
5. Dean, P.M. (editor), *Molecular Similarity in Drug Design*, Chapman and Hall, Glasgow (1995).
6. Carhart, R.E., Smith, D.H. and Venkataraghavan, R., *J. Chem. Inf. Comput. Sci.*, 25 (1985) 64.
7. Willett, P., Winterman, V. and Bawden, D., *J. Chem. Inf. Comput. Sci.*, 26 (1986) 36.
8. Willett, P., *Similarity and Clustering in Chemical Information Systems*, Research Studies Press, Letchworth (1987).
9. Drayton, S.K., Edwards, K., Jewell, N.E., Turner, D.B., Wild, D.J., Willett, P., Wright, P.M. and Simmons, K., *Internet J. Chem.* at URL http://www.ijc.com/articles/1998v1/37/.
10. Ginn, C.M.R., Willett, P. and Bradshaw, J., Submitted for publication.
11. Dean, P.M. and Perkins, T.D.J., In Martin. Y.C. and Willett, P. (editors), *Designing Bioactive Molecules: Three-Dimensional Techniques and Applications*, American Chemical Society, Washington (1998).
12. Kubinyi, H., Folkers, G. & Martin, Y.C. (editors), *3D QSAR in Drug Design*, Kluwer/ESCOM, Leiden (1998).
13. Carbó, R., Leyda, L. and Arnau, M., *Int. J. Quant. Chem.*, 17 (1980)1185.
14. Manaut, F., Sanz, F., Jose, J. and Milesi, M., *J. Comput.-Aid. Mol. Design*, 5 (1991) 371.
15. Richard, A.M., *J. Comput. Chem.*, 12 (1991) 959.
16. Good, A.C., Hodgkin, E.E. and Richards, W.G., *J. Chem. Inf. Comput. Sci.*, 32 (1992) 188.
17. Good, A.C. and Richards, W.G., *J. Chem. Inf. Comput. Sci.*, 33 (1993) 112.
18. Petke, J.D. (1993) *J. Comput. Chem.*, 14 (1993) 928.
19. Mestres, J., Rohrer, D.C. and Maggiora, G.M., *J. Comput.-Aid. Mol. Design*, 13 (1999) 79.
20. Turner, D.B., Willett, P., Ferguson, A. and Heritage, T.W., *SAR QSAR Environ. Res.*, 3 (1995) 101.
21. Thorner, D.A., Willett, P., Wright, P.M. and Taylor, R., *J. Comput.-Aid. Mol. Design*, 11 (1997) 163.
22. Wild, D.J. and Willett, P., *J. Chem. Inf. Comput. Sci.*, 36 (1996) 159.
23. Thorner, D.A., Wild, D.J., Willett, P. and Wright, P.M., *J. Chem. Inf. Comput. Sci.*, 36 (1996) 900.
24. Lipinski, C.A., *Ann. Reports Med. Chem.*, 21 (1986) 283.
25. Gaillard, P., Carrupt, P., Testa, B. and Boudon, A., *J. Comput.-Aid. Mol. Design*, 8 (1994) 83.
26. Croizet, F., Dubost, J.P., Langlois, M.H. and Audrey, E., *Quant. Struct.-Activ. Relat.*, 10 (1991) 211.
27. The *World Drug Index* database is available from Derwent Information at URL http://www.derwent.co.uk
28. Kearsley, S.K., Sallamack, S., Fluder, E.M., Andose, J.D., Mosley, R.T. and Sheridan, R.P., *J. Chem. Inf. Comput. Sci.*, 36 (1996) 118.
29. UNITY is available from Tripos Inc. at http://www.tripos.com
30. Pepperrell, C.A., Willett, P. and Taylor, R., *Tetrahed. Comp. Methodol.*, 3 (1990) 575.
31. Briem, H. and Kuntz, I.D., *J. Med. Chem.*, 39 (1996) 3401.
32. Turner, D.B., Tyrrell, S.M. and Willett, P., *J. Chem. Inf. Comput. Sci.*, 37, (1997) 18.

33. The BIOSTER database is available from Synopsys Systems at URL http://www.synopsys.co.uk/
34. Gillet, V.J., Schuffenhauer, A. and Willett, P., Submitted for publication.
35. Hall, D.L., *Mathematical Techniques in Multisensor Data Fusion*, Artech House, Northwood MA (1992).
36. Belkin, N.J., Kantor, P., Fox, E.A. and Shaw, J.B., *Inf. Proc. Manag.*, 31 (1995) 431.
37. Ginn, C.M.R., Turner, D.B., Willett, P., Ferguson, A.M. and Heritage, T.W., *J. Chem. Inf. Comput. Sci.*, 37 (1997) 23.
38. Kahn, S.D., In Schleyer, P.v.R., Allinger, N.L., Clark, T., Gasteiger, J., Kollman, P.A., Schaefer, H.F. and Schreiner, P.R. (editors), *Encyclopedia of Computational Chemistry*, John Wiley, Chichester (1998).
39. Molecular Simulations Inc. is at URL http://www.msi.com
40. ChemX products are available from Oxford Molecular Limited at URL http://www.oxmol.co.uk
41. Daylight Chemical Information Systems Inc. is at URL http://www.daylight.com
42. Bradshaw, J., at URL http://www.daylight.com/meetings/mug97/Bradshaw/MUG97/tv_tversky.html

Chapter 5

Topological Pharmacophore Description of Chemical Structures using MAB-Force-Field-Derived Data and Corresponding Similarity Measures

Paul R. Gerber
Pharmaceutical Research and Development, F. Hoffmann-La Roche, Basel, Switzerland

1. INTRODUCTION

The characterization of chemical structures in terms of their potential pharmacophoric action is a central issue in the computer-aided search for new leads in drug discovery. The number of proposed schemes is large [1-3], and the difficulties in finding relevant measures to judge the merits of any method make it hard to discriminate among them. A second problem consists of the huge size of many databases that one would wish to mine for possible drug candidates. This size problem imposes significant restrictions on the complexity of possible algorithms for searching [4]. Correspondingly, there is a span of models in which a trade between computational speed and accuracy (or relevance to a particular purpose) is emphasized in various ways.

The characterization of pharmacophoric properties of chemical structures consists of two components: a description in terms of a relevant scheme of any single structure, and a prescription of how to compare any pair of structures. The description attributes to each structure a set of values, such as

Fundamentals of Molecular Similarity, Edited by Carbó-Dorca *et al.*
Kluwer Academic/Plenum Publishers, New York 2001

counts of substructure elements, physical parameters, geometrical quantities, etc. The similarity measure combines the two value sets of a pair of structures to yield a value (normally between 0 and 1) which describes the similarity of the pair. A value of 0 indicates that the two structures have nothing in common, while the value 1 tells us that the two structures are identical with respect to the chosen description.

An example of a bit-vector type description is the concept of fingerprints as used in the DAYLIGHT software [5]. Fingerprints essentially monitor the occurrence of linear structural elements of various types. For such a bit-vector description, a natural similarity measure is the Tanimoto index. This type of characterization is specially designed to handle large datasets at high speed and is neither restricted to nor optimally suited for pharmacophoric applications.

A second example is the pharmacophoric similarity concept [6] as implemented in the software Moloc [7]. Here full atomic coordinates and simple pharmacophore properties of the atoms are used as descriptors. The similarity measure is a rather involved function of these quantities. It compares the position and possibly directionality of polar atoms, and also accounts for volume overlap. In addition, this similarity function is maximized by varying the rigid body superposition of the pair of structures. Clearly, this type of comparison is considerably more demanding computationally than logical operations on bit-vectors, and, correspondingly, such an analysis is restricted to small sets of structures (up to a few thousand in the most favorable cases), but is highly relevant for pharmacophoric purposes. In this chapter, a new type of characterization is proposed, which attempts to retain as much as possible of the pharmacophoric accuracy but avoids three-dimensional description with its rather expensive computational treatment [8].

2. TOPOLOGICAL PHARMACOPHORE DESCRIPTION (TPR)

The new type of description uses a set of centers with pharmacophoric properties, which we will call agons, and a topological distance matrix between the agons. Agons are classified into hydrogen binders and hydrophobics.

In the most detailed case each atom is an agon by itself. The basis for assigning pharmacophoric properties is the analysis of the structure within the force field MAB [9,10] of the modeling program Moloc [7]. An atom is a hydrogen binder, when it has a hydrogen-bond donor- or acceptor strength above a corresponding donor or acceptor strength threshold, θ_d or θ_a, otherwise it is classified as hydrophobic. However, we introduce an additional parameter ν_h with the effect that if a hydrophobic atom has at least ν_h hydrogen binders as neighbors, it is discarded as hydrophobic ($\nu_h = 0$ has no effect).

Topological distances are also derived from MAB by taking the reference bond distance between bonded atoms, and the sum of these for the shortest path between two atoms that are not directly connected. This single-atom description is in general much too detailed for a speedy comparison.

In order to reduce the complexity of description the hydrophobic atoms are first grouped together by complete-linkage clustering with respect to the topological distance matrix as obtained by removing the Hydrogen binders and possibly, for nonzero ν_h, neighbors to them from the structure. By choosing an appropriate clustering level, as specified by a critical distance ρ_p, the number of hydrophobic agons can be drastically reduced. An agon is then characterized by its extension as determined by the average mutual distance between all pairs in a cluster. Figure 1 illustrates how the number of clusters decreases with increasing value of ρ_p for the example structure of valium. Increasing the ρ_p–values through 3, 5, 7, to 12 leads to 9, 5, 3, and 2 clusters respectively.

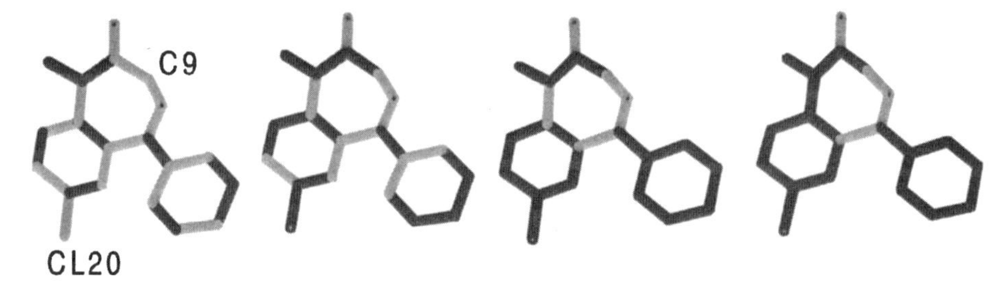

Figure 1. Dependence of hydrophobic cluster-size on the range parameter ρ_p. When going from left to right through the pictures of the molecule valium the parameter assumed the values 3, 5, 7, and 12 respectively, leading to 9, 5, 3, and 2 clusters respectively. Clusters are indicated in red color (dark). In the leftmost structure the two atoms carrying a label are clusters by their own.

Furthermore, a minimal cluster size σ_p may be specified, which excludes hydrophobic clusters with at most σ_p atoms.

A further reduction in the number of agons can be achieved by clustering the hydrogen-binding atoms, this time using the full distance matrix and a distinct clustering level. This level is also specified by a critical distance, ρ_h. Donor and acceptor strengths of such a cluster are obtained by taking the square root of the sum of squares of the corresponding atomic properties within the cluster. The extension of such clusters is calculated as in the case of hydrophobic agons.

The distance matrix for the clustered agons consists now of distance values between pairs of cluster. The distance between two clusters is simply obtained by averaging the distance values for all distinct pairs of atoms, which can be formed by taking one atom from each cluster.

By going to increasingly coarse clustering the level of description decreases, i.e., the amount of data per structure decreases. This reduction can be quite substantial, although the computational effort to produce the data remains essentially unchanged.

2.1 Parameters for Pharmacophore Specification

Six parameters determine the level of topological description, namely

- θ_d = H-bond donor threshold (1.3), h
- θ_a = H-bond acceptor threshold (1.3), a
- ρ_h = H-binder critical cluster distance (3.5), b
- ν_h = minimal number of H-binder neighbors for exclusion as hydrophob (2), q
- ρ_p = hydrophob critical cluster distance (10), d
- σ_p = upper size-limit for exclusion of hydrophobic clusters (1), c

Numbers in brackets give the default values in the program Mtprgn (see below), which calculates topological pharmacophores for a list of structures given as SMILES codes. The letters at the end are the qualifiers needed to modify the values in this program. Thus we may characterize a set of tpr's by the six numbers of the above list. A default-set is characterized by (1.3, 1.3, 3.5, 2, 10, 1).

3. SIMILARITY MEASURES

The calculation of the similarity between two topological pharmacophores consists of two essential steps. The first one is to pair up the agons of the pharmacophores, the second one to calculate a similarity value for the given pairing. Each pairing consists of the same number of agon pairs and is subject to the condition that only agons of equal type (hydrogen binders or hydrophobs) can make a pair. Thus, the number of H-binder (hydrophob) pairs is equal to the smaller of the two numbers of H-binder (hydrophob) agons of the two pharmacophores. Because the number of different pairings increases in a factorial fashion with the number of agons, a limitation in this number is usually unavoidable. Combining, for example, a (4,3)-pharmacophore (4 Hydrogen binders, 3 hydrophobs) with a (2,4)-

pharmacophore yields 12*24 = 288 different pairings. The reduction by clustering of the previous section becomes often a necessity, if one wants to avoid that many of the structures cannot be treated because of excessive computational effort. Alternatively, one could think of using approximate graph theoretical methods to speed up the similarity calculation [11].

For a given pairing the similarity value is calculated as follows. For each pair of agons a weight is calculated. For a hydrophobic pair this weight has the form

$$P_p = 1 + e_1 * e_2$$

where e is the extension of an agon. For H-binders the weight reads

$$P_h = w_d * D_1 * D_2 + w_a * A_1 * A_2$$

where D and A are donor and acceptor strengths respectively of an agon, and w_d and w_a fix the relative weight of donor and acceptor properties with respect to hydrophobic ones. Finally, the weights P are combined with distance data to yield the similarity value

$$S = \sum_{k}^{pairs} P_k^{\,2} + \sum_{k<l}^{pairs} P_k * P_l * g(d_{kl,1}, d_{kl,2})$$

The second (double) sum runs over all combinations of distinct agon pairs, while the weight function, g, takes into account that the distance values d_{kl} have in general nonequal values in pharmacophores 1 and 2. For this function we took the form

$$g(d_1, d_2) = \{1 - (d_1 - d_2)^2 / W_g\}^2, \quad \text{if} \quad (d_1 - d_2)^2 < W_g, \quad \text{else } g = 0.$$

This represents a finite-range bell-shaped function for the difference of the two arguments. The maximum value of one is assumed for equal argument values, and the width is given by the square root of the parameter W_g. This quantity is the sum of squares of agon-type dependent width parameters δ. For H-binders δ has a constant value δ_h, while for hydrophobs we take

$$\delta_p = \sqrt{P_p + \Delta_p^{\,2}}$$

where P_p is the hydrophobic agon weight, defined above, and Δ_p is a bare width for hydrophobs. This combination makes sense because P_p can be seen as a quadratic extension of the agon.

Finally, when a positive width-increase parameter, δ_d, is specified, the definition of W_g by the sum of squares of δ's is augmented by multiplication with a factor of the form

$$f_W = 1 + (d_m / \delta_d)^2$$

This factor provides the possibility to set less rigorous bounds for agons at large (exceeding δ_d) topological distances.

For each pairing of agons a value S is calculated. The pairing with the maximum value of S yields the best superposition of agons. The corresponding S-value yields, after normalization the final similarity value for the two pharmacophores.

Normalization is achieved by dividing the maximum S-value by the geometric mean of the two self-similarity values of the two pharmacophores k and l:

$$s_{kl} = S_{kl} / \sqrt{S_{kk} * S_{ll}}$$

Clearly, the self-similarity value is directly obtained for the identical agon pairing. All other pairings yield smaller S-values.

Alternative normalization rules may be adequate depending on the nature of the problem. Thus instead of taking the total self-similarity value of either pharmacophore, one can envisage to just consider the self-similarity value of the minimal subpharmacophore made up only from the agons occurring in the actual optimal pairing. This leads to a total of four similarity values, either of which may be useful in a particular problem. Thus we call:

- s^f = full similarity = total self-similarity taken for both pharmacophores,
- s^b = sub-similarity = minimal self-similarity taken for first pharmacophore,
- s^s = super-similarity = minimal self-similarity taken for second pharmacophore,
- s^p = partial similarity = minimal self-similarity taken for both pharmacophores.

s^b and s^s make only sense if pharmacophores k and l occur in a nonsymmetric context, e.g., when l is a target pharmacophore and k runs through a database. For diversity assessment of a database these nonsymmetric similarity measures make little sense.

3.1 Parameters for Similarity Calculations

The parameters specifying a similarity calculation are:

- w_d = relative weight of donor property with respect to hydrophobic (1), d
- w_a= relative weight of acceptor property with respect to hydrophobic (1), a
- δ_h= width for H-binders (2), h
- Δ_p= basic width for hydrophobs (2), p
- δ_d= critical distance for width increase (0), w

Numbers in brackets are default values of the program Mtprsml, while the letters are the qualifiers used to modify these values.

4. PHARMACOPHORE REPRESENTATION OF A DATABASE

For the structures in a database it may be useful to calculate and store topological pharmacophore representations for several reasons:

The degree of diversity among the structures as determined with the given similarity function may be of interest in order to select minimal divers subsets, i.e., a small subsets of structures representing a high degree of the inherent diversity.

The similarity measure may also help to obtain an indication, whether two databases differ in pharmacophoric content, and if so, which selection of one database would optimally complement the other.

Furthermore, the structures of the database may be ranked with respect to their similarity with a given structure, or more generally with respect to the average similarity with several pharmacophors. The given structure may be a known active pharmacophore or possibly a substrate. The ranking yields an indication as to which structures of the database are alternative candidates for a desired pharmacophoric effect. Such a ranking provides a prioritization of structures for screening of databases of available compounds.

As a technical remark we may add that, while it is possible to represent in general most of the structures in a database as topological pharmacophores, some these may turn out to have a large number of agons. In subsequent similarity calculations a threshold in this number must often be imposed for efficiency reasons, such that for practical purposes there is always some loss of structures.

As an example we have calculated pharmacophore representations of the available compounds from the Roche inhouse database. We obtained some 180,000 structures. The calculation of topological pharmacophore data took some 12 hours of CPU-time on MIPS-R10000-processors which corresponds to a speed of four structures per second. Using several processors in parallel, a database of one million entries can be converted over night.

A characteristic of such a representation of structures as pharmacophores is the distribution of agon numbers for the structures. The following table shows this distribution for the (1.3, 1.3, 3.5, 2, 10, 1)-representation of the above mentioned Roche structures.

This table shows that (2, 2) pharmacophores occur most frequently, namely 26733 times.

The following shorthand representation indicates, which fraction of all structures is taken into account when a limit on the number of agons per pharmacophore for similarity calculations is imposed. For this case, the representation is:

Limit	5	6	7	8	9	10
% of database	58.1	75.2	84.6	90.2	93.6	95.5

The more fine-grained (1.3, 1.3, 2, 2, 7, 1)-representation of the same database, in which critical clustering distances for H-binders and hydrophobs are reduced to 2 and 7 respectively, yields the following characterization:

Limit	5	6	7	8	9	10
% of database	22.9	42.5	59.7	72.0	80.5	85.9

It is obvious, that restriction to a maximum of eight agons per pharmacophore, as can usually be recommended for reasonable speed, only catches 72% of the database in this latter representation, in contrast to the previous case, where over 90% would be treated. An illustration for the difference in detailing between these two descriptions is, that amide units are taken as a single H-binder agon in the coarse-grained description, while carbonyl and nitrogen are separate agons in the fine-grained case.

Table 1. Distribution of compounds of the Roche database among pharmacophores with h hydrogen-binder agons (rows) and p hydrophobic agons (columns) in a (1.3, 1.3, 3.5, 2, 10, 1)-representation.

H\p	0	1	2	3	4	5	6	7	8	9	more	total
0	0	378	1071	687	292	134	41	13	6	1	5	2628
1	38	5228	13730	7922	2001	472	125	29	11	7	3	29566
2	100	9771	26733	17489	5774	1825	441	148	39	24	7	62351
3	134	5560	18398	14755	6116	2339	873	277	83	34	25	48594
4	76	1946	6660	6627	4145	1911	559	203	75	64	36	22302
5	14	447	1803	2437	1854	999	417	104	28	23	13	8139
6	3	136	626	1105	1058	802	339	142	46	24	18	4299
7	3	24	261	615	647	451	287	116	42	13	11	2470
8	0	9	61	202	413	276	147	84	53	14	13	1272
9	0	2	28	83	186	143	143	62	70	32	12	761
More	0	0	18	79	235	394	219	231	187	142	708	2213
Total	368	23501	69389	52001	22721	9746	3591	1409	640	378	851	184595

4.1 Ionizable Groups

The hydrogen-binder properties, D and A, of an agon containing basic or acidic groups, depend very strongly on the state of protonation of that group. This state, in turn, is determined by the pK_a of the group and by the actual pH. Since pK_a values are not generally known for the molecules of a whole database, we utilized the program pkalc [12] to estimate the pK_a values of the various groups. This program gives calculated values as well as an indication whether a group is basic or acidic, and which atom takes or releases a proton. Before calculating pharmacophore data, basic groups were protonated whenever their calculated pK_a value was higher than the assumed pH (taken to be 7), and acidic groups with pK_a smaller than pH were deprotonated. The computational resources needed for this protonation calculation are an order of magnitude larger than for calculating the pharmacophore data, and the CPU-times quoted above do not include this part. Thus, it saves much CPU-time if protonation of the various groups is already known.

5. LEAD PHARMACOPHORES AND RANKING OF A DATABASE

Pharmacophore representations of structures have lead-character in projects where the underlying structures are leads or active compounds. An advantage of a lead pharmacophore is that it is not tied to chemical entities but rather to pharmacophoric properties. Thus, a lead pharmacophore may be able to uncover structural leads with novel chemical groups. Furthermore, a pharmacophoric lead may also be generated from a substrate or even from a transition state analog of a substrate.

To illustrate ranking, we have taken the (1.3, 1.3, 3.5, 2, 10, 1)-representation the, above-mentioned database of available Roche structures. The lead pharmacophore was generated from the peptide sequence, D-Phe-Pro-Arg, for short fpr, which is well known as a thrombin inhibitor, and may also be considered a substrate model. The terminal amine- and carboxylate groups of this peptide were replaced by methyl groups in order to avoid spurious agons in the lead pharmacophore. From this structure the (1.3, 1.3, 3.5, 2, 10, 1)-pharmacophore was generated, which turned out to carry three H-binders and three hydrophobs. Ranking was made with the default similarity measure.

In order to judge this ranking, we used an inhouse database of thrombin inhibitors. For simplicity, every structure that was a member of this thrombin database was considered a hit. Figures 2 and 3 show the percentages of hits among all thrombin inhibitors versus the ranked enumeration of the full database.

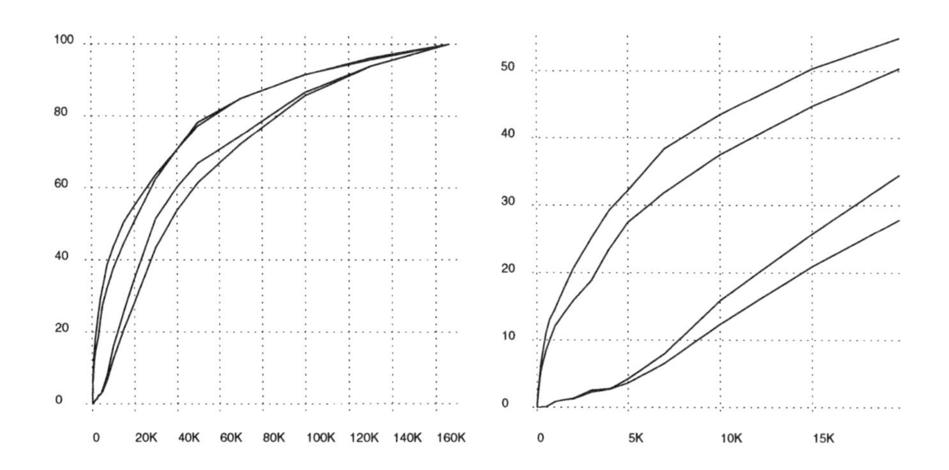

Figure 2, 3. Percentages of hits among all thrombin inhibitors are shown versus the ranked enumeration of the full database. The four curves correspond to the different types of normalization. From top to bottom the curves result from s^b-, s^f-, s^p-, and s^s-normalization, respectively. The arrangement of database structures is according to decreasing similarity values, separately so for each normalization, and thus, the same abscissa value corresponds in general to different structures for the four curves. Figure 3 is a close-up of Figure 2 for the region near the origin.

It is evident that the hit-rate for the best-ranked structures is substantially enhanced over the rate encountered by random sampling. Using subpharmacophore similarity, s^b, the 500 best-ranked structures contain 123 hits as compared to a total of 166359 structures of suitable tpr representation, which include 1115 structures of the thrombin database. This corresponds to an almost 40-fold enhancement of the hit-rate over random selection.

It must be kept in mind that some of the thrombin inhibitors have a binding mode, which makes use of additional features of the protein, not occupied by D-Phe-Pro-Arg, such that not all of the 1115 structures would be properly represented by our lead pharmacophore. This may also account for the rather steady increase in hit number for larger ranking values, as seen in the left-hand graph.

Furthermore, one can envisage ranking virtual structural databases, of which pharmacophore representations can be generated with justifiable computational effort. Such virtual databases originate e.g. from an enumeration of products in combinatorial chemistry. However, if large sets of substituents are considered for each possible substitution site, the number of product structures may be prohibitively large to envisage a pharmacophoric evaluation. In such cases the educt sets may be subjected to previous diversity analysis in order to select a subset of maximal pharmacophoric diversity and manageable size. Such a diversity analysis can, in addition to the methods mentioned in the Introduction, also be performed within the framework of topological pharmacophores as we will show later on.

As a further possible application one can envisage to look for binding motives in a database. For the case of ATP binding proteins, for example, inhibitor motives to substitute the adenine may be extracted from a database by prioritizing it with the topological pharmacophore of N-methyl-adenine itself. For this case it may be advisable to use a rather fine-grained description level in order to separate all H-binder groups. With a (1.3, 1.3, 2, 2, 7, 1)-representation we obtained a five-agon (4, 1) pharmacophore. Clearly, the database should be described on a similar level of detail. Although this leads to many pharmacophores with high agon numbers, this is not really a problem for the prioritization because pharmacophores with many agons will yield a small similarity value with the Adenine pharmacophore just because of the non-matching size. Quite generally, it is a feature of the used full-similarity measure, that a pair of structures of different size obtains a low similarity value.

The rating has been made for the MEDCHEM97 database as provided by the DAYLIGT software company [5]. From the 33000 structures in this database 27000 survived under the condition that pharmacophores of more than eight agons were omitted.

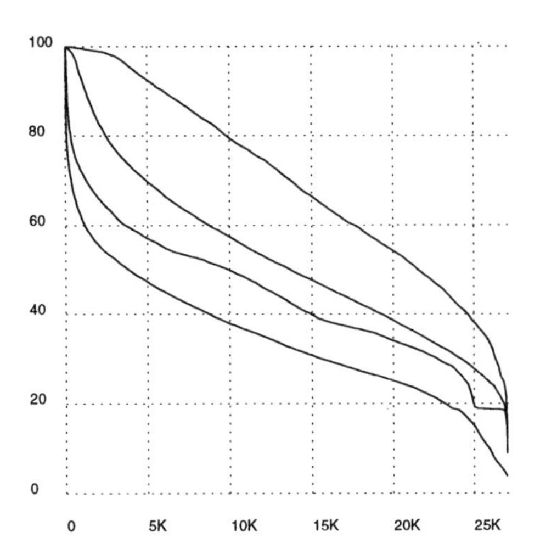

Figure 4. Similarity values of the structures of the Medchem97 database N-methyl-adenine. All pharmacophores were generated in a (1.3, 1.3, 2, 2, 7, 1)-representation. The arrangement of the structures of the database is according to decreasing similarity value to N-methyl-adenine. The four curves from bottom to top correspond to from s^f-, s^b-, s^s-, and s^p-normalization, respectively. The selection of 99 structures corresponds to the s^f-normalization of the bottom curve.

Figure 4 shows the similarity values of the structures in the sequence obtained from the similarity calculation against the N-methyl-adenine pharmacophore. All four normalizations are shown. However, we restrict the discussion to the full similarity case, s_f. It is obvious that only a narrow selection of structures reaches similarity values of 0.8 or higher. This value is also a limit where a sharp decrease over some hundred structures crosses over to a range with a somewhat more moderate slope, comprising some 2000 structures. After that the similarity value decreases more slowly over almost the whole database. A final range of again steeper decrease is found for the most hydrophobic structures. Of course, there is no consideration of chemical or pharmacokinetic aspects included in such a selection. The 99 most similar structures are displayed in Figure 5.

Since adenine itself is generally to small to provide good binding to the proteins, the purpose of such an investigation is mainly to obtain a set of pre-rated proposals to replace adenine by alternative groups.

6. DIVERSITY ANALYSIS

The proposed similarity measure can also be used to assess the diversity with respect to pharmacophoric properties of a database of structures.

However, the speed with which similarity values can be calculated sets a technical limit of a few thousand entries per database. Because the computational effort grows with the square of the number of entries, this limit is not likely to improve quickly. One can either calculate a full similarity matrix, which can be subjected to hierarchical clustering methods, or just a limited set of nearest neighbours per structure, suited for partitional clustering methods.

Figure 5. The 99 structures out of the MEDCHEM97 database, which are most similar to N-methyl-adenine. They have been aligned by a rigid-body matching procedure, which superimposes the centroid coordinates of corresponding agons as good as possible by means of a standard rigid-body-match procedure. The structures have then been spread apart in their most flat plane such that similarity values decrease from left to right within each row and then from top to bottom among rows. The target N-methyl adenine itself is also a member of this database (similarity value one) and is displayed at the top left-hand corner.

As an illustration we analyse the set of 99 adenine analogs shown above. Figure 7 shows the clustering tree of the similarity matrix, and illustrates that the set consists of rather similar molecules (similarities above .5) as one would expect from its mode of generation.

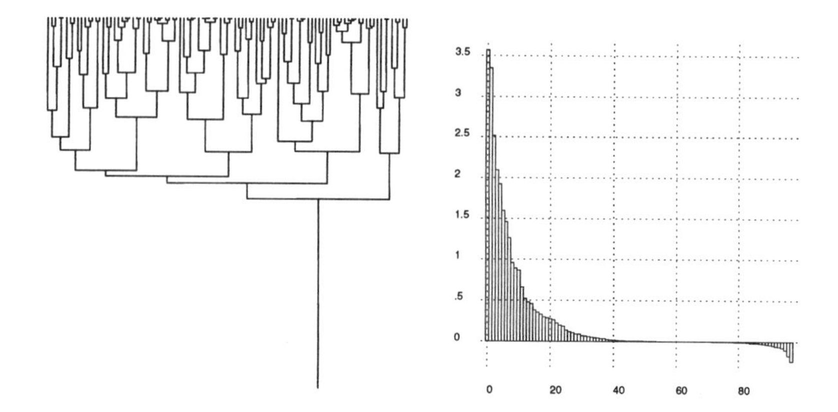

Figure 6, left. Hierarchical tree for complete linkage clustering of the similarity matrix the set of 99 adenine analogs shown above. While the similarity values against N-methyl-adenine are all above 0.77, similarities among the set reach lower values. Nevertheless, the set still consists of rather similar molecules (similarities above .5) as one would expect from its mode of generation.

Figure 7, right. Eigenvalues of the distance matrix, the elements of which are d_{ij}, which are related to the similarity matrix elements s_{ij} through $d_{ij} = 1 - s_{ij}$. The single large negative eigenvalue (trace(d_{ij}) = 0) has been omitted. Every eigenvalue of about one or higher represents a separate additional diversity component. Evidently, only some twelve independent components occur in this set. This means there is a redundancy factor of about 8 in the set of 99 structures.

While clustering is a useful way to group structures together; it is more problematic to use the diversity matrix as a tool to find absolute measures for characterizing the diversity of a database. Given a pharmacophore description and a similarity measure, a diversity assessment can readily be made. Figure 7 shows the eigenvalues of the distance matrix ($d_{ij} = 1 - s_{ij}$). This representation allows for estimation of the minimal number of compounds, necessary to represent the full diversity content of the set. For every eigenvalue of about one or larger represents a diversity component. In the present case a subset of about a dozen structures would be sufficient to represent the diversity of the whole, or equivalently, there is a redundancy factor of about 8 in the set of 99 structures. However, it must be kept in mind that the most difficult question remains to be whether description and similarity measure are adequate to the required purpose of judging the set or database.

7. SOFTWARE IMPLEMENTATION

The topological pharmacophore concept, as described above, has been implemented in the modeling software package Moloc (for a description see http://www.Moloc.ch). The auxiliary program Mtprgn generates the topological pharmacophore description for a whole database in batch mode. The program Mtprsml performs various types of similarity calculations for given tpr-representations. Furthermore, some of the mentioned features, such as examination of atom assignments to agons, generation of topological pharmacophores from single structures or from 3-dimensional pharmacophores, or superposition of structures onto a target, can be addressed directly from various menus within Moloc itself.

REFERENCES

1. Weininger, D. *J. Chem. Inf. Comput. Sci.* **1988**, *28*, 31-36.
2. Greene, J.; Kahn, S.; Savoj, H.; Sprague, P.; Teig, S. *J.Chem.Inf.Comput.Sci.* **1994**, *34*, 1297-1308.
3. Boehm, H. J. *J.Comput.Aided Mol.Des.* **1992**, *6*, 61-78.
4. Downs, G. M.; Barnard, J. M. *J. Inf. Comput. Sci.* **1997**, *37*, 59-61.
5. Daylight Chemical Information Systems Inc. *www.Daylight.com.*
6. Gerber, P. R. *Roche Internal Report # B174895* **1995**.
7. Mueller, K.; Ammann, H. J.; Doran, D. M.; Gerber, P. R.; Gubernator, K.; Schrepfer, G. *Bull.Soc.Chim.Belg.* **1988**, *97*, 655-667.
8. Willet, P. *J. Mol. Recog.* **1995**, *8*, 290-303.
9. Gerber, P. R.; Muller, K. *J. Comput.-Aided Mol. Design* **1995**, *9*, 251-268.
10. Gerber, P. R. *J. Comput.-Aided Mol. Design* **1998**, *12*, 37-51.
11. Rarey, M.; Dixon, J. S. *J. Comput.-Aided Mol. Design* **1998**, *12*, 471-490.
12. CompuDrug, *www.Compudrug.hu.*

Chapter 6

Dissimilarity Measures: Introducing a Novel Methodology

Guido Sello
Dipartimento di Chimica Organica e Industriale, Universita' degli Studi di Milano

1. INTRODUCTION

In recent years the concept of similarity has been mathematically transformed into a practical tool for compound comparison. The consequent development of many methods for similarity calculation have permitted the application of the correlated similarity measures to many chemical fields [2-8]. Successively, because of the challenging task posed by the combinatorial generation of large databases of compounds researchers became interested in the development of methodologies to select subsets of diverse compounds from those databases. In this perspective, many systems for diversity evaluation have been developed and applied mainly to databases analysis [9-13]. Nevertheless, the relatively recent introduction of the dissimilarity use in compound comparison still permits the addition of new methodologies.

Similarity and dissimilarity could be separately defined; however, we can also define dissimilarity (D) as a function of similarity (S) and consider this a good working definition. We can thus measure, or evaluate, molecular similarity and state that compounds showing scarce or no similarity are dissimilar. (For example, we can define D = 1-S; or D = 1/S). In any case, we need to calculate one or more molecular descriptors and to develop a method that by their use can evaluate similarity. One problem that still remains concerns the significance of the measure when comparing very diverse molecules. For the sake of clarity, we can consider a toy example. If we compare two hydrocarbons, e.g. butane X and hexadecane Y, to a heterocyclic compound, e.g. purine Z, and we use the number and type of

Fundamentals of Molecular Similarity, Edited by Carbó-Dorca *et al.*
Kluwer Academic/Plenum Publishers, New York 2001

functional groups as our molecular descriptor, the solution is that X is similar to Y, X is diverse to Z, and Y is diverse to Z. But if we would like to order the compounds we cannot order X and Y with respect to Z.

A different approach is nevertheless possible. We can develop a methodology that assures a result that is always different from zero. In addition, we can decide that the calculated values are contained at most in two or three magnitude orders. This way, we can conserve the fundamental method of similarity measure and put all the approximations in the method of combining the single measures. Recently, we developed a new approach in this perspective [14]. It compares two molecules through the use of an a priori defined comparison set of structures that is used to connect the two current molecules. The two compounds whose similarity is needed are compared with the molecules in the predefined set and the most similar reference molecules are selected. As all the compounds in the set possess a pairwise similarity measure, it is possible to combine the results into one value.

The methodology that we are going to present is a further step towards the development of a reliable approach to the measure of dissimilarity, or, better, to the measure of the similarity of dissimilar compounds.

1.1 Background

The development of our new methodology requires the choice of a structural descriptor. Indeed, any structural descriptor that has a precise mathematical definition can be used because one of the principal aims of the procedure is just the preservation of the descriptor meaning. Nevertheless, we are going to use the descriptor we commonly use: atom electronic energy. Its calculation is made through our standard method [15]; thus, the calculated values are completely comparable to our preceding applications. Atom electronic energy is calculated by our methodology. The energy is derived from the equation that correlates the electronic energy to the chemical potential: $E = \int \mu \, dn$ where E is the energy, μ is the chemical potential and dn is the electron variation. Because μ depends on the molecular structure as a function of the electron distribution, the atomic energy value is connected to the atom environment. Consequently, it is possible the calculation of the perturbation of the atomic energy comparing the energy of the atom in the molecule with that of the isolated atom. The result is a measure of the importance (weight) of each atom in each structure, as defined by the following equation (1)

$$W_i = | E_{tot} - E_{(tot-i)} | \tag{1}$$

where E_{tot} is the energy of the complete structure and $E_{(tot-i)}$ is the energy of the structure from which atom (i) has been eliminated; the energies come from the formula $E = \Sigma_i E_i$ where E_i is the energy of atom (i) [16].

Using the calculated Ws the compounds are canonical numerated and their comparison is made by means of the obtained canonical sequences [17]. Atoms with similar W are inserted into sets of connected similar atoms and the operation is repeated saving only the best result. Finally, a similarity index is calculated weighting the atoms member of similar structural fragments against the total structures. It is important to note that both the energy calculation and the comparison procedure can be performed in two different ways: either topologically [15], i.e. considering the molecules as graphs and consequently ignoring everything correlated with structure geometry; or in 3D space [17], i.e. considering the molecules as objects formed by atoms positioned in space and consequently considering spatial relationships and the influence of the molecular configuration and conformation. In addition, the methodology we are going to introduce can be, in principle, used with any descriptor, if it is capable of quantitatively describing molecular modifications.

For the sake of simplicity, we are going to use the topological scheme in the present paper.

1.2 The method

Differently from our previous approach [14] we are now proposing a system that is based on the following assumption: during the application of the comparison scheme it is possible to change a structure, initially dissimilar with respect to the other term of the comparison, into a new compound that is sufficiently similar to it. At the end of the transformation, or during its carrying out, we can measure the transformation cost and accordingly modify the calculated similarity measure.

A brief example can help understanding the procedure. The compounds used for the comparison are: naphthalene 1 and decalin 2 (Fig. 1). At the beginning their atoms are all energetically different, thus their similarity is null. Then, we begin changing one of the structures: for example 1. The first change is the reduction of one double bond: 1 is transformed to 1,2-dihydronaphthalene 3; but similarity is still null. Second change: reduction of a second double bond: 3 is transformed to tetralin 4. At this point the similarity of 2 and 4 is greater than zero. At the end we will obtain, in further three steps, decalin and the similarity is now equal to one. The final result is that in order to get the best similarity we had to travel five steps; i.e. the cost is five transformation steps.

This represents the general scheme: the stepwise transformation of the structures into the most similar educts, the measure of their similarity, the calculation of the transformation cost, and the calculation of a combined value representative of the similarity of the starting compounds. Consequently, we have to decide: a) what kind of transformations are allowed; b) when the changes must be considered sufficient; c) what calculation can be used to evaluate the cost.

In order to solve the first problem (what transformations) we have to look further inside the similarity calculation procedure. The comparison between two compounds is performed in three actions: first, the two compounds are canonical sequenced using the atomic energies and connections; second, the obtained sequences are compared and similar connected atoms are selected

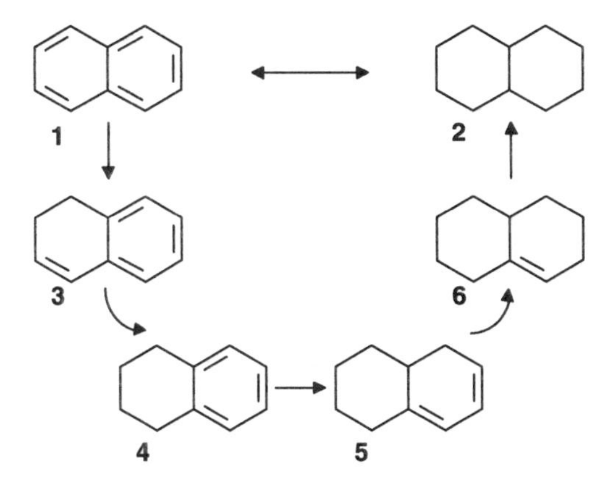

Figure 1. Comparison of naphthalene and decalin. The transformation of naphthalene

and inserted into similarity sets; third, the best result (i.e. that giving the biggest similarity sets) is selected. At the end of the procedure we obtain one or more sets containing the similar atoms ordered in connected fragments. Because our aim is the search at the minimal cost for the best similarity between our compounds, it is obvious that the similarity sets must be preserved. This means that any atoms that is member of a similarity set must be preserved in its integrity, i.e. without changing its energy. Consequently, all the functional groups that contain a preserved atom, different from carbon, are preserved. In addition, all the skeletal atoms are preserved to guarantee the integrity of the structure backbone. Taking into account those cautions, the transformation procedure changes, one by one, all the functional groups into saturated carbons [18]; vacant valences are saturated by hydrogens. The comparison sketched in Figure 1 can serve as an example.

The procedure is straightforward and finishes as soon as all the dissimilar functional groups have been eliminated (thus answering the question: when to stop the procedure). However, there are still two questions that remain to be answered:

a) is the procedure insensitive to the order of the group removal?
b) how many of the calculated similarities must be considered?

The elimination of functional groups is absolutely random, therefore there is the probability that the order can influence the result. One possibility could be to analyse all the possible order permutations to be confident that the selected order is the best. Obviously, such an approach is unacceptable because the number of permutations would be very high. A second possibility can be a random selection of permutations just to check that the calculated values are insensitive to the order; but, again, we risk to perform many calculations without improving the reliability of the result. The third possibility is based on the following assumption: if an atom energy is changed, the atom has been affected by the structure modification. Thus, we sum all the energy variations that record the structure variations that affected a particular atom. If the atom becomes part of a similarity set then it contributes to the similarity gain; its contribution is weighted by the number of the suffered variations. In this mode, we can be relatively sure that the calculated contributions do not depend on the order of FG modifications. We cannot prove this assumption, but we can add some considerations to its validity.

First, in our calculation scheme we use a definition of functional groups that is based on atomic energy (and not, for example, on atom and bond types) that depends on the interactions of all the atoms in the molecule. As a consequence, the perturbations produced by the structural modifications are completely correlated with the final structure and do not depend on the travelled path. This assures the same transformation cost for the same final point.

Second, in the case of different final points in two different comparisons the hypothetical greater/smaller energy cost is weighted against the greater/smaller number of transformation steps. Here again, because the cost is expressed in variation energy terms it is impossible to have a greater cost in a smaller number of steps. These considerations are true in all the similarity calculations that use descriptors of fundamental aspects of the molecules. In the Result and Discussion part we will also present the results obtained artificially changing the order of functional group elimination (see Fig. 8).

The second question is easier to address. In fact, it is obvious to consider all the similar structures that show an increase in the reciprocal similarities with respect to the starting compounds. It remains to develop a calculation scheme to transform these similarities into a quantity.

Thus, the last point to address concerns the calculation scheme. We have available the similarity indexes calculated for each structure pair and they are standard indexes. This means that for each structure pair we have a reliable, in our approach, calculated similarity value. But the problem is to find a way to combine those values into a single final value that can be representative of the similarity of the starting compounds through their transformations. We assign a prevalent weight to the similarity index of the unchanged structures. All the other indexes have a similar contribution to the final value. To assure the correct contribution the energy variations are compared to the energy of the starting molecule and correlated with the number of transformation steps that are the steps in which an atom has its energy changed, because the variation is greater the greater their number. The resulting equation is:

$$SI_f = SI^0 + \Sigma_i (SI_p) \quad (2)$$

where SI_f is the final similarity index, SI^0 is the starting similarity index, and the sum over (i) is determined by the number of similarity increases of atom (i). More,

$$SI_p = \Sigma_i \, FACT / VD$$

$$FACT = (SI_{curr} - SI^0) / NVA$$

$$VD = NS + 1$$

where FACT is the weighted contribution of each atom to the similarity in the current step, SI_{curr} is the current similarity index, NVA is the number of changed atoms, NS is the number of steps in which atom(i) is varied, and VD is the effective number of steps.

It is clear that the system we developed is simple and straightforward. But there is an objection that we cannot neglect: as the triangular relation is not valid in the calculation of similarity, it can be impossible to guarantee the validity of the SI_f. Indeed, we can add some considerations:

a) First, all calculated intermediate SIs maintain their intrinsic validity.
b) Second, the transformations of the structures are always very slow; as a consequence, the similarity between the parent and the son structures is always guaranteed.

c) Third, the use of proportional contributions makes the direct similarities (i.e. those of the starting compounds) dominant.

We believe that the discussion of some examples will help in corroborating our assumptions.

2. RESULTS AND DISCUSSION

2.1 Simple examples

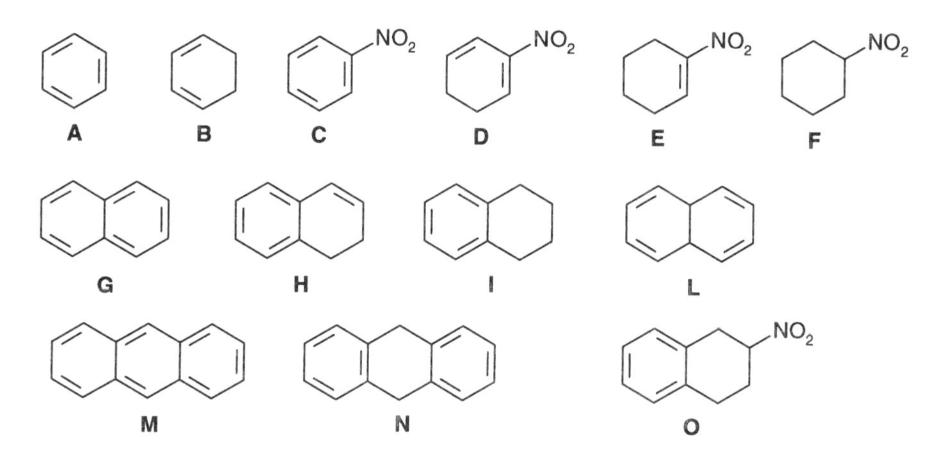

Figure 2. Structures in set 1

As the first test set we selected the set containing the compounds sketched in Fig. 2. They are 13 simple cyclic hydrocarbons at different level of unsaturation, some of them containing a nitro group. In this case we are comparing potentially similar compounds, but, as we shall see, their similarity can be even null. This example clearly shows that the calculated values correctly represent structure similarity. In Table 1 it is possible to observe the similarity values that have been calculated by our procedure. Not all the structures have been compared because we would only like to point to some aspects.

Table 1. Calculated similarity indexes of compounds in set 1

A-G	A-H	A-I	A-L	A-M	A-O	B-C	B-G	C-D	C-E	C-F	G-M	G-N	I-L
0.750	0.750	0.750	0.265	0.600	0.632	0.472	0.258	0.667	0.611	0.563	0.833	0.623	0.450

The first remark concerns the stability of the values when comparing the same compound (A) to a set of correlated compounds (G, H, I). The expected result confirms that the calculated similarity is insensitive to small changes. However, it changes with molecular size (compound M) and with energetically important variations (compounds L and O). A similar situation is present when comparing compound G with compounds M and N.

The second point shows the ordering of the compounds with respect to a reference. Compounds D, E, and F, are correctly ordered and grouped together by comparison with compound C; on the contrary, compound B is evidently more different (see B-C and C-D).

2.2 Similar complex structures

A second set of structures is reported in Fig. 3. The compounds are all part of a common biogenetic family and thus they are similar enough to allow for their direct comparison. Nevertheless, in Table 2 are compared the similarity values calculated using our three different schemes using compound A as a reference: direct comparison (DC), comparison through a predefined set (PSC), and comparison through molecular transformations (MTC). We can observe that the values show a similar compound orderings, excluding A5, A6, A15, A17, for PSC, and A6, A10, A19, for MTC. In addition, MTC and DC are better correlated than PSC and DC. It is obvious that PSC is much faster than MTC; thus, the choice of PSC or MTC depends on the current need.

Table 2. Calculated similarity indexes of compounds in set 2

	A1	A2	A3	A4	A5	A6	A7
DC[a]	0.610	0.558	0.353	0.555	0.649	0.154	0.372
PSC[b]	0.550	0.440	0.270	0.490	0.470	0.230	0.375
MTC[c]	0.598	0.613	0.376	0.590	0.716	0.356	0.404

	A8	A9	A10	A11	A12	A13	A14
DC[a]	0.616	0.450	0.595	0.389	0.455	0.488	0.545
PSC[b]	0.530	0.400	0.540	0.310	0.350	0.420	0.510
MTC[c]	0.616	0.461	0.827	0.414	0.478	0.503	0.527

	A15	A16	A17	A18	A19
DC[a]	0.766	0.353	0.714	0.585	0.543
PSC[b]	0.440	0.290	0.550	0.497	0.560
MTC[c]	0.766	0.409	0.704	0.634	0.688

[a] Similarity index calculated by Direct Comparison.
[b] Similarity index calculated through Predefined Set Comparison.
[c] Similarity index calculated through Molecular Transformation Comparison.

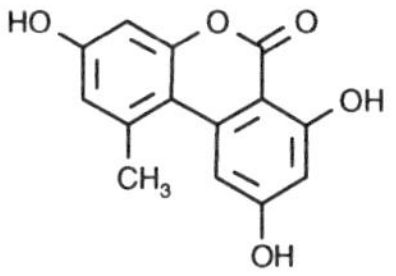

A 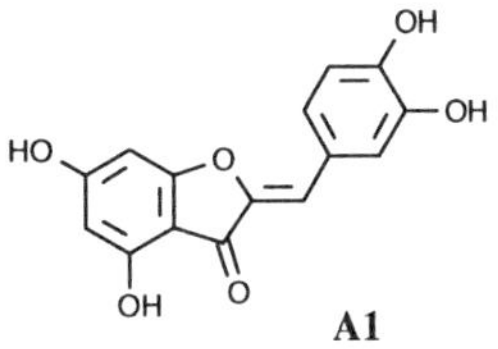 A1 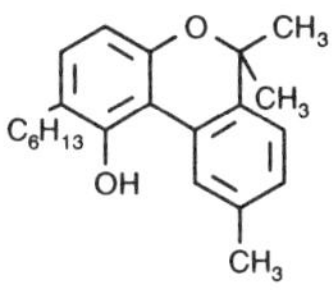 A2

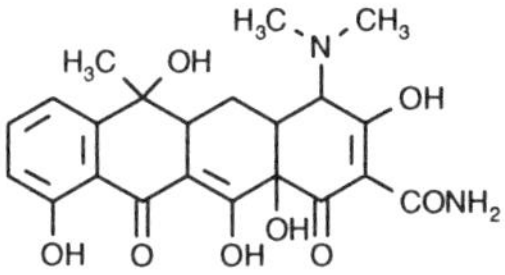

A3 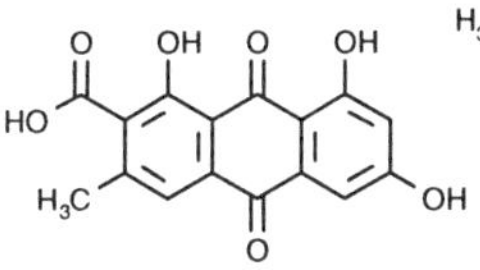 A4 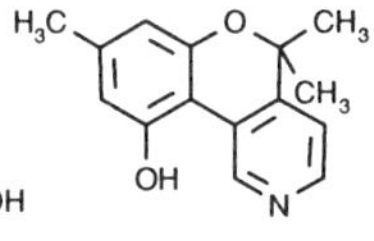 A5

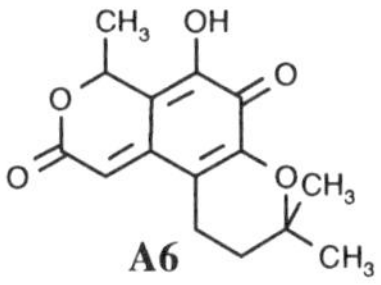

A6 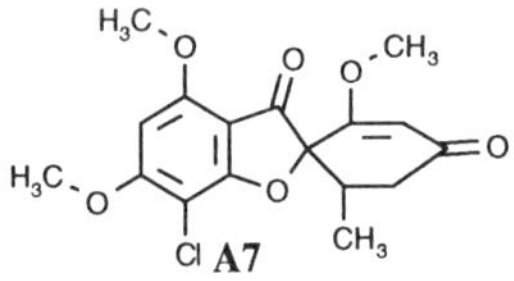 A7 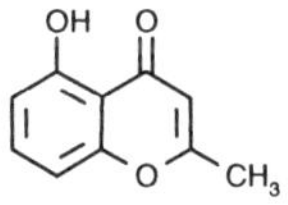 A8

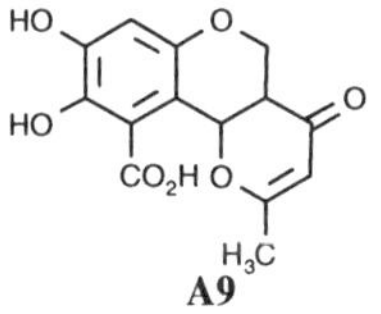

A9 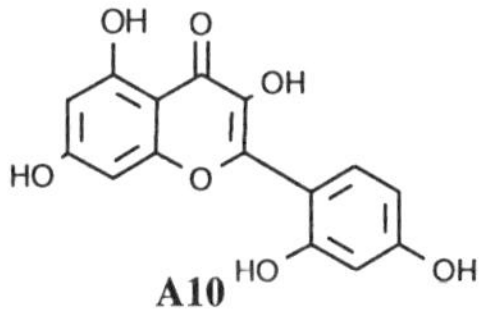 A10 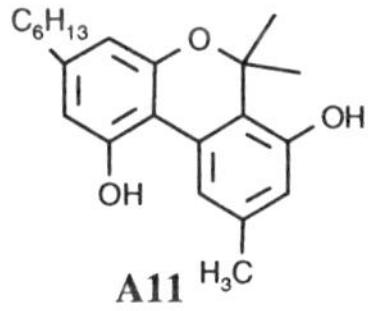 A11

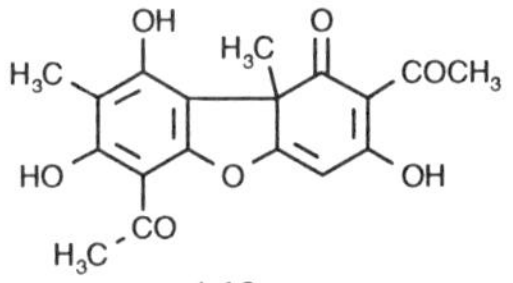

A12 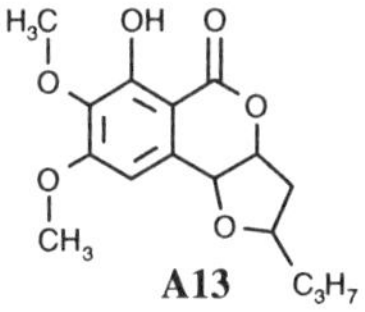 A13 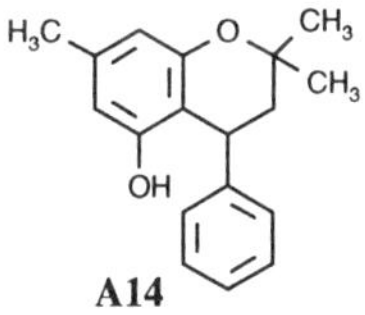 A14

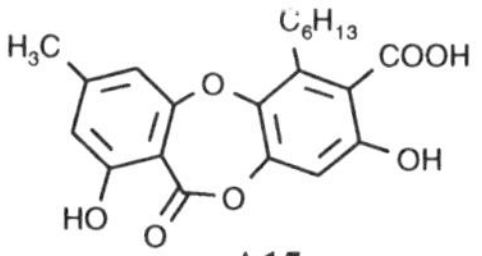

A15 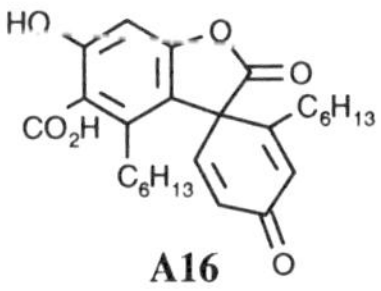 A16 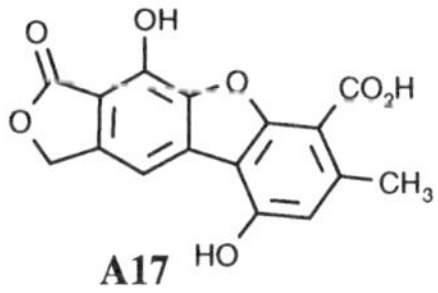 A17

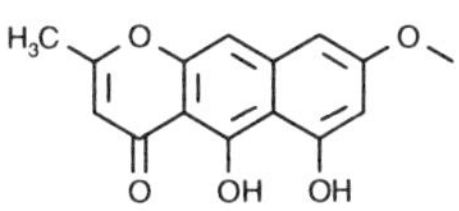

A18 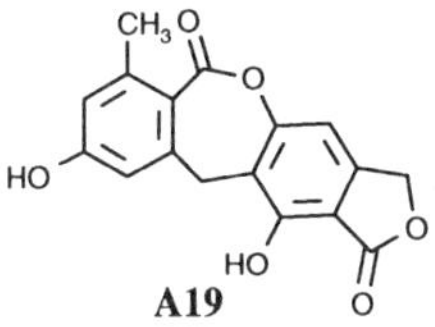 A19

Figure 3. Structures in set 2

2.3 Other examples

After the previous checks on the coherence of the procedure when comparing simple and similar compounds, we analyse a set containing very different compounds with the aim of completing the discussion. This set is sketched in Figures 4 and 5; it contains 13 structures chosen in a list of organic compounds that cover many structural variations. The size of the compounds is most often the same, thus reducing the influence of size-dependency on the similarity values.

In Table 3 are reported the values of the similarity indexes calculated using equation 2; they range from 0.064 to 0.494 and are representative of many different situations. There are 11 indexes that are initially equal to 0, whereas the greatest value for an initial index is 0.489. The most interesting result is that, after all the transformations, no index is still null showing that the procedure is capable of calculating an index for all the pairwise comparisons. The goodness of the calculated values is, however, a different issue. As usual it is not easy to affirm that two compounds are more similar than other two; here, in addition, we are comparing compounds that are enough different and their similarity is extremely difficult to be assessed.

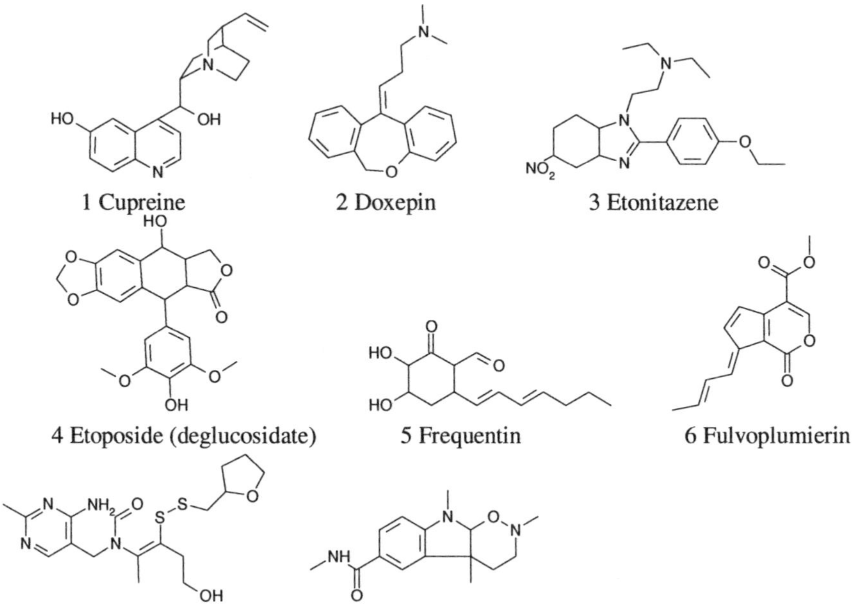

1 Cupreine 2 Doxepin 3 Etonitazene

4 Etoposide (deglucosidate) 5 Frequentin 6 Fulvoplumierin

7 Fursultiamine 8 Geneserine

Figure 4. Structures in set 3; part I

Nevertheless we will try to critically look at the numbers.

First, we are going to discuss the absolute values. The most similar pair of compounds is the 6-9 pair (0.494, they have as a common substructure an α,β,γ,δ-unsaturated carbonyl), whereas the most dissimilar pairs are the 3-10 (0.064, they have no similar substructures) and 10-11 (0.116) pairs; this result is overall quite reasonable. In addition, for each compound it is possible to select its most similar / dissimilar partner: 1-2 / 1-6; 1-2 / 2-10; 3-2 / 3-10; 4-2 / 4-5; 5-6 / 5-8; 6-9 / 6-1; 7-13 / 7-10; 8-13 / 8-5; 9-6 / 9-3; 10-12 / 10-3; 11-13 / 11-10; 12-6 / 12-7; 13-11 / 13-10. The result is qualitatively very interesting because looking at the corresponding structure pairs we have the impression that the selection has been very good.

A second comment that we can do concerns the selection of the most and the least representative compounds in the set and we are going to use the index mean as a measure; they are compounds 13 (0.394) and 10 (0.214), respectively. 10 seems to be particularly well chosen as the least representative, but both of them have mean value well apart from the others.

A third point that we can discuss concerns the influence of the transformations on the index values. In Tables 4-12 are reported the values of the initial indexes (i.e. before doing any transformation), of the final indexes (i.e. at the end of the transformation), of the calculated indexes (i.e. the initial index modified by the contribution of the transformations), and of the percent influence of the modification.

Table 3. Calculated similarity indexes of compounds in set 3

1-2	1-3	1-4	1-5	1-6	1-7	1-8	1-9	1-10	1-11	1-12	1-13	2-3	2-4
0.409	0.329	0.344	0.304	0.138	0.261	0.305	0.203	0.189	0.326	0.244	0.335	0.398	0.400
2-5	2-6	2-7	2-8	2-9	2-10	2-11	2-12	2-13	3-4	3-5	3-6	3-7	3-8
0.242	0.297	0.285	0.319	0.136	0.131	0.391	0.354	0.372	0.281	0.190	0.152	0.268	0.269
3-9	3-10	3-11	3-12	3-13	4-5	4-6	4-7	4-8	4-9	4-10	4-11	4-12	4-13
0.122	0.064	0.372	0.228	0.342	0.221	0.228	0.280	0.268	0.332	0.264	0.232	0.304	0.426
5-6	5-7	5-8	5-9	5-10	5-11	5-12	5-13	6-7	6-8	6-9	6-10	6-11	6-12
0.389	0.245	0.128	0.288	0.250	0.315	0.356	0.388	0.239	0.239	0.494	0.317	0.307	0.420
6-13	7-8	7-9	7-10	7-11	7-12	7-13	8-9	8-10	8-11	8-12	8-13	9-10	9-11
0.382	0.320	0.198	0.143	0.310	0.207	0.356	0.206	0.255	0.282	0.238	0.385	0.319	0.241
9-12	9-13	10-11	10-12	10-13	11-12	11-13	12-13						
0.257	0.316	0.116	0.330	0.187	0.217	0.460	0.242						
M1[a]	M2	M3	M4	M5	M6	M7	M8	M9	M10	M11	M12	M13	
0.282	0.311	0.251	0.298	0.276	0.300	0.259	0.268	0.259	0.214	0.297	0.283	0.349	

[a] Indexes in this row are the calculated means.

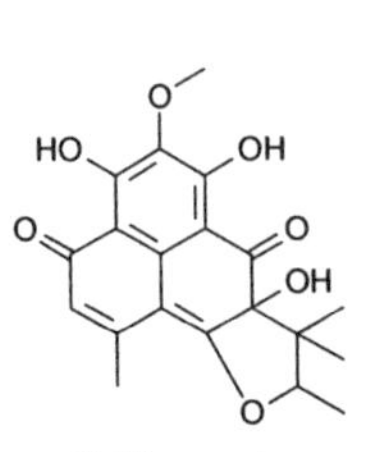

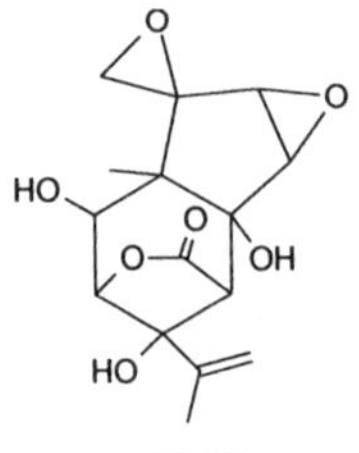

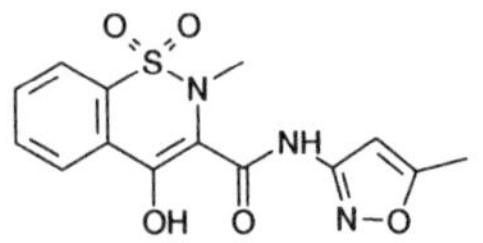

9 Herqueinone

10 Hyenanchin

11 Isoxicam

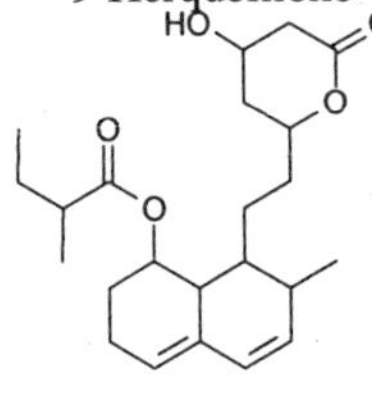

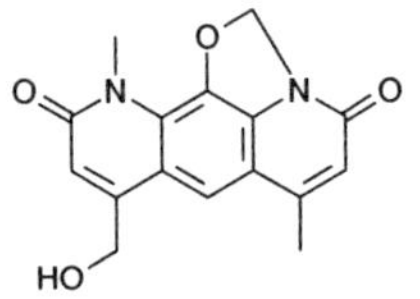

12 Mevastin

13 Nybomycin

Figure 5. Structures in set 3; part II

The influence of the modifications have very different weight; it can represent the whole index (when the initial similarity is null), or it can be null (when the initial and the final indexes are equal), or it can differently change the index (between 1.0 and 65.5 %). The changes are usually greater when the initial similarity is small, but this is not always the case (e.g. compare 6-12 and 3-12).

Table 4. Initial, final, and calculated indexes and their percent variation

	1-2	1-3	1-4	1-5	1-6	1-7	1-8	1-9	1-10	1-11	1-12	1-13
initial	0.409	0.263	0.344	0.276	0	0.245	0.305	0.12	0.189	0.261	0.196	0.267
final	0.409	0.374	0.344	0.457	0.337	0.285	0.305	0.324	0.212	0.39	0.292	0.45
SI_f	0.409	0.329	0.344	0.304	0.138	0.261	0.305	0.203	0.189	0.326	0.244	0.335
Δ	0	0.066	0	0.028	0.138	0.016	0	0.083	0	0.065	0.048	0.068
Δ %	0	20.1	0	9.2	N/A	6.1	0	40.9	0	19.9	19.7	20.3

Table 5. Initial, final, and calculated indexes and their percent variation

	2-3	2-4	2-5	2-6	2-7	2-8	2-9	2-10	2-11	2-12	2-13
initial	0.385	0.4	0	0	0.285	0.319	0	0	0.391	0.122	0.372
final	0.41	0.4	0.511	0.648	0.298	0.319	0.364	0.41	0.403	0.609	0.381
SI_f	0.398	0.4	0.242	0.297	0.285	0.319	0.136	0.131	0.391	0.354	0.372
Δ	0.013	0	0.242	0.297	0	0	0.136	0.131	0	0.232	0
Δ %	3.3	0	N/A	N/A	0	0	N/A	N/A	0	65.5	0

Table 6. Initial, final, and calculated indexes and their percent variation

	3-4	3-5	3-6	3-7	3-8	3-9	3-10	3-11	3-12	3-13	14-15
initial	0.281	0	0	0.222	0.245	0	0	0.372	0.143	0.342	0
final	0.286	0.372	0.284	0.314	0.298	0.245	0.186	0.387	0.342	0.342	0.283
SI_f	0.281	0.19	0.152	0.268	0.269	0.122	0.064	0.372	0.228	0.342	0.283
Δ	0	0.19	0.152	0.046	0.024	0.122	0.064	0	0.085	0	0.283
Δ %	0	N/A	N/A	17.2	8.9	N/A	N/A	0	37.3	0	N/A

Table 7. Initial, final, and calculated indexes and their percent variation

	4-5	4-6	4-7	4-8	4-9	4-10	4-11	4-12	4-13
initial	0.17	0.17	0.244	0.268	0.143	0.196	0.196	0.246	0.335
final	0.311	0.452	0.322	0.268	0.667	0.372	0.292	0.404	0.526
SI_f	0.221	0.228	0.28	0.268	0.332	0.264	0.232	0.304	0.426
Δ	0.051	0.058	0.036	0	0.189	0.068	0.036	0.058	0.091
Δ %	23.1	25.4	12.9	0	56.9	25.8	15.5	19.1	21.4

Table 8. Initial, final, and calculated indexes and their percent variation

	5-6	5-7	5-8	5-9	5-10	5-11	5-12	5-13
initial	0.278	0.193	0	0.285	0.25	0.195	0.356	0.283
final	0.516	0.3	0.256	0.422	0.278	0.424	0.367	0.492
SI_f	0.389	0.245	0.128	0.288	0.25	0.315	0.356	0.388
Δ	0.111	0.052	0.128	0.003	0	0.12	0	0.105
Δ %	28.5	21.2	N/A	1	0	38.1	0	27.1

The final values are always well greater than zero, sometimes the increment is in the order of 0.2-0.5, demonstrating that the structures are changed in the right direction.

Table 9. Initial, final, and calculated indexes and their percent variation

	6-7	6-8	6-9	6-10	6-11	6-12	6-13
initial	0.136	0.154	0.489	0.25	0.195	0.246	0.292
final	0.359	0.324	0.513	0.647	0.4	0.603	0.595
SI_f	0.239	0.239	0.494	0.317	0.307	0.42	0.383
Δ	0.103	0.085	0.005	0.067	0.112	0.174	0.091
Δ %	43.1	35.6	1	21.1	36.5	41.4	23.8

Table 10. Initial, final, and calculated indexes and their percent variation

	7-8	7-9	7-10	7-11	7-12	7-13
initial	0.298	0.113	0.125	0.294	0.185	0.356
final	0.335	0.283	0.244	0.337	0.28	0.356
SI_f	0.32	0.198	0.143	0.31	0.207	0.356
Δ	0.022	0.085	0.018	0.016	0.022	0
Δ %	6.9	42.9	12.6	5.2	10.6	0

Table 11. Initial, final, and calculated indexes and their percent variation

	8-9	8-10	8-11	8-12	8-13	9-10	9-11	9-12	9-13
initial	0.177	0.14	0.273	0.122	0.372	0.238	0.2	0.257	0.274
final	0.311	0.292	0.335	0.267	0.429	0.464	0.364	0.395	0.375
SI_f	0.206	0.255	0.282	0.238	0.385	0.319	0.241	0.257	0.316
Δ	0.029	0.111	0.009	0.116	0.013	0.081	0.041	0	0.042
Δ %	14.1	45.1	3.2	48.7	3.4	25.4	17	0	13.3

Table 12. Initial, final, and calculated indexes and their percent variation

	10-11	10-12	10-13	11-12	11-13	12-13
initial	0	0.305	0.136	0.157	0.444	0.2
final	0.244	0.397	0.27	0.331	0.486	0.327
SI_f	0.116	0.33	0.187	0.217	0.46	0.242
Δ	0.116	0.025	0.051	0.06	0.016	0.042
Δ %	N/A	7.6	27.3	27.6	3.5	17.4

The calculation permits some operations that would have not been possible, otherwise. For example, the pairs 2-5, 2-6, 2-9, and 2-10, can be ordered at the end of the procedure, operation that is impossible at the beginning because all the similarity indexes were null. A similar chance is present for the pairs 3-5, 3-6, 3-9, and 3-10. On the contrary, the pairs 6-7 and 6-8 have exactly the same calculated index, whilst the pairs 4-10 and 4-11 that had equal SIs at the beginning reach different end points.

In conclusion, we can affirm that the procedure gives the expected result, i.e. the possibility to calculate a similarity index even for very different compounds.

2.4 Structure modifications

As the final point in this discussion it can be interesting to look at the modifications that the structures suffer during the procedure application. To this end, we are going to use two structure pairs 14-15 and 1-6 (Figures 4 and 6).

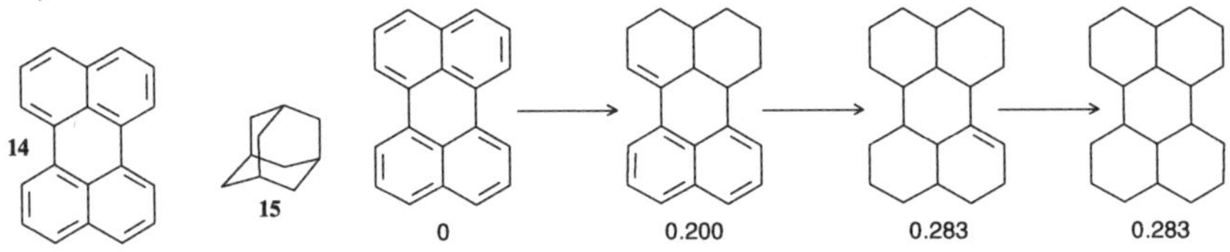

14 15 0 0.200 0.283 0.283

Figure 6. Structures 14 and 15 and their comparison scheme

The first pair is formed by a completely aromatic and completely saturated compounds. Their initial index is null. The modifications concern compound 14 because it is the only one that has functional groups. After four steps the index has become equal to 0.200 and after five steps more it becomes equal to 0.283. It is worth noting that the theoretical maximum value would be 0.667. The calculated index is equal to 0.110 and it is in agreement with expectations: the two structures are not very similar, indeed.

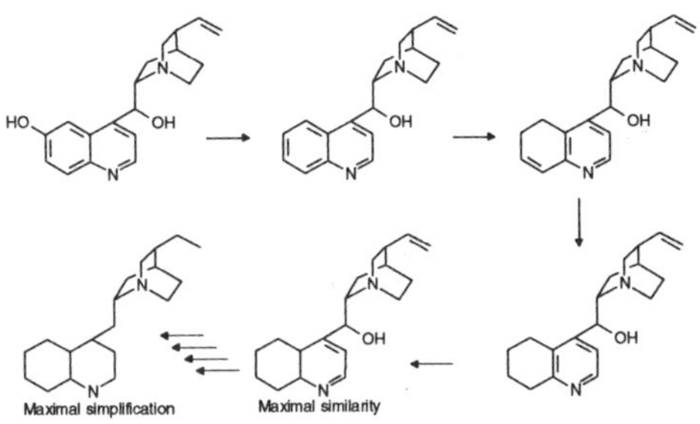

HO OH OH OH OH OH

Maximal simplification Maximal similarity

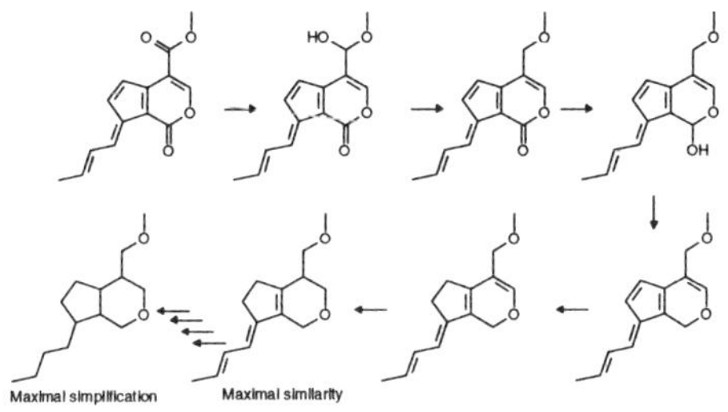

Maximal simplification Maximal similarity

Figure 7. The transformation of compounds 1 and 6. Two different end points can be identified

The second pair is part of the previously discussed set. In this case, it is interesting to note that the final structures have still some functional groups that remain unchanged because they are either part of the similar substructures, or part of the carbon skeleton. Compounds 1 and 6 reach two diverse final points: the point of maximal similarity and the point of maximal simplification (Figure 7). The final index will take into account only the first point because all the subsequent modifications are unimportant for what concerns the structure comparison. For the sake of completeness, we also analysed all the pairs in the third set changing the numbering scheme. The calculated indexes were always very close to those reported here with changes at the third digit. In addition, we artificially modified the pair C-E (Figure 2) in two different orders as shown in Fig. 8. The calculated SI_f value is always equal to 0.648 despite the different number of modifications.

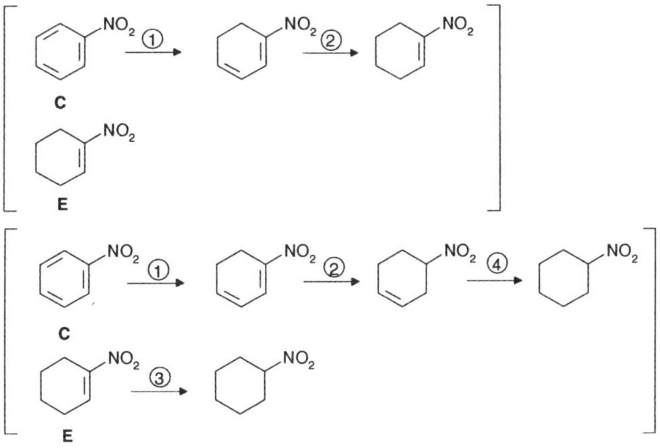

Figure 8. Two diverse ways to compare compounds C and E

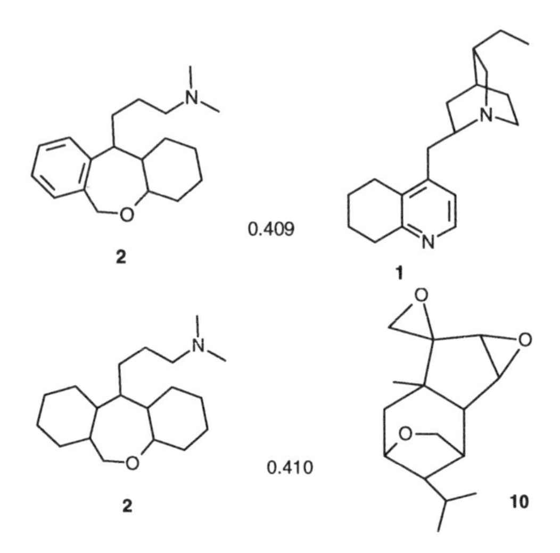

Figure 9. Comparisons of compounds 1 and 2, and 2 and 10. Compound 2 is differently modified depending on the partner

Another point is the different level of modification that a structure undergoes depending on the comparison. For example, we can look at the two pairs 1-2 and 2-10 (Fig. 9); compound 2 is differently modified because the initial similarity sets are different. Even if at the end of the transformations the two comparisons give two calculated similarity indexes in the same range (0.409 and 0.410), in the first case the similarity is present since the beginning, whilst in the second case it has been reached through many passages and this fact is reflected by the SI_f index values (0.409 and 0.131).

3. CONCLUSION

The presented methodology for measuring similarity between dissimilar compounds would like to contribute to the achievements that the use of similarity permitted in the field of molecule comparison. Even taking into account the possibility that the methodology couldn't completely solve the problem, we think that it represents, together with our previous approach, a strong challenge for everybody interested in assessing the very chance to compare compounds that are not similar with the aim of calculating a measure of their similarity in a congruent measuring space.

The application of the present approach has exactly the same limits as all other applications in the field of similarity; consequently, we are confident on the possibility of its profitable use.

ACKNOWLEDGEMENT

Partial financial support by the Consiglio Nazionale delle Ricerche, and by the Ministero dell'Universita' e della Ricerca Scientifica e Tecnologica, is gratefully acknowledged.

REFERENCES AND NOTES

1. Dedicated to professor E.J. Corey on occasion of his 70th birthday
2. *Concepts and Applications of Molecular Similarity*. M.A. Johnson, G.M. Maggiora, eds.: Wiley Interscience, (1990) , New York
3. *Molecular Similarity and Reactivity: from Quantum Chemical to Phenomenological Approaches*. R. Carbo', ed.: Kluwer Academic Publishers, (1995), Dordrecht
4. *Advances in Molecular Similarity*. R. Carbo'-Dorca, P.G. Mezey, eds.: JAI Press Inc., (1996), London
5. P.A. Bath, A.R. Poirrette, P. Willett, F.H. Allen, *J. Chem. Inf. Comput. Sci.* 1994, *34*, 141
6. P.N. Judson, *J. Chem. Inf. Comput. Sci.* 1994, *34*, 148
7. For some commercial packages using molecular similarity: a) G. Grethe, T.E. Moock, *J. Chem. Inf. Comput. Sci.* 1990, 30, 511. b) G. Grethe, W.D. Hounshell. In *Chem. Struct. 2 Proc. Int. Conf. 2nd 1990.* W.A. Warr, ed.: Springer-Verlag, (1993), Berlin, p.399
8. R. Benigni, C. Andreoli, A. Giuliani, *Environ. Mol. Mutagen.* 1994, *24*, 208
9. M.J. McGregor, P.V. Pallai, *J. Chem. Inf. Comput. Sci.* 1997, *37*, 443
10. D.K. Agrafiotis, *J. Chem. Inf. Comput. Sci.* 1997, *37*, 841
11. E.J. Martin, J.M. Blaney, M.A. Siani, D.C. Spellmeyer, A.K. Wong, W.H. Moos, *J. Med.Chem.* 1995, *38*, 1431
12. R.A. Lewis, J.S. Mason, I.M. McLay, *J. Chem. Inf. Comput. Sci.* 1997, *37*, 599
13. V.J. Gillet, P. Willet, J. Bradshaw, *J. Chem. Inf. Comput. Sci.* 1997, *37*, 731
14. Sello, G., *J. Chem. Inf. Comput. Sci.* 1998, *38*, 691
15. B. Leoni, G. Sello. In *Molecular Similarity and Reactivity: from Quantum Chemical to Phenomenological Approaches,* R. Carbo' ed.: Kluwer Academic Publishers, (1995), Dordrecht, p.267
16. L. Baumer, G. Sello, *J. Chem. Inf. Comput Sci.* 1992, *32*, 125
17. G. Sello, M. Termini. In *Advances in Molecular Similarity,* R. Carbo'-Dorca and P.G. Mezey eds.: JAI Press Inc., (1996), London, p. 213
18. J. Gasteiger, W.D. Ihlenfeldt, R. Fick, J.R. Rose, *J. Chem. Inf. Comput Sci.* 1992, *32*, 700

Chapter 7

Quantum-Mechanical Theory of Atoms in Molecules: A Relativistic Formulation

Jerzy Cioslowski
Department of Chemistry and CSIT, Florida State University, Tallahassee, FL 32306-3006, USA

Jacek Karwowski
Instytut Fizyki, Universytet Mikołaja Kopernika, Grudziądzka 5, 87-100 Toruń, Poland

1. INTRODUCTION

Analysis of chemical and physical phenomena often calls for partitioning of electronic properties into contributions from atoms, bonds, and molecular fragments [1]. Such a partitioning requires a theoretical prescription for discerning atoms in molecules (AIMs) that is not provided by the conventional formulation of quantum mechanics. However, AIMs can be rigorously defined as open subsystems with properties that satisfy well-known quantum-mechanical relationships, such as the virial and Ehrenfest theorems. A nonrelativistic theory of AIMs that incorporates such a definition has been formulated by Bader [2]. Central to this theory is the generalization of Schwinger's principle of stationary action [3] that results in the identification of AIMs as unions of nuclei and disjoint domains in the Cartesian space (called atomic basins) bordered by surfaces of zero flux in the electron density gradient [2,4-9].

In this chapter, we arrive at a relativistic definition of AIMs by investigating the Hamilton's principle applied to the Lagrangian density of a system with a finite boundary. Although not as general as the principle of

Fundamentals of Molecular Similarity, Edited by Carbó-Dorca *et al.*
Kluwer Academic/Plenum Publishers, New York 2001

stationary action, such an approach does not only furnish a relativistic condition for the domain of the proper quantum system but also sheds new light upon the conceptual basis of the nonrelativistic theory of AIMs.

2. THEORY

The following notation is employed throughout the text. The spacetime coordinates are denoted by x_μ, $\mu = 1, 2, 3, 4$, where $x_4 \equiv ict$ and c is the velocity of light. The coordinates of the position vector $\vec{r}$ are written as x_j, $j = 1, 2, 3$. Since the wavefunction Ψ may in general consist of several components, the symbol $\Psi^\dagger$ rather than Ψ^* is used for its conjugate. The derivatives of the wavefunction (and those of its components) are denoted by appropriate subscripts, namely

$$\Psi_\mu \equiv \frac{\partial\Psi}{\partial x_\mu}, \quad \Psi_j \equiv \frac{\partial\Psi}{\partial x_j}, \quad \Psi_t \equiv \frac{\partial\Psi}{\partial t} . \tag{1}$$

Summation over repeated indices is implied (the Einstein's convention). Although the presence of only one electron is assumed for the sake of simplicity, a generalization to a many-electron case is straightforward.

2.1 Hamilton's Principle

The Hamilton's principle, which greatly facilitates theoretical analysis of diverse natural phenomena, may be formulated in terms of the Lagrangian density L and the action integral W [10]. The Lagrangian density, which contains complete information about the physical system and the theoretical model employed in its description, depends upon the field amplitude $\Psi(\vec{r},t)$ and its space and time derivatives. In the case of a molecule with clamped nuclei, $\Psi(\vec{r},t)$ is simply its electronic wavefunction. The action integral pertaining to a system that is contained within the volume $\Omega(t)$ and evolves between two fixed time points $t_1 < t_2$ is given by

$$W[\Psi] = \int_{t_1}^{t_2} dt \int_{\Omega(t)} \mathrm{L}\, d^3x . \tag{2}$$

Equations of motion (EOMs) that describe the temporal evolution of an isolated system follow from the Hamilton's principle, according to which [10]

$$\delta\, W[\Psi]=0 \tag{3}$$

for all variations $\delta\,\Psi$ such that

$$\delta\,\Psi(\vec{r},t_1)\Big|_{\vec{r}\in\Omega}=\delta\,\Psi(\vec{r},t_2)\Big|_{\vec{r}\in\Omega}=\delta\,\Psi^\dagger(\vec{r},t_1)\Big|_{\vec{r}\in\Omega}=\delta\,\Psi^\dagger(\vec{r},t_2)\Big|_{\vec{r}\in\Omega}=0\,. \tag{4}$$

The condition (3) establishes several relations between the form of L and the structure of the respective EOMs:

- First, if the EOMs are to be linear then L has to be a bilinear function of Ψ and its derivatives.
- Second, if the EOMs are to be Galileo- or Lorentz-invariant then L has to transform accordingly.
- Third, the maximum degree of derivatives of Ψ in L cannot be higher than the degree of the corresponding EOMs.

In light of the above discussion, one may assume without any loss of generality that L depends only on Ψ, $\Psi^\dagger$, and their first derivatives, i.e.

$$\mathrm{L}=\mathrm{L}\left(\Psi,\Psi_\mu,\Psi^\dagger,\Psi^\dagger_\mu\right). \tag{5}$$

For such L, the variation of W is given by

$$\delta\, W[\Psi]=\int_{t_1}^{t_2}dt\int_{\Omega(t)}\delta\,\mathrm{L}\,d^3x+\int_{t_1}^{t_2}dt\int_{\delta\Omega(t)}\mathrm{L}\,d^3x\,, \tag{6}$$

where

$$\delta\,\mathrm{L}=\frac{\partial \mathrm{L}}{\partial\Psi}\delta\,\Psi+\frac{\partial \mathrm{L}}{\partial\Psi_\mu}\delta\,\Psi_\mu+\delta\,\Psi^\dagger\frac{\partial \mathrm{L}}{\partial\Psi^\dagger}+\delta\,\Psi^\dagger_\mu\frac{\partial \mathrm{L}}{\partial\Psi^\dagger_\mu}\,.$$

Since

$$\frac{\partial \mathrm{L}}{\partial\Psi_\mu}\delta\,\Psi_\mu=\frac{\partial}{\partial x_\mu}\left(\frac{\partial \mathrm{L}}{\partial\Psi_\mu}\delta\,\Psi\right)-\frac{\partial}{\partial x_\mu}\left(\frac{\partial \mathrm{L}}{\partial\Psi_\mu}\right)\delta\,\Psi\,, \tag{7}$$

and

$$\int_{t_1}^{t_2} dt \int_{\Omega(t)} \frac{\partial}{\partial x_\mu} \left(\frac{\partial \mathrm{L}}{\partial \Psi_\mu} \, \delta \, \Psi \right) d^3x =$$
$$\int_{t_1}^{t_2} dt \int_{\Omega(t)} \frac{\partial \mathrm{L}}{\partial \Psi_t} \, \delta \, \Psi \, \Bigg|_{t_1}^{t_2} d^3x + \int_{t_1}^{t_2} dt \int_{S(t)} \frac{\partial \mathrm{L}}{\partial \Psi_j} \, \delta \, \Psi \, ds_j \ , \tag{8}$$

where $S(t) \in \Omega(t)$ is the boundary of $\Omega(t)$ and $d\vec{s}$ is the oriented surface element of S, Eq. (6) may be rewritten as

$$\delta W[\Psi] =$$
$$\int_{t_1}^{t_2} dt \int_{\Omega(t)} \left[\left(\frac{\partial \mathrm{L}}{\partial \Psi} - \frac{\partial}{\partial x_\mu} \frac{\partial \mathrm{L}}{\partial \Psi_\mu} \right) \delta \, \Psi + \delta \, \Psi_\mu^\dagger \left(\frac{\partial}{\partial \Psi^\dagger} - \frac{\partial}{\partial x_\mu} \frac{\partial \mathrm{L}}{\partial \Psi_\mu^\dagger} \right) \right] d^3x$$
$$+ \int_{t_1}^{t_2} dt \int_{S(t)} \left(\frac{\partial \mathrm{L}}{\partial \Psi_j} \, \delta \, \Psi + \delta \, \Psi^\dagger \frac{\partial \mathrm{L}}{\partial \Psi_j^\dagger} \right) ds_j + \int_{t_1}^{t_2} dt \int_{\delta\Omega(t)} \mathrm{L} \, d^3x \tag{9}$$
$$+ \int_{\delta\Omega(t)} \left(\frac{\partial \mathrm{L}}{\partial \Psi_t} \, \delta \, \Psi + \delta \, \Psi^\dagger \frac{\partial \mathrm{L}}{\partial \Psi_t^\dagger} \right) \Bigg|_{t_1}^{t_2} d^3x \ .$$

Due to the boundary conditions imposed by Eqs. (4), the last integral in the rhs of Eq. (9) vanishes. In the case of an isolated system, where $\delta\Psi$ vanishes on S and $\delta\Omega = 0$, neither the second nor the third integral contributes to the rhs of Eq. (9), and the Hamilton's principle yields the following EOMs:

$$\frac{\partial \mathrm{L}}{\partial \Psi} - \frac{\partial}{\partial x_\mu} \frac{\partial \mathrm{L}}{\partial \Psi_\mu} = 0 \quad \text{and} \quad \frac{\partial \mathrm{L}}{\partial \Psi^\dagger} - \frac{\partial}{\partial x_\mu} \frac{\partial \mathrm{L}}{\partial \Psi_\mu^\dagger} = 0 \ . \tag{10}$$

Inspection of Eqs.(10) immediately leads to the conclusion that only Lagrangian densities with the structure

$$\mathrm{L}\left(\Psi, \Psi_\mu, \Psi^\dagger, \Psi_\mu^\dagger\right) = \frac{i\hbar}{2}\left(\Psi^\dagger \Psi_t - \Psi_t^\dagger \Psi\right) + \tilde{\mathrm{L}}\left(\Psi, \Psi_j, \Psi^\dagger, \Psi_j^\dagger\right) \tag{11}$$

are compatible with EOMs of the standard Hamiltonian form, i.e.

$$i\hbar \Psi_t = \hat{H} \Psi, \qquad \hat{H} = \hat{H}^\dagger \ . \tag{12}$$

For such L,

$$\hat{H}\Psi = \frac{\partial}{\partial x_j}\frac{\partial \tilde{L}}{\partial \Psi_j^\dagger} - \frac{\partial \tilde{L}}{\partial \Psi^\dagger}. \tag{13}$$

2.2 Definition of a Proper Quantum Subsystem

The domain $\overline{\Omega}(t) \in \Omega(t)$ of a proper quantum subsystem is defined by the requirement that the Hamilton's principle [Eq.(3)] carries over to the action integral [Eq.(2)] defined with $\overline{\Omega}$ rather than Ω [2,4-9]. Since EOMs hold for any $\vec{r} \in \overline{\Omega}$, this definition imposes the condition [compare Eq. (9)]

$$\int_{t_1}^{t_2} dt \left[\int_{\bar{S}(t)} \left(\frac{\partial L}{\partial \Psi_j} \delta\Psi + \delta\Psi^\dagger \frac{\partial L}{\partial \Psi_j^\dagger} \right) ds_j + \int_{\delta\bar{\Omega}(t)} L d^3x \right] = 0 , \tag{14}$$

upon the boundary: $S(t)$ of $\Omega(t)$.

In the nonrelativistic case, the Schrödinger Hamiltonian

$$\hat{H} = -\frac{\hbar^2}{2m}\hat{\nabla}^2 + \hat{V} \tag{15}$$

obtains from [compare Eq. (13)][1]

$$\tilde{L} = -\frac{\hbar^2}{2m}\Psi_j^\dagger \Psi_j - V\Psi^\dagger\Psi, \tag{16}$$

where V is the external potential and Ψ is a scalar (one-component) wavefunction [10]. For Ψ^0 and $\Psi^{0\dagger}$ that satisfy the corresponding EOMs,

$$L\left(\Psi^0, \Psi^0_\mu, \Psi^{0\dagger}, \Psi^{0\dagger}_\mu\right) = -\frac{\hbar}{4m}\nabla^2\left(\Psi^{0\dagger}\Psi^0\right) \tag{17}$$

and Eq. (14) becomes

$$-\frac{\hbar^2}{4m}\int_{t_1}^{t_2} \delta \left[\int_{\bar{\Omega}(t)} \nabla^2\left(\Psi^{0\dagger}\Psi^0\right) d^3x \right] dt = 0 . \tag{18}$$

Since

$$\int_{\bar{\Omega}(t)} \nabla^2 \left(\Psi^\dagger \Psi\right) d^3x = \int_{\bar{S}(t)} \vec{\nabla}\rho \cdot d\vec{s} = 0 \ , \tag{19}$$

where $\rho = \Psi^\dagger \Psi$ is the probability density, Eq.(18) is satisfied when

$$\vec{\nabla}\rho \cdot d\vec{s} = 0 \quad \text{for all} \quad \vec{r} \in \bar{S} \ . \tag{20}$$

Consequently, Eq. (20) constitutes the basis of the quantum-mechanical definition of AIMs as regions in space bordered by surfaces of zero flux in the electron density gradient [2,4-9].

2.3 Relativistic Generalization

In the case of an electron moving in the field of a stationary potential V, in the absence of a vector potential, the Dirac Hamiltonian of relativistic quantum mechanics,

$$\hat{H} = -ic\hbar\alpha_j \frac{\partial}{\partial x_j} + mc^2\beta + \hat{V} \ , \tag{21}$$

follows from [compare Eq. (13)][2]

$$\tilde{L} = \frac{ic\hbar}{2} \left(\Psi^\dagger \alpha_j \Psi_{,j} - \Psi^\dagger_{,j} \alpha_j \Psi\right) - \Psi^\dagger \left(mc^2\beta + V\right) \Psi \ , \tag{22}$$

where

$$\alpha_j = \begin{pmatrix} 0 & \sigma_j \\ \sigma_j & 0 \end{pmatrix}, \quad \beta = \begin{pmatrix} I & 0 \\ 0 & -I \end{pmatrix}, \tag{23}$$

σ_j are the 2×2 Pauli matrices, I is a 2×2 unit matrix and the wavefunction Ψ is a four-component spinor [10].

For Ψ^0 and $\Psi^{0\dagger}$ that satisfy the corresponding EOMs,

$$\mathrm{L}\left(\Psi^0, \Psi^0_\mu, \Psi^{0\dagger}, \Psi^{0\dagger}_\mu\right) = 0, \tag{24}$$

and Eq. (14) becomes an identity. In other words, $\tilde{\mathrm{L}}$ given by Eq. (22) does not impose any constraints upon the domain $\overline{\Omega}$ of a proper quantum subsystem. This apparent inconsistency with the nonrelativistic case discussed in the previous section of this chapter, to which the present result should reduce at the limit of $c \to \infty$, can be resolved as follows: Let

$$\Psi = \begin{pmatrix} G \\ F \end{pmatrix}, \tag{25}$$

where G and F are, respectively, the large and small components of Ψ. In this representation, Eqs. (11) and (22) yield

$$\mathrm{L} = \frac{i\hbar}{2}\left(G^\dagger G_t + F^\dagger F_t - G_t^\dagger G - F_t^\dagger F\right) + \tilde{\mathrm{L}}, \tag{26}$$

where

$$\begin{aligned} \tilde{\mathrm{L}} = {} & \frac{ic\hbar}{2}\left(G^\dagger \sigma_j F_j + F^\dagger \sigma_j G_j - G_j^\dagger \sigma_j F - F_j^\dagger \sigma_j G\right) \\ & - G^\dagger G\left(mc^2 + V\right) + F^\dagger F\left(mc^2 - V\right) . \end{aligned} \tag{27}$$

The corresponding EOMs for G and F read

$$\begin{aligned} i\hbar G_t &= -ic\hbar\sigma_j F_j + \left(mc^2 + V\right) G , \\ i\hbar F_t &= -ic\hbar\sigma_j G_j - \left(mc^2 - V\right) F . \end{aligned} \tag{28}$$

As the nonrelativistic limit is approached, $|F| << |G|$ and $|E - mc^2| << mc^2$, where E is the energy of a stationary state. Consequently, at this limit

$$F_t = \frac{E}{i\hbar} F \to \frac{mc^2}{i\hbar} F , \tag{29}$$

$$\frac{i\hbar}{2}\left(F^\dagger F_t - F_t^\dagger F\right) \to mc^2 F^\dagger F , \tag{30}$$

and

$$\left(mc^2+V\right) F^\dagger F \to mc^2 F^\dagger F \,. \tag{31}$$

After substitution into Eqs. (26) and (27), these asymptotic expressions produce

$$\begin{aligned} \mathrm{L} \to & \frac{i\hbar}{2} \left(G^\dagger G_t - G_t^\dagger G\right) \\ & + \frac{ic\hbar}{2} \left(G^\dagger \sigma_j F_j + F^\dagger \sigma_j G_j - G_j^\dagger \sigma_j F - F_j^\dagger \sigma_j G\right) \\ & - G^\dagger G \left(mc^2+V\right) + 2mc^2 F^\dagger F \,. \end{aligned} \tag{32}$$

The corresponding EOMs,

$$i\hbar G_t = -ic\hbar \sigma_j F_j + \left(mc^2+V\right) G, \quad 0 = ic\hbar \sigma_j G_j + 2mc^2 F \,, \tag{33}$$

are known as the Levy-Leblond equations [11]. Elimination of the small component from the first of these equations affords an EOM for the large component that reads

$$i\hbar G_t = -\frac{\hbar^2}{2m} \nabla^2 G + \left(mc^2+V\right) G \,. \tag{34}$$

By the same token, the nonrelativistic Lagrangian density

$$\mathrm{L} = \frac{i\hbar}{2} \left(G^\dagger G_t - G_t^\dagger G\right) - \frac{\hbar^2}{2m} G_j^\dagger G_j - \left(mc^2+V\right) G^\dagger G + \frac{\hbar^2}{4m} \nabla^2 \left(G^\dagger G\right) \tag{35}$$

is obtained with the help of the identities

$$(\vec{\sigma}\, \nabla)\, (\vec{\sigma}\, \nabla)\, G = \nabla^2 G \,, \tag{36}$$

$$\left[\nabla (\vec{\sigma}\, \nabla G)^\dagger\right] \vec{\sigma}\, G = \left(\nabla^2 G^\dagger\right) G \,, \tag{37}$$

and

$$(\vec{\sigma}\,\nabla G)^{\dagger}(\vec{\sigma}\,\nabla)\,G = (\nabla G)^{\dagger}\,\nabla G + i\vec{\sigma}\left[(\nabla G)^{\dagger}\times(\nabla G)\right]. \tag{38}$$

This Lagrangian density differs from its nonrelativistic counterpart given by Eqs. (11) and (16) by a trivial energy shift of $-mc^2$ and the presence of an expression that involves $\nabla^2(G^{\dagger}G)$. This term does not contribute to the action integral of an isolated system and thus does not alter the nonrelativistic EOMs. It cancels out the analogous contribution in Eq. (17), explaining the origin of the rhs of Eq. (24). Therefore, if the nonrelativistic equality (17) is to be recovered as $c \to \infty$, the relativistic $\tilde{\mathrm{L}}$ given by Eq. (22) has to be augmented with a heuristic term Υ that does not contribute to W of an isolated system, and reduces to the rhs of Eq. (17) in the case of a stationary state at the nonrelativistic limit.

In order not to contribute to W of an isolated system, Υ has to be of the form

$$\Upsilon = \frac{\partial J_{\mu}}{\partial x_{\mu}}, \tag{39}$$

where J_{μ} is a vector that either vanishes on S or is perpendicular to it [3]. If we restrict our choice to J_{μ} bilinear in Ψ and Ψ then either $J_{\mu} = \Psi\gamma_{\mu}\Psi$ or $J_{\mu} = \Psi\gamma_{\mu}\gamma_{\nu}\Psi a_{\nu}$, where $\gamma_j = -i\beta\alpha_j$, $\gamma_4 = \beta$, $\Psi = \Psi\gamma_4$, and a_{ν} is a Ψ-independent vector. The first of these two choices leads to Υ vanishing everywhere within Ω due to the continuity equation [10]. In the second choice, for a stationary state,

$$\Upsilon \sim \vec{\nabla}\left[\bar{\Psi}\left(\vec{\Sigma}\times\vec{a} - a_4\vec{\alpha}\right)\Psi\right], \tag{40}$$

where

$$\vec{\Sigma} = \begin{pmatrix} \vec{\sigma} & 0 \\ 0 & \vec{\sigma} \end{pmatrix}.$$

Since a_{ν} should not favor any particular direction in the coordinate space, the only choice is $a_{\nu}=(0, 0, 0, a_4)$. Then, setting $a_4 = i\dfrac{c\hbar}{2}a$, we get

$$\Upsilon = \frac{ic\hbar}{2}\vec{\nabla}\left(\Psi^{\dagger}\vec{\alpha}\ \beta\Psi a\right). \tag{41}$$

At the nonrelativistic limit Eq. (41) yields

$$\Upsilon = -\frac{\hbar^2}{4m}\nabla^2\left(G^{\dagger}Ga\right) \tag{42}$$

i.e., with $a = 1$, it is equal to the rhs of Eq. (17). Thus, since

$$\int_{\bar{\Omega}(t)}\left(\Psi^{\dagger}\vec{\alpha}\ \beta\ \Psi\right)d^3x = \int_{\bar{S}(t)}\left(\Psi^{\dagger}\vec{\alpha}\ \beta\ \Psi\right)\cdot d\vec{s} = 0\,, \tag{43}$$

the relativistic counterpart of the condition (20) reads

$$\left(\Psi^{\dagger}\vec{\alpha}\ \beta\ \Psi\right)\cdot d\vec{s} = 0 \quad \text{for all} \quad \vec{r}\in\bar{S}\,. \tag{44}$$

In other words, the relativistic AIMs are delineated by surfaces of zero flux in $\Psi^{\dagger}\alpha\beta\Psi$.

Let us note that, in principle, a may be chosen in a rather arbitrary way. In particular it may be set equal to the stationary nuclear potential V or to any other scalar function of the electronic coordinates[3]. If we allow for the second derivatives of Ψ in the Lagrangian density, then the most natural choice $J_\mu = -\frac{\hbar^2}{4m}\frac{\partial}{\partial x_\mu}\left(\bar{\Psi}\ \Psi\right)$ yields the same nonrelativistic limit as Eq. (41), however the relativistic AIMs would be defined by

$$\vec{\nabla}\left(\Psi^{\dagger}\beta\ \Psi\right)\cdot d\vec{s} = 0 \quad \text{for all} \quad \vec{r}\in\bar{S}\,. \tag{45}$$

rather than by the condition (44). Then, as we see, a "natural" definition of AIMs is far from being unique.

3. CONCLUSIONS

Careful analysis of the relationship between the relativistic Lagrangian density and its nonrelativistic counterpart elucidates certain new aspects of the quantum-mechanical theory of AIMs. In the nonrelativistic case, the partitioning of quantum system into proper subsystems follows from the properties of the Lagrangian density [Eq. (11)] with $\tilde{L}$ given by Eq. (16). Although this form of the Lagrangian density is the simplest one that yields the Schrödinger EOMs, it does not obtain at the nonrelativistic limit from the simplest L associated with the Dirac EOMs (which does not furnish constraints upon the domain $\overline{\Omega}$ of a proper quantum subsystem). Consequently, the relativistic definition of AIMs requires the addition of a heuristic term to the conventional Lagrangian density. The form of this term can be easily identified, yet such a definition of AIMs lacks unambiguous physical interpretation.

The findings of the present study are of a dual nature. On one hand, the relativistic generalization of the concept of AIMs is bound to aid quantum chemists in their understanding of electronic structures of molecules containing heavy elements. On the other hand, the arbitrariness in the expression for the Lagrangian density and its consequence to the theory of proper quantum subsystems raise questions about the uniqueness of a rigorous definition of AIMs.

ACKNOWLEDGEMENTS

This work was supported by the National Science Foundation under the grant CHE-9632706 and by the Polish KBN under the project No. 2 P03B 126 14. The authors are indebted to S. Dembinski for helpful discussions.

NOTES

[1] This is the simplest expression for L that yields the Schrödinger EOMs.

[2] This is the simplest expression for L that yields the Dirac EOMs.

[3] In the case of a clamped-nuclei molecule, the coordinate system is always connected with the fixed frame of the nuclei. Therefore the nuclear potential *V* may be considered as an external scalar potential rather than as the fourth component of the relativistic four-potential.

REFERENCES

1. J. Cioslowski, in *Encyclopedia of Computational Chemistry*, Wiley, New York (1998).
2. R. F. W. Bader, *Atoms in Molecules: A Quantum Theory*, Clarendon Press, Oxford (1990).
3. J. Schwinger, *Phys. Rev.*, **82**, 914 (1951).
4. S. Srebrenik and R. F. W. Bader, *J. Chem. Phys.*, **63**, 3945 (1975).
5. R. F. W. Bader, S. Srebrenik, and T. T. Nguyen-Dang, *J. Chem. Phys.*, **68**, 3680 (1978).
6. R. F. W. Bader, *J. Chem. Phys.*, **73**, 2871 (1980).

7. R. F. W. Bader and T. T. Nguyen-Dang, *Adv. Quantum Chem.*, **14**, 63 (1981).
8. R. F. W. Bader, *J. Chem. Phys.*, **91**, 6989 (1989).
9. R. F. W. Bader, *Phys. Rev.*, **B 49**, 13348 (1994).
10. J. J. Sakurai, *Advanced Quantum Mechanics*, Addison-Wesley, Redwood City (1987).
11. J. M. Levy-Leblond, *J. Commun. Math. Phys.*, **6**, 286 (1967).

Chapter 8

Topological Similarity of Molecules and the Consequences of the Holographic Electron Density Theorem, an Extension of the Hohenberg-Kohn Theorem

Paul G. Mezey

Institute for Advanced Study, Collegium Budapest, 1014 Budapest, I. Szentháromság u. 2, Hungary and Mathematical Chemistry Research Unit, Department of Chemistry and Department of Mathematics and Statistics. University of Saskatchewan, 110 Science Place, Saskatoon, SK, Canada, S7N 5C9. Tel. 1 306 966 4661; Fax 1 306 966 4730; e-mail mezey@sask.usask.ca

1. INTRODUCTION

The principle of "Geometrical Similarity as Topological Equivalence" (the GSTE Principle) has been proposed and applied to the problems of molecular similarity in several studies (see, e.g. refs. 1-3 and references therein). The basic idea behind this principle is simple: similarity between two geometrical constructions can be expressed as a topological equivalence within a suitable chosen topology.

Originally introduced within the geometrical context, the Resolution-Based Similarity Measures (RBSM) are also based on a simple idea: if it requires a higher resolution to distinguish two objects, than these two objects are more similar that two objects distinguishable already at a lower resolution [4]. Within the context of topology, topological resolution can be interpreted as a particular level within a hierarchy of weaker and stronger topologies introduced into the same space. By finding the weakest topology that already distinguishes a given pair of objects, a topological similarity measure can be defined [5]. This approach is a direct generalization of the geometrically defined Resolution-Based Similarity Measures [1,4], and is

Fundamentals of Molecular Similarity, Edited by Carbó-Dorca *et al.*
Kluwer Academic/Plenum Publishers, New York 2001

properly referred to as Topological Resolution-Based Similarity Measures [5].

In the field of molecular electron density computations and related fields of molecular modeling, several advances have been achieved in recent years, some of these in local representations of molecules and macromolecular systems. In particular, advances in macromolecular electron density computations have special importance in molecular engineering and pharmaceutical drug design. Due to these advances, it has become possible to compute high quality (in fact, *ab initio* quality) electron densities, as well as many other properties of macromolecules, even those containing thousands of atoms. For example, the new, *ab initio* quality quantum chemical computational techniques are applicable to proteins [2]. Using these new computational techniques, it is now possible to calculate the forces acting on individual atomic nuclei in various distorted conformation of a macromolecule [6,7], for example, to calculate the forces involved in the internal motions and the folding processes of proteins [2].

The fundamental approach leading to these developments is the Additive, Fuzzy Density Fragmentation (AFDF) scheme applied to the electron density clouds of molecules [2], where the individual fuzzy electron density fragments are generated from accurate, "custom-made" quantum chemical representations of smaller molecules. These smaller molecules, called "parent molecules", include a "coordination shell" representing correctly all the local interactions about each fuzzy density fragment within any desired finite range of the actual macromolecule. The earlier numerical method [8,9,10] is now replaced by an analytical technique based on macromolecular density matrices [6,7]

The study of the theoretical foundation of these approaches has lead to a somewhat surprising theoretical development concerning pieces of fuzzy, boundaryless electron density clouds of molecules. The so-called "Holographic Electron Density Theorem", proven recently, states that any nonzero-volume piece of a fuzzy, boundaryless molecular electron density cloud, in a non-degenerate ground state, contains the complete information about the complete electron density cloud, hence, about the entire molecule [11,12]. This theorem provides a stronger statement than the Hohenberg-Kohn Theorem [13] based on the complete electron density. According to the Holographic Electron Density Theorem, the complete electron density is not required; all molecular properties are already fully determined by any, nonzero volume piece of a non-degenerate ground state electron density cloud [11,12]. The Holographic Electron Density Theorem provides a new tool to the growing field of Density Functional Theory [14].

Accordingly, any small, positive-volume piece of a molecular electron density can serve in principle as a tool for establishing correlations with complex molecular behaviour, such as relative toxicities [15]. This result reconfirms and provides a theoretical foundation of the long-established practice of drawing conclusions concerning the global behaviour of molecules based on some local features of some of the functional groups within the molecule.

One should point out the fundamental differences between the local electron density pieces used in the holographic electron density theorem and those used in the AFDF computational techniques. On the one hand, the holographic electron density theorem establishes a relation between any finite, nonzero-volume electron density piece of some boundary surface and the complete, boundaryless, fuzzy molecular electron density. On the other hand, the local density pieces used in the new AFDF computational methods are themselves fuzzy, boundaryless fragment densities. One important connection between the two approaches is provided by the theorem, by giving a justification for the claim that the local interactions between fuzzy electron density fragments are properly represented by the AFDF methods. Within any, suitably chosen coordination shell selected within the parent molecules used in the generation of the fuzzy fragments, the local interactions are influencing the central part of the electron density cloud of the fragment.

One of the recent applications of local electron density analysis has involved the analysis of local chirality and its relation to measurable chiral effects [16].

2. GLOBAL AND LOCAL MOLECULAR PROPERTIES, GLOBAL AND LOCAL SIMILARITIES AND THEIR INTERRELATIONS

On molecular level, all properties of molecules are fully determined by the electron density charge cloud surrounding the atomic nuclei present in the molecule. A molecule contains only positively charged atomic nuclei and the negatively charged electron density cloud. Consequently, there is simply no other material present in a molecule that could possibly contain any additional information; the complete information about all molecular properties is present in the electron density cloud. Since the electronic charge cloud becomes dense in the vicinity of various nuclei to a different degree, it is possible to unambiguously identify each nucleus. As a result, the electronic charge cloud alone already contains all information about all properties of the molecule.

A more formal statement of essence of this idea is given by the celebrated Hohenberg-Kohn theorem [13], the fundamental relation of density functional theory [14]. The Hohenberg-Kohn theorem states that the non-degenerate ground state electron density fully determines the energy of the molecule [13]. It can be easily seen that all other ground state properties of the molecule are also determined by the electron density [14]; in fact, the Hohenberg-Kohn theorem implies that the molecular electron density determines all the essential properties of the molecular Hamiltonian.

This could hardly be otherwise: the molecular electron density fully determines the nuclear charges and locations, and besides nuclei, there is only the electron density cloud in a molecule. Clearly, the matter present within the object must fully identify the object, consequently, the electron density, by default, must carry all the information about the molecule.

Of course, one may try to associate molecular properties with local molecular regions. For example, it is commonly expected that molecules containing common functional groups, that is, local regions that are described by a common bonding pattern at the level of structural formulas, are exhibiting similar chemical properties, for example, they undergo similar chemical reactions. The electron density contributions of these local regions also carry information, and it is of some interest to study how much information is present in the electron density of the local regions as compared to the information content of the electron density of the complete molecule.

An intuitively plausible, yet from some perspective, a surprising relation holds between the corresponding local and global information contents, as expressed by a recently proven theorem, the Holographic Electron Density Theorem [11,12]. This theorem states the following: *Any small, positive volume local region of a complete, boundaryless non-degenerate electronic ground state electron density cloud of a molecule contains the complete information about the entire electron density.* This implies that any such local region of the electron density cloud completely determines all properties of the molecule [11,12]. The Holographic Electron Density Theorem represents a considerable improvement over the Hohenberg-Kohn theorem [13] that relies on the complete electron density of the molecule. In fact, there is no need for the complete electron density to determine the molecular energy and other properties; according to the Holographic Electron Density Theorem, already any small, nonzero volume piece of the electron density cloud is sufficient to determine all molecular properties [11,12].

Accordingly, local molecular regions, such as individual functional groups are always influenced in a fundamental way by the rest of the molecule. This influence is characteristic to such an extent that the induced local changes in the fragment fully identify the molecule that contains the fragment, that is, the electron density of the (possibly very small, but positive volume) fragment contains the entire molecular information.

The Holographic Electron Density Theorem has fundamental implications with respect to the interrelations between local and global symmetry and chirality [11,16], and also with respect to advanced molecular modeling [15]. In the study of a series of molecules with similar pharmacological activity, the holographic electron density theorem justifies the approach that relies on local molecular regions. Even if these local regions are not directly involved in the primary interactions triggering the given type of pharmacological activity, nevertheless, they must contain the complete molecular information. Consequently, (at least in principle) it should be possible to find correlations between the experimentally determined descriptors of the pharmacologically relevant biochemical processes and some of the local features of the electron density clouds and [15]. The similarities in the global biochemical behaviours of these molecules must be reflected in some similarities in their local electron density clouds [15].

3. SHAPE CHARACTERIZATION OF LOCAL MOLECULAR REGIONS

The Shape Group Methods have been originally introduced for the shape characterization of three-dimensional, fuzzy sets embedded in the ordinary 3D Euclidean space E^3, such as molecular electron density clouds. The same methodology is applicable to local molecular electron density components, as long as they are also represented as fuzzy objects. In the molecular context, the algebraic-topological Shape Groups are the homology groups of level sets of electron densities truncated according to some curvature or some other conditions. For a detailed review of the fundamentals and typical applications of Shape Groups the reader may consult refs. [1-3]; here only a brief summary of the essential aspects, relevant to the interplay of local and global electron density properties will be reviewed.

The molecular electron density function is denoted by $\rho(K,\mathbf{r})$, where K is the nuclear configuration specified by some nuclear coordinates, and $\mathbf{r}$ is the three-dimensional position vector. The function $\rho(K,\mathbf{r})$ can be represented by an infinite family of its level sets, called molecular isodensity contour (MIDCO) surfaces, denoted by G(K,a). Here K is the nuclear configuration, as before, and a is the density threshold that may take values from the $[0, \infty)$ interval.

A MIDCO level set G(K,a) is defined as:

$$G(K,a) = \{ \mathbf{r} : \rho(K,\mathbf{r}) = a \}. \tag{1}$$

In the most often employed application of the shape group method, the local curvatures of each MIDCO G(K,a) are compared to a range of reference curvatures b. This leads to the classification of all points **r** along each MIDCO G(K,a) according to the local curvatures. At a point **r** the MIDCO G(K,a) is either:

a) locally convex relative to the chosen reference curvature b; in this case we say that the point **r** belongs to a domain of type $D_2(b)$, or
b) locally of the saddle type relative to reference curvature b; in this case we say that the point **r** belongs to a domain of type $D_1(b)$, or
c) locally concave relative to curvature b; in this case we say that the point **r** belongs to a domain of type $D_0(b)$.

The index μ of each domain $D_\mu(b)$ is called its curvature type. For each curvature type μ relative to a reference curvature value b, the domains $D_\mu(b)$ of each MIDCO G(K,a) generate a pattern P(K,a,b) of domains on the MIDCO surface G(K,a). The topology of these patterns P(K,a,b) provides a shape description.

By selecting a curvature type μ relative to a reference curvature value b, we may assume that the corresponding domains $D_\mu(b)$ are removed from the MIDCO surface G(K,a), leading to a truncated object $G_\mu(K,a,b)$ with a certain set of holes.

The Shape Groups $H^k_\mu(K,a,b)$ of the molecule are the algebraic-topological homology groups of the truncated MIDCO surfaces $G_\mu(K,a,b)$.

For three-dimensional fuzzy objects, such as the molecular electron densities, there are three classes of Shape Groups: the zero-, one- and two-dimensional Shape Groups $H^k_\mu(K,a,b)$. In these notations, the dimension of the Shape Group is indicated by the index k.

The Shape Groups $H^k_\mu(K,a,b)$ are invariant within certain small ranges the nuclear configurations K, the electron density threshold values a, and reference curvature values b; as a result, for each molecule there are only a finite number of different shape groups $H^k_\mu(K,a,b)$ [1,2].

The Shape Groups are partially characterized by a set of integers, such as the ranks the respective homology groups which are called the Betti numbers and are denoted by $b^0_\mu(a,b)$, $b^1_\mu(a,b)$, and $b^2_\mu(a,b)$, for the zero-, one-, and two-dimensional Shape Groups, $H^0_\mu(a,b)$, $H^1_\mu(a,b)$, and $H^2_\mu(a,b)$, respectively. Using this partial characterization, a finite number of these integers generate a discretized representation of the shape of the fuzzy electron density cloud of the molecule.

Indeed, one of the simplest choices of discretization is offered by the zero-, one-, and two-dimensional Betti numbers, by $b^0_\mu(a,b)$, $b^1_\mu(a,b)$, and $b^2_\mu(a,b)$, that is, by the ranks of the zero-, one-, and two-dimensional Shape

Groups, $H^0_\mu(a,b)$, $H^1_\mu(a,b)$, and $H^2_\mu(a,b)$, respectively. The sets of Betti numbers determined for the entire range of MIDCOs G(K,a) of the molecule M and for the appropriate range of curvature parameters b, generate a set of numerical shape descriptors.

In most application to date, some computational convenience was achieved by having the actual reference curvatures b and density thresholds a sampled only at finite number, n_a and n_b values, respectively. Also for computational convenience, in most applications these thresholds are changed by constant increments in a logarithmic scale.

This results in a representation of the distribution of various values of Betti numbers $b^p_\mu(a,b)$ as a function of the density threshold a and curvature parameter b using discretized (a,b)-maps. The (a,b)-maps are represented numerically in terms of shape matrices $\mathbf{M}^{(a,b)}$ with elements encoding the values of Betti numbers. These shape matrices, regarded as numerical shape codes for the molecules, provide a concise shape representation.

The total number of elements in each shape code (shape matrix $\mathbf{M}^{(a,b)}$) is

$$t = n_a n_b. \tag{2}$$

Based on these shape codes $\mathbf{M}^{(a,b)}$, a simple shape-similarity measure is obtained. Let us assume that A and B refer to either two molecules, or to two fuzzy (AFDF) molecular fragments, then the quantity

$$m(A,B) = m[\mathbf{M}^{(a,b),A}, \mathbf{M}^{(a,b),B}] \tag{3}$$

is defined as the number of matches between corresponding elements of the two shape codes $\mathbf{M}^{(a,b),A}$ and $\mathbf{M}^{(a,b),B}$.

The quantity:

$$s(A,B) = m(A,B) \,/\, t = m[\mathbf{M}^{(a,b),A}, \mathbf{M}^{(a,b),B}] \,/\, t \tag{4}$$

defines a shape-similarity measure, where s(A,B), describes the degree of similarity between the global shape group representations of the two molecules or of the two fuzzy molecular electron density fragments A and B.

Due to the discretization process of the shape information, performed in the course of the computation of the topological shape codes, the numerical comparison of shape codes is a much simpler process than any direct comparisons of the fuzzy electron density clouds. By contrast, in most direct density comparisons of two molecules or two molecular fragments, an optimization, or at least a standardization of the mutual orientation and superposition of the two molecules or molecular fragments is required before a meaningful comparison. One advantage of the shape-similarity evaluation of two molecules or two molecular fragments based on their shape codes is the fact that no such optimum superposition is required.

4. THE ROLE OF TOPOLOGICAL RESOLUTION

Within a geometrical framework, resolution-based similarity measures introduced earlier [4] expressed a natural representation of the perceived level of similarity between objects by the measurable level of (geometrical) resolution that allows the detection of differences between the objects.

Within a topological framework, similarity measures based on topological resolution [5] are also expected to represent the perceived similarity by determining the level within a hierarchy of topologies, where this level is the minimum level that distinguishes the two objects.

One may formulate a hierarchy of topologies relying on subbases. (Note that, some of the fundamental topological concepts and tools used in this section are discussed in greater detail in ref. [1]). A simple relation between generating subbases and topologies is relevant to the approach of topological resolution. Consider two generating subbases S_1 and S_2 containing families of subsets from an underlying space X, where the inclusion relation

$$S_2 \supset S_1 \tag{5}$$

holds. As a consequence, the corresponding topologies generated by these two subbases are not only comparable, but also topology $\mathbf{T}_2$ is necessarily finer than topology $\mathbf{T}_1$:

$$\mathbf{T}_2 \supset \mathbf{T}_1. \tag{6}$$

If $\mathbf{T}$ is a family of topologies $\mathbf{T}_i$ in the underlying set X,

$$\mathbf{T} = \{\mathbf{T}_1, \mathbf{T}_2, \ldots \mathbf{T}_i, \mathbf{T}_{i+1}, \ldots\}, \tag{7}$$

where these topologies $\mathbf{T}_i$ are fully ordered by the stronger-weaker relation,

$$\mathbf{T}_{i+1} \supset \mathbf{T}_i \tag{8}$$

for any two indices i and i+1, then the corresponding topological spaces are also fully ordered, that is,

$$(X,\mathbf{T}_{i+1}) \supset (X,\mathbf{T}_i) \qquad (9)$$

for every index-pair i and i+1.

Consider the similarity analysis of molecules, and assume that these topologies describe some of the essential attributes of the molecular electron density cloud. The choice of family **T** of topologies is discriminative for the given problem, if from the ordered sequence of topological spaces

$$\ldots \supset (X,\mathbf{T}_{i+1}) \supset (X,\mathbf{T}_i) \ldots \supset (X,\mathbf{T}_2) \supset (X, \mathbf{T}_1) \qquad (10)$$

it is possible to select at least one that provides distinction among the molecules. Alternative choices of topologies can serve for selecting different levels of topological resolutions, either to increase or decrease discrimination.

The relative level of topological resolution is defined in terms of the finer-cruder relations, and relation (9) implies that the topological space $(X,\mathbf{T}_{i+1})$ gives a higher topological resolution than the topological space $(X,\mathbf{T}_i)$ for the *same* underlying space X.

The topological space $(X,\mathbf{T}_{i+1})$ describes more or at least as much detail of the underlying space X than the topological space $(X,\mathbf{T}_i)$, as implied by relation (9). If in addition to relation (9), the constraint

$$(X,\mathbf{T}_{i+1}) \neq (X,\mathbf{T}_i) \qquad (11)$$

also holds, that is, if the strict inclusion relation applies in (9), than the discrimination is stronger. In this case, one can find functions f that maps X onto itself and is $\mathbf{T}_i$ –continuous but not $\mathbf{T}_{i+1}$– continuous. Consequently, a topological description $(X,\mathbf{T}_{i+1})$ with a higher topological resolution provides more information, than a description $(X,\mathbf{T}_i)$ with a topological resolution of lower level, where the latter may not be fine enough to describe some details.

5. A GENERALIZATION OF THE GSTE PRINCIPLE: TOPOLOGICAL SIMILARITY AS WEAKER TOPOLOGICAL EQUIVALENCE (TSWTE)

One important characteristic of a hierarchy of topological spaces given by the sequence of inclusion relations (10) is the relative roles they play with respect to resolution. If in this series the metric topology is included as one with extreme topological resolution, then a mere similarity present at the level of the metric topology may correspond to a topological equivalence in a topological space with a weaker topology. In this context, the strongest topology that already provides a topological equivalence may be taken as representing the level of topological resolution describing the degree of similarity.

This approach can be generalized as follows. Instead of addressing the characterization of similarity at the level of extreme resolution, that in the case of a suitable metric corresponds to geometrical similarity, one may consider the problem of topological similarity that corresponds to topological equivalence at some lower level of the hierarchy of the sequence (10). If the task is the characterization of topological similarity objects within the underlying space X at some intermediate topological resolution represented by the level i of topological space $(X,\mathbf{T}_i)$, then one may determine the level j of the strongest topology $(X,\mathbf{T}_j)$ that already provides a topological equivalence for these objects, where the relation

$$(X,\mathbf{T}_i) \supset (X,\mathbf{T}_j) \qquad (12)$$

applies.

In fact, topological similarity at some level within the hierarchy becomes a weaker topological equivalence at some other level of the hierarchy with a weaker topology within the sequence. The strongest of these weak topologies represents a characteristic level for the given similarity. This approach, Topological Similarity as Weaker Topological Equivalence (TSWTE) becomes the ordinary Geometrical Similarity as Topological Equivalence (GSTE) approach if the topological space $(X,\mathbf{T}_i)$ happens to be one with $\mathbf{T}_i$ the metric topology suitable for the geometrical representation of the objects within the underlying space X. Consequently, the TSWTE characterization obtained is indeed a generalization of the GSTE approach.

6. CONCLUSIONS AND SUMMARY

This generalization of the Geometrical Similarity as Topological Equivalence (GSTE) approach to the Topological Similarity as Weaker Topological Equivalence (TSWTE) approach allows one to decide the level of relevant resolution at the outset, possibly ignoring unimportant detail, and then, by describing similarities in terms of the remaining, already significant detail, one can find the characteristic topological level at which equivalence is obtained. The difference $i - j$ in relation (12) is a simple indication of the degree of topological similarity. The local electron density descriptors themselves can be taken at various levels of resolution; in fact, local electron density contributions of various sizes may be taken to form defining subbases for a hierarchy of topologies. In this approach, finer or cruder details of the molecular electron density cloud may be chosen to reflect similarities, and the differences can be characterized by the Topological Similarity as Weaker Topological Equivalence (TSWTE) approach.

REFERENCES

1. Mezey, P.G. (1993) Shape in Chemistry: An Introduction to Molecular Shape and Topology, VCH Publishers, New York.
2. Mezey, P.G. (1997) Quantum Chemistry of Macromolecular Shape, Internat. Rev. Phys. Chem., **16**, 361-388.
3. Mezey, P.G. (1997) Shape in Quantum Chemistry. In Conceptual Trends in Quantum Chemistry, Vol. 3, Calais. J.-L.; Kryachko, E.S. (eds.), Kluwer Academic Publ., Dordrecht, The Netherlands, pp 519-550.
4. Mezey, P.G. (1991) The Degree of Similarity of Three-Dimensional Bodies; Applications to Molecular Shapes, J. Math. Chem., **7**, 39-49.
5. Mezey, P.G. (1999) Shape-Similarity Relations Based on Topological Resolution, J. Math. Chem., in press.
6. Mezey, P.G. (1995) Macromolecular Density Matrices and Electron Densities with Adjustable Nuclear Geometries, J. Math. Chem., **18**, 141-168.
7. Mezey, P.G. (1997) Quantum Similarity Measures and Löwdin's Transform for Approximate Density Matrices and Macromolecular Forces, Int. J. Quantum Chem., 63, 39-48.
8. Walker, P.D., and Mezey, P.G. (1994) Ab initio Quality Electron Densities for Proteins: A MEDLA Approach, J. Am. Chem. Soc., **116**, 12022-12032.
9. Walker, P.D., and Mezey, P.G. (1994) Realistic, Detailed Images of Proteins and Tertiary Structure Elements: Ab Initio Quality Electron Density Calculations for Bovine Insulin, Can J. Chem., **72**, 2531-2536.
10. Walker, P.D., and Mezey, P.G. (1995) A New Computational Microscope for Molecules: High Resolution MEDLA Images of Taxol and HIV-1 Protease, Using Additive Electron Density Fragmentation Principles and Fuzzy Set Methods, J. Math. Chem., **17**, 203-234.
11. Mezey, P.G. (1998) Generalized Chirality and Symmetry Deficiency, J. Math. Chem., **23**, 65-84.
12. Mezey, P.G. (1999) The Holographic Electron Density Theorem and Quantum Similarity Measures, Mol. Phys., **96**, 169-178.
13. Hohenberg, P., and Kohn, W. (1964) Inhomogeneous electron gas. Phys. Rev. **136**, B864-B871.
14. Levy, M. (1996) Elementary Concepts in Density Functional Theory. In Recent Developments and Applications of Modern Density Functional Theory, Theoretical and Computational Chemistry, Vol. 4; J.M. Seminario, Ed.; Elsevier Science B.V.: Amsterdam, pp 3 - 24.
15. Mezey, P.G. (1999) Holographic Electron Density Shape Theorem and Its Role in Drug Design and Toxicological Risk Assessment, J. Chem. Inf. Comput. Sci., **39**, 224-230.

16. Mezey, P.G., Ponec, R., Amat, L., and Carbo-Dorca, R. (1999) Quantum Similarity Approach to the Characterization of Molecular Chirality, Enantiomers, in press.

Chapter 9

Quantum Chemical Reactivity: Beyond the Study of Small Molecules

J. M. Bofill
Departament de Química Orgànica i Centre Especial de Recerca en Química Teòrica, Universitat de Barcelona, Martí i Franquès 1, E-08028 Barcelona, Catalunya, Spain. E-mail: jmbofill@qo.ub.es

J. M. Anglada
Institut d'Investigacions Químiques i Ambientals. C.I.D. - C.S.I.C., Jordi Girona Salgado 18 - 26, E-08034 Barcelona, Catalunya, Spain. E-mail: anglada@qteor.cid.csic.es

E. Besalú
Institut de Química Computacional i Departament de Química, Universitat de Girona, Campus de Montilivi, E-17071 Girona, Catalunya, Spain. E-mail: emili@iqc.udg.es

R. Crehuet
Institut d'Investigacions Químiques i Ambientals. C.I.D. - C.S.I.C., Jordi Girona Salgado 18 - 26, E-08034 Barcelona, Catalunya, Spain.

1. INTRODUCTION

It is well known the paramount importance attached to the quantum mechanical methods related to the transition states (TS) or minimal energy surfaces localisation. In this field, many efforts are done to apply such techniques to medium, and large, sized molecules. The main goal is to obtain molecular descriptions of such systems within the highest possible level of accuracy.

Fundamentals of Molecular Similarity, Edited by Carbó-Dorca *et al.*
Kluwer Academic/Plenum Publishers, New York 2001

This chapter tries to present a suitable framework scheme and thus has been structured into three parts. In the first one, a specific mathematical modelisation of the Bell-Evans-Polanyi principle (BEP) is proposed to locate approximate transition structures of elementary reactions. The main idea consists of representing the adiabatic energy surface associated to any chemical reaction as a combination of three quadratic energy surfaces: two of them associated to the reactants and the products and the third one to the crossing point energy. The obtained results, not only serve to approximate transition structures and the corresponding Hessian matrix, but also to analyse the transition in terms of the contribution weights of the reactants and products. In the second part, a special implementation of a Newton-Raphson method to locate TS is outlined for a large number of variables. Its description is based on a novel way to implement the Powell formula for updating Hessian matrices. Finally, the last part of this chapter describes how to apply the Gaussian wave packet (GWP) method to semiclassical dynamic studies. The full content of this work encompasses the recent interests to the authors related to the possibility to study molecular systems of medium or large size.

2. PREDICTION OF APPROXIMATE TRANSITION STATES BY THE BELL-EVANS-POLANYI PRINCIPLE

As stated in reference [1], useful chemical models to describe a TS have been described by several authors. In brief, a TS is nothing more than the minimum energy point of the intersection line defined by the potential energy surfaces (PES) associated to the reactants and products. It is common to define a parameter that, in some way, measures the similarity of the TS to reactants and products on a scale that goes from 0 to 1. Thus, in this context, a TS can be viewed as hybrid of the reactants and products structures.

2.1 Classical approach

Within the crude model described in the previous paragraph, the potential surface along the reaction coordinate is approximated by a linear combination of two harmonic potentials associated to the reactants and products [2]. Usually, the PES of the reactants and products are expanded quadratically around the stationary points associated to the corresponding equilibrium geometries. The TS occurs at the intersection minima of the two parabolas. This is the diabatic surface model of TS based on the intersection of two surfaces [3-8]. These ideas can be expressed mathematically using a Lagrangian function:

$$\begin{cases} L(\mathbf{q},\lambda) = w_R V_{RR}(\mathbf{q}) + w_P V_{PP}(\mathbf{q}) + \lambda\left(V_{RR}(\mathbf{q}) - V_{PP}(\mathbf{q})\right) \\ V_{RR}(\mathbf{q}) - V_{PP}(\mathbf{q}) = 0 \end{cases} \tag{1}$$

where $\mathbf{q}$ are the internal coordinates describing a reaction path leading from reactants to products, V_{RR} and V_{PP} are the harmonic PES for reactants and products, respectively, λ is a Lagrangian multiplier, and w_R and w_P are two weights such that $w_R + w_P = 1$.

Based on equation (1) many algorithms exist to locate TS in an approximated way. These are described in the work of Robb et al.[9,10], Jensen[11,12], Ruedenberg and Sun[13]. Some examples or applications were reported by Pross and Shaik[14-19], Warshel[20-26], and Kim and Hynes[27], among others.

From the equation (1) we obtain the extrema as a solution of the set of equations:

$$\begin{cases} \nabla_q L(\mathbf{q},\lambda) = w_R \mathbf{g}_R + w_P \mathbf{g}_P + \lambda(\mathbf{g}_R - \mathbf{g}_P) = 0 \\ \nabla_\lambda L(\mathbf{q},\lambda) = V_{RR}(\mathbf{q}) - V_{PP}(\mathbf{q}) = 0 \end{cases} \tag{2}$$

And from the above equation it is straightforward to have

$$\mathbf{g}_R = \frac{\lambda - w_P}{\lambda + w_R}\mathbf{g}_P = \xi \mathbf{g}_P \tag{3}$$

Since the crossing point should be near the TS, in this point, the two gradients $\mathbf{g}_R$ and $\mathbf{g}_P$ should be antiparallels, that is $\xi<0$. Using this characteristic and the fact that the condition $w_R + w_P = 1$ must be fulfilled, the domain of the Lagrangian multiplier can be found:

$$\xi = \frac{\lambda - w_P}{\lambda + w_R} = \frac{\lambda + w_R - 1}{\lambda + w_R} = 1 - \frac{1}{\lambda + w_R} < 0 \tag{4}$$

and

$$-w_R < \lambda < w_P. \tag{5}$$

Moreover, ξ can be transformed in a useful reaction coordinate: since $\xi=0$ at $\mathbf{q}_R$ and $\xi=-\infty$ at $\mathbf{q}_P$, we define

$$0 < rc = \frac{\xi}{\xi - 1} < 1 \tag{6}$$

and the value of 0 can be associated to the reactants while the value of rc=1 to the products.

2.2 The extended PES description and formalism

The hybrid TS model described, sometimes presents unbalanced or inappropriate bond breaking-bond formation characteristics. This is because this model doesn't the resonance energy terms take into account. One goal of the approach presented here is to introduce the resonance effects in some way to improve the quality of the TS description and the appropriate bond breaking and bond formation features. The approach proposed here will consider this characteristic in an approximate and computationally inexpensive way [1].

In the standard BEP model, the PES attached wave function is obtained from the superposition of two state electronic wave functions:

$$\Psi_{PES} = c_R \Phi_R + c_P \Phi_P \tag{7}$$

where Φ_R and Φ_P are the wave functions that describe the electronic structure of the reactants and the products, respectively. In this context, the ground state Born-Oppenheimer PES, i.e. the adiabatic PES, corresponds to the lower root of a 2x2 secular equation, V_{adi}:

$$V_{adi}(\mathbf{q}) = \frac{V_{RR}(\mathbf{q}) + V_{PP}(\mathbf{q})}{2} - \left[\left(\frac{V_{RR}(\mathbf{q}) - V_{PP}(\mathbf{q})}{2}\right)^2 + V_{RP}^2(\mathbf{q})\right]^{1/2}, \tag{8}$$

where the matrix elements are defined as

$$\begin{cases} V_{RR}(\mathbf{q}) = \langle \Phi_R | H_{ele} | \Phi_R \rangle \\ V_{PP}(\mathbf{q}) = \langle \Phi_P | H_{ele} | \Phi_P \rangle \\ V_{RP}(\mathbf{q}) = V_{PR}(\mathbf{q}) = \langle \Phi_R | H_{ele} | \Phi_P \rangle \end{cases} \tag{9}$$

and H_{ele} is the electronic Hamiltonian that depends on the nuclear coordinates. Consequently, V_{RR}, V_{PP}, V_{RP} and V_{adi} also depend on the set of nuclear coordinates, as clearly shown in the notation.

A possibility is to change the Ψ_{PES} given in equation (7) by using the following more general expression

$$\Psi_{PES} = c_R\Phi_R + c_P\Phi_P + c_I\Phi_I \tag{10}$$

where Φ_I represents a set of intermediate (typically ionic) configurations. This expression can be understood as an extension of generalization of the Pross and Shaik model [14-19]. Using the wave function of equation (10) one obtains a 3x3 secular equation. The questions which arises are how to recover the BEP description of a PES and how to introduce the intermediate configurations in the TS region using the BEP principle. A possible answer is found when working with Löwdin's partitioning technique [28,29]. In this case it will be demonstrated how to recast again a 2x2 secular equation as required by the BEP model.

The secular equation resulting of the wave function proposed in equation (10) is

$$\begin{pmatrix} \langle\Phi_R|H_{ele}|\Phi_R\rangle & \langle\Phi_R|H_{ele}|\Phi_P\rangle & \langle\Phi_R|H_{ele}|\Phi_I\rangle \\ \langle\Phi_P|H_{ele}|\Phi_R\rangle & \langle\Phi_P|H_{ele}|\Phi_P\rangle & \langle\Phi_P|H_{ele}|\Phi_I\rangle \\ \langle\Phi_I|H_{ele}|\Phi_R\rangle & \langle\Phi_I|H_{ele}|\Phi_P\rangle & \langle\Phi_I|H_{ele}|\Phi_I\rangle \end{pmatrix}\begin{pmatrix} c_R \\ c_P \\ c_I \end{pmatrix} = V_{adi}\begin{pmatrix} c_R \\ c_P \\ c_I \end{pmatrix} \tag{11}$$

and following Löwdin, it can be partitioned in the following way

$$\begin{cases} \begin{pmatrix} \langle\Phi_R|H_{ele}|\Phi_R\rangle & \langle\Phi_R|H_{ele}|\Phi_P\rangle \\ \langle\Phi_P|H_{ele}|\Phi_R\rangle & \langle\Phi_P|H_{ele}|\Phi_P\rangle \end{pmatrix}\begin{pmatrix} c_R \\ c_P \end{pmatrix} + \begin{pmatrix} \langle\Phi_R|H_{ele}|\Phi_I\rangle \\ \langle\Phi_P|H_{ele}|\Phi_I\rangle \end{pmatrix} c_I = V_{adi}\begin{pmatrix} c_R \\ c_P \end{pmatrix} \\ \left(\langle\Phi_I|H_{ele}|\Phi_R\rangle \quad \langle\Phi_I|H_{ele}|\Phi_P\rangle\right)\begin{pmatrix} c_R \\ c_P \end{pmatrix} + \left(\langle\Phi_I|H_{ele}|\Phi_I\rangle\right) c_I = V_{adi} c_I \end{cases} \tag{12}$$

From equation (12) we get

$$c_I = -\left(\langle\Phi_I|H_{ele}|\Phi_I\rangle - V_{adi}\right)^{-1}\left(\langle\Phi_I|H_{ele}|\Phi_R\rangle \quad \langle\Phi_I|H_{ele}|\Phi_P\rangle\right)\begin{pmatrix} c_R \\ c_P \end{pmatrix} \tag{13}$$

Substituting c_I from equation (13) in equation (12) we obtain

$$\begin{pmatrix} V_{RR}'(\mathbf{q}) & V_{RP}'(\mathbf{q}) \\ V_{PR}'(\mathbf{q}) & V_{PP}'(\mathbf{q}) \end{pmatrix} \begin{pmatrix} c_R \\ c_P \end{pmatrix} = V_{adi} \begin{pmatrix} c_R \\ c_P \end{pmatrix} \tag{14}$$

Equation (14) can be seen as the 2x2 secular equation form of the 3x3 secular equation (11). From equation (14) it is obvious that the elements $V_{RR}'(\mathbf{q})$, $V_{PP}'(\mathbf{q})$ and $V_{RP}'(\mathbf{q})$ have the generic form:

$$V_{JK}'(\mathbf{q}) = \langle \Phi_J | H_{ele} | \Phi_K \rangle - \langle \Phi_J | H_{ele} | \Phi_I \rangle \frac{\langle \Phi_I | H_{ele} | \Phi_K \rangle}{\langle \Phi_I | H_{ele} | \Phi_I \rangle - V_{adi}} \quad , \quad \forall\, J,K = R,P \tag{15}$$

Rather than compute the terms V_{JK}' using equation (15), a simplification can be made. Since H_{ele} depends on the coordinate $\mathbf{q}$ we take $V_{RR}'(\mathbf{q})$ as the quadratic expansion of V_{adi} around $\mathbf{q}_R$, the reactant equilibrium geometry; $V_{PP}'(\mathbf{q})$ as the quadratic expansion of V_{adi} around $\mathbf{q}_P$, the product equilibrium geometry and $V_{RP}'(\mathbf{q})=V_{PR}'(\mathbf{q})$ as a *type* of quadratic expansion of V_{adi} around the crossing point geometry, $\mathbf{q}_{CP}$. In this context, the optimal points are denoted by an asterisk: $V_{RR}'(\mathbf{q}_R)=V_{RR}^*$, $V_{PP}'(\mathbf{q}_P)=V_{PP}^*$.

The important and crucial part of the BEP model is the resonance matrix element, $V_{RP}'(\mathbf{q})$. It is not obvious how this matrix element should be approximated. In this chapter the approximation of Chang and Miller [30] to evaluate the $V_{RP}'(\mathbf{q})$ matrix element is employed.

From the above equations, it is possible to obtain an expression for the BEP adiabatic PES, V_{adi}. The corresponding gradient and Hessian expression of V_{adi} in internal coordinates, $\mathbf{q}$, are given in ref. [1,31]. With this knowledge, it is straightforward to employ any type of Newton-Raphson (NR) algorithm to locate a TS geometry in the proposed BEP adiabatic PES. The authors recommend the approaches described in ref. [32-42].

2.3 Algorithmic procedure

The following is a schematic description of the procedure that finds the TS of the proposed BEP adiabatic PES model [1], some numerical results obtained by this algorithm are reported in ref. [31]. Briefly, the algorithm is:

1. Given $\mathbf{q}_R$, V_{RR}^*, $\mathbf{H}_R$, $\mathbf{q}_P$, V_{PP}^*, and $\mathbf{H}_P$ find the minimum energy crossing point, $\mathbf{q}_{CP}$, of the intersection of the two quadratic PES, V_{RR}' and V_{PP}'. Since we can take any value within the condition $w_R + w_P = 1$, the basic choice is to consider $w_R = w_P = 1/2$.
2. Compute the value of the real PES at $\mathbf{q}_{CP}$ and take this as V_{adi}^0.
3. Depending on the difficulty to compute the gradient vector and Hessian matrix of the real PES, two choices are available:

- 3A. If the gradient and the Hessian matrix are easy to be evaluated from the real PES at $\mathbf{q}_{CP}$, take these as $\mathbf{g}_{adi}{}^{0}$ and $\mathbf{H}_{adi}{}^{0}$. Make sure that $\mathbf{H}_{adi}{}^{0}$ possesses the desired spectra.
- 3B. If the gradient and the Hessian matrix are difficult to be evaluated then the gradient vector $\mathbf{g}_{adi}{}^{0}$ is approximated as $w_R\mathbf{g}_R + w_P\mathbf{g}_P$ where $\mathbf{g}_R$ and $\mathbf{g}_P$ are the gradient vectors of the quadratic potentials V_{RR}' and V_{PP}' at $\mathbf{q}_{CP}$ respectively and the Hessian matrix $\mathbf{H}_{adi}{}^{0}$ as $w_R\,\mathbf{H}_R + w_P\,\mathbf{H}_P$. If we consider the case in which the weights w_R and w_P are 1/2 and we take into account the fact that, at the point $\mathbf{q}_{CP}$, $V_{RR}{}^{0} = V_{PP}{}^{0}$, and then the $\mathbf{b}_0$ vector is equal to the zero vector and the $\mathbf{B}_0$ matrix takes the following form

$$\mathbf{B}_0 = \frac{1}{V_{RR}^{0} - V_{adi}^{0}}\left(\mathbf{H}_R + \mathbf{H}_P - 2\mathbf{H}_{adi}^{0}\right) - \frac{1}{2}\frac{1}{\left(V_{RR}^{0} - V_{adi}^{0}\right)^2}\left(\mathbf{g}_R^{0} - \mathbf{g}_P^{0}\right)\left(\mathbf{g}_R^{0} - \mathbf{g}_P^{0}\right)^{T}, \tag{16}$$

the square of the resonance energy matrix term takes the following form

$$\left(V_{RP}'\right)^2 = A_0 \exp\left(\frac{1}{2}\Delta\mathbf{q}_0^{T}\mathbf{B}_0\Delta\mathbf{q}_0\right) \tag{17}$$

4. Using a NR algorithm, locate the TS in the BEP surface.

The present algorithm is computationally inexpensive, except in step 2 and option 3A of step 3. The option 3B of step 3 needs some comments. This option should be taken very carefully. A necessary but not sufficient criteria to be considered in order to see if this option can be used, consists in regarding the largest value of the all square overlaps between the vector $N(\mathbf{g}_R\text{-}\mathbf{g}_P)$ and the set of eigenvectors $\{\mathbf{v}_i\}_{i=1}{}^{n}$ of the matrix $w_R\,\mathbf{H}_R + w_P\,\mathbf{H}_P$, we get

$$s = \max\left\{\left[\,\mathbf{v}_i^{T}\left(N\left(\mathbf{g}_R^{0} - \mathbf{g}_P^{0}\right)\right)\right]^2,\ i = 1,\ldots,n\right\} \tag{18}$$

Note that the s scalar is defined within the interval $0<s\leq1$. If the s parameter is small, for example lower than 0.5, then option 3B should not be used. It is important to emphasize that the option 3A should be used in

general rather than option 3B since this last option assumes a set of strong approximations.

The converged TS geometry takes into account the resonance effects and consequently is much closer to the real TS geometry than the minimum energy crossing point, $\mathbf{q}_{CP}$, geometry.

One can conclude that by using the described algorithm, the TS geometry found and the corresponding Hessian matrix evaluated are both good initial geometry and Hessian to start the optimization of the geometry TS on the real PES using a NR type algorithm [43,44]. Several application examples of the procedure given here are described in ref. [31].

3. A NOVEL METHOD TO IMPLEMENT THE POWELL FORMULA FOR UPDATING HESSIAN MATRICES RELATED TO TRANSITION STRUCTURES: A NEWTON-RAPHSON ALGORITHMIC IMPLEMENTATION

Newton and quasi-Newton-Raphson methods are attractive because they are rapidly convergent to a local stationary point. An important characteristic of this kind of procedures is the high demanding of storage memory. Consequently, it can not be used for higher demanding problems. For example, it is well known that for a problem of n variables the storage of the Hessian matrix requires $O(n^2)$ memory locations. In quasi-Newton-Raphson methods the Hessian matrix is updated at each iteration. For the optimization of saddle points it has been proposed to use the Powell formula to update the Hessian matrix at each iteration.

One can to apply this proposed algorithm, by modifying a NR procedure and adapting it to deal with a big number of variables. Here, presented is a way to update the Hessian matrix according to the Powell expression but only storing two vectors of length n at each iteration [45].

3.1 The new Powell formula implementation

The Powell representation formula can be written as

$$\mathbf{B}_{i+1} = \mathbf{B}_i + \mathbf{E}_i = \mathbf{B}_i + \mathbf{j}_i \mathbf{u}_i^T + \mathbf{u}_i \mathbf{j}_i^T - \left(\mathbf{d}_i^T \mathbf{j}_i\right) \mathbf{u}_i \mathbf{u}_i^T \quad i = 0, 1, \ldots \tag{19}$$

where

$$\begin{cases} \mathbf{j}_i = \mathbf{y}_i - \mathbf{B}_i \mathbf{d}_i \\ \mathbf{y}_i = \mathbf{g}_{i+1} - \mathbf{g}_i \\ \mathbf{d}_i = \mathbf{x}_{i+1} - \mathbf{x}_i \\ \mathbf{u}_i = \dfrac{\mathbf{M}_i \mathbf{d}_i}{\mathbf{d}_i^T \mathbf{M}_i \mathbf{d}_i} \end{cases} \tag{20}$$

and $\mathbf{M}_i$ is a symmetric and positive definite matrix. In Powell's update formula, the matrices $\mathbf{M}_0 = \mathbf{M}_1 = ... = \mathbf{M}_i = ... = \mathbf{I}$, being $\mathbf{I}$ the unit matrix.

In the approach presented here, Powell's formula is rewritten as

$$\begin{aligned} \mathbf{B}_{k+1} = \mathbf{B}_0 + \sum_{i=0}^{k} \mathbf{E}_i = \mathbf{B}_0 \\ + \sum_{i=0}^{k} \left[\mathbf{j}_i \mathbf{u}_i^T + \mathbf{u}_i \mathbf{j}_i^T - \left(\mathbf{d}_i^T \mathbf{j}_i \right) \mathbf{u}_i \mathbf{u}_i^T \right] \quad k = 0, 1, ... \end{aligned} \tag{21}$$

This formula contains the information of all previous iterations. At the iteration k in a standard quasi-Newton-Raphson the matrix $\mathbf{B}_{k+1}$ is updated by the matrix $\mathbf{B}_0$ and the sets of the pair vectors $\{\mathbf{j}_i, \mathbf{u}_i\}_{i=0}^{k}$ and the scalars $\{\mathbf{d}_i^T \mathbf{j}_i\}_{i=0}^{k}$. It is well known that, in this algorithm, given some n-dimensional vector $\mathbf{v}$, it is important to respect the computation of the matrix-vector product $\mathbf{B}_{k+1}\mathbf{v}$. Within the formalism presented here, this product can be evaluated as:

$$\mathbf{B}_{k+1} \mathbf{v} = \mathbf{B}_0 \mathbf{v} + \sum_{i=0}^{k} \left[\mathbf{j}_i \mathbf{u}_i^T \mathbf{v} + \mathbf{u}_i \mathbf{j}_i^T \mathbf{v} - \left(\mathbf{d}_i^T \mathrm{j}_i \right) \mathbf{u}_i \mathbf{u}_i^T \mathbf{v} \right] \quad k = 0, 1, ... \tag{22}$$

An efficient algorithm for evaluating the previous product can be outlined as follows:

1. Compute $\mathbf{B}_0\mathbf{v}$ and store it in a temporal vector, say $\mathbf{b}$.
2. Using the pair of vectors $\{\mathbf{j}_i, \mathbf{u}_i\}_{i=0}^{k}$ and the scalars $\{\mathbf{d}_i^T \mathbf{j}_i\}_{i=0}^{k}$ do:
   Loop from i = 0 to k.
   2.1. Compute the scalar products $\mathbf{u}_i^T \mathbf{v}$ and $\mathbf{j}_i^T \mathbf{v}$.
   2.2. Compute $\mathbf{b} \leftarrow \mathbf{b} + \mathbf{j}_i \mathbf{u}_i^T \mathbf{v} + \mathbf{u}_i (\mathbf{j}_i^T \mathbf{v} - \mathbf{j}_i^T \mathbf{d}_i \mathbf{u}_i^T \mathbf{v})$.
   End of loop.

The computation of this vector requires about 4kn multiplications. Note that at each iteration, two vectors of length n are stored. Consequently if $\mathbf{B}_0$

is selected to be diagonal, then at the iteration k the amount of memory needed using this formula is 2nk + n.

3.2 A quasi-Newton-Raphson method for several variables

Here, is described an algorithm useful for saddle point search with limited memory resources or for problems of high dimension. The strategy consists of never storing any full matrix and working only with a reduced number of eventually large vectors:

1. Give $\mathbf{B}_0$, the vectors $\{\mathbf{j}_i, \mathbf{u}_i\}_{i=0}^{k-1}$ and the scalars $\{\mathbf{d}_i^T\mathbf{j}_i\}_{i=0}^{k-1}$. It is assumed that $\mathbf{B}_0$ is not complete but it possesses the necessary correct structure in order to locate the desired transition structure. It can be assumed that the most relevant terms in the matrix are stored in a small diagonal box, as shown in the following picture:

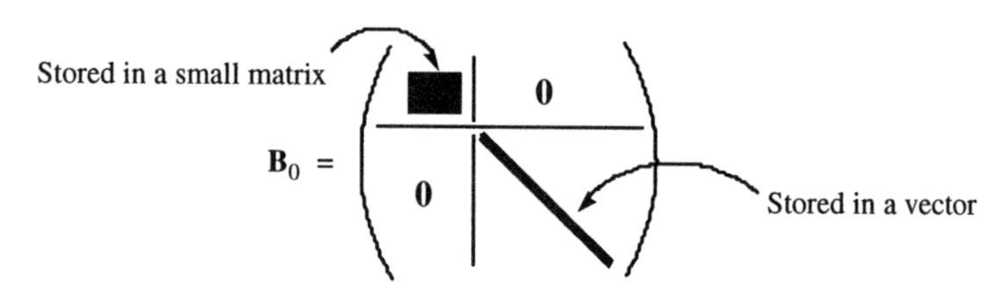

2. Obtain the second eigenpair of the eigenvalue equation

$$\mathbf{B}_k^a \mathbf{v}_\nu^{(k)} = \begin{bmatrix} 0 & \mathbf{g}_k^T \\ \mathbf{g}_k & \mathbf{B}_k \end{bmatrix} \mathbf{v}_\nu^{(k)} = \begin{bmatrix} 0 & \mathbf{g}_k^T \\ \mathbf{g}_k & \mathbf{B}_k \end{bmatrix} \begin{pmatrix} v_{1,\nu}^{(k)} \\ \mathbf{v}_\nu^{'(k)} \end{pmatrix} = \lambda_\nu^{(k)} \mathbf{v}_\nu^{(k)} \quad \forall \nu = 1,\ldots,n+1 \qquad (23)$$

by a Lanczos-Davidson type algorithm [46-48]. At each Lanczos-Davidson type iteration the vector $\mathbf{B}_k\mathbf{v}_\nu^{(k)}$ (ν=1,2) is evaluated by using the algorithm described above. The new geometry is given by

$$\mathbf{x}_{k+1} = \mathbf{x}_k + \frac{1}{v_{1,2}^{(k)}} \mathbf{v}_2^{'(k)} \qquad (24)$$

3. At the new $\mathbf{x}_{k+1}$, compute the energy $E(\mathbf{x}_{k+1})$ and gradient $\mathbf{g}_{k+1}$. Test for convergence: if $\|\mathbf{g}_{k+1}\| < \varepsilon$, stop ($\| \cdot \|$ defines Euclidean norm).
4. Compute the vectors $\mathbf{u}_k$ and $\mathbf{B}_k\mathbf{d}_k$. For the latter one use, the algorithm described above taking care to store the vector $\mathbf{j}_k$ and the scalar $\mathbf{d}_k^T\mathbf{j}_k$.

5. Go to 1.

In general, this method demands more iterations [45] than does the classical one. Also, this number of iterations depends on the correct selection of the active part of the initial guess for the approximated Hessian matrix. Nevertheless, the method has the advantage that it can be implemented in limited memory resources systems. It is important to remember that the coupling of the formulation given here together with the augmented Hesssian method diagonalized by a Lanczos-type algorithm does not need to store any full matrix.

4. THE GAUSSIAN WAVE PACKET METHOD APPLIED TO SEMICLASSICAL DYNAMIC STUDIES

This section presents an approach related to the Gaussian wave packet (GWP) method applied to semiclassical dynamics [49].

4.1 Time evolution description

In quantum mechanics the state of the system at any time t, is represented by the wave function, $\Psi(\mathbf{x},t)$, and the time evolution is given by the Schrödinger equation

$$H\Psi(\mathbf{x},t) = i\hbar \frac{\partial \Psi(\mathbf{x},t)}{\partial t}, \tag{25}$$

where H is the Hamiltonian matrix and **x** are the set of the cartesian system of coordinates.

Following Dirac [50] and Heller [51] we take as wave function a GWP

$$\Psi(\mathbf{x},t) \rightarrow \Psi(\mathbf{x}|\mathbf{C}(t),\mathbf{p}(t),\mathbf{x}_0(t),c(t)) = \exp\left\{\frac{i}{\hbar}\left[\frac{1}{2}(\mathbf{x}-\mathbf{x}_0(t))^T\mathbf{C}(t)(\mathbf{x}-\mathbf{x}_0(t)) + \mathbf{p}^T(t)(\mathbf{x}-\mathbf{x}_0(t)) + c(t)\right]\right\} \tag{26}$$

where $\mathbf{x}_0(t)$ and $\mathbf{p}(t)$ are the expectation values of the cartesian position and momentum operators of the wave function. The $\mathbf{x}_0(t)$ and $\mathbf{p}(t)$ are vectors and $\mathbf{C}(t)$ and c(t) are complex matrix and scalar functions of time respectively.

Since we are interested for molecular systems of middle or big size we introduce the normal coordinates, **Q**, in the usual way by diagonalizing the mass-weighted Hessian matrix,

$$\begin{cases} \mathbf{Q}(t) = \mathbf{L}^T \mathbf{M}^{1/2} \Delta\mathbf{x}(t) \\ \mathbf{G}_0 = \mathbf{L}^T \mathbf{M}^{-1/2} \mathbf{g}_0 \\ \mathbf{l}_0 = \mathbf{L}^T \mathbf{M}^{-1/2} \mathbf{B}_0 \mathbf{M}^{-1/2} \mathbf{L} \\ \mathbf{P}(t) = \mathbf{L}^T \mathbf{M}^{-1/2} \mathbf{p}(t) \end{cases} \tag{27}$$

where **L** is an orhtogonal matrix. Using these expressions the GWP can be written as

$$\Psi(\mathbf{Q}(t)|\mathbf{A}(t), \mathbf{P}(t), c(t)) = \exp\left\{ \frac{i}{\hbar}\left[\frac{1}{2}(\mathbf{Q}(t))^T \mathbf{A}(t)\mathbf{Q}(t) + \mathbf{P}^T(t)\mathbf{Q}(t) + c(t) \right] \right\} \tag{28}$$

The Hamiltonian governing the motion in normal coordinates is

$$\begin{aligned} H = -\frac{\hbar^2}{2}\Delta_Q + V_t(\mathbf{x}_0(t) + \mathbf{Q}(t)) = -\frac{\hbar^2}{2}\sum_{k=1}^{3N-6}\frac{\partial^2}{\partial Q_k^2} \\ + V_t(\mathbf{x}_0(t)) + \sum_{k=1}^{3N-6}(G_0)_k Q_k(t) + \frac{1}{2}\sum_{k=1}^{3N-6}(l_0)_k (Q_k(t))^2 \end{aligned} \tag{29}$$

Inserting this Hamiltonian and the GWP in the Schrödinger equation gives the following set of equations

$$\begin{cases} A_k^2(t) + (l_0)_k = -\dfrac{dA_k(t)}{dt} & ; \quad k = 1,\ldots,3N-6 \\ A_k(t)P_k(t) + (G_0)_k = -A_k(t)\dfrac{dQ_k(t)}{dt} - \dfrac{dP_k(t)}{dt} & ; \quad k = 1,\ldots,3N-6 \\ \dfrac{dc(t)}{dt} = -V_t(\mathbf{x}_0) - \sum_{k=1}^{3N-6} P_k(t)\dfrac{dQ_k(t)}{dt} - \dfrac{1}{2}\sum_{k=1}^{3N-6} P_k^2(t) + i\dfrac{\hbar}{2}\sum_{k=1}^{3N-6} A_k(t) & \end{cases} \tag{30}$$

The complex matrix **A**(t) controls the Q-space spread of Gaussian-Wave-Packet, and the $x_\mu x_\nu$ correlation in the x-space. In addition, in the normal coordinates representation, this matrix is diagonal.

Since each element of $A_k(t) = B_k(t) + i\,D_k(t)$ and the complex scalar $c(t) = \delta(t) + i\gamma(t)$ then it is found that

$$\begin{cases} A_k^2(t) + (l_0)_k = -\dfrac{dA_k(t)}{dt} & ;\quad k = 1,\ldots,3N-6 \\ (G_0)_k + \dfrac{dP_k(t)}{dt} = 0 & ;\quad k = 1,\ldots,3N-6 \\ P_k(t) + \dfrac{dQ_k(t)}{dt} = 0 & ;\quad k = 1,\ldots,3N-6 \\ \dfrac{d\delta(t)}{dt} = -E_c - \sum_{k=1}^{3N-6} P_k(t)\dfrac{dQ_k(t)}{dt} - \dfrac{\hbar}{2}\sum_{k=1}^{3N-6} D_k(t) & \\ \dfrac{d\gamma(t)}{dt} = \dfrac{\hbar}{2}\sum_{k=1}^{3N-6} B_k(t) & \end{cases} \tag{31}$$

A solution of the above set of differential equations is

$$\begin{aligned} & D_k(t) = (l_0)_k^{1/2} \\ & P_k(t) = P_k(t_0) - (G_0)_k\,\Delta t \\ & Q_k(t) = Q_k(t_0) - P_k(t_0)\Delta t + \frac{1}{2}(G_0)_k\,\Delta t^2 \\ & \delta(t) = \delta(t_0) - \Delta t\left(E_{GWP} - \sum_{k=1}^{3N-6} P_k^2(t_0)\right) \\ & \qquad\qquad - \Delta t^2 \sum_{k=1}^{3N-6} P_k(t_0)(G_0)_k + \frac{1}{3}\Delta t^3 \sum_{k=1}^{3N-6} (G_0)_k^2 \\ & \gamma(t) = \gamma(t_0) \end{aligned} \tag{32}$$

Using these results the GWP takes the following structure

$$\begin{aligned} \Psi(\mathbf{Q}(t)|\mathbf{D}(t),\mathbf{P}(t),\delta(t)) = & \left(\prod_{k=1}^{3N-6} \frac{(l_0)_k^{1/2}}{\pi\hbar}\right)^{1/4} \\ & \exp\left(-\frac{1}{2\hbar}\sum_{k=1}^{3N-6} (l_0)_k^{1/2} (Q_k(t))^2\right) \exp\left[\frac{i}{\hbar}\left(\sum_{k=1}^{3N-6} P_k(t) Q_k(t) + \delta(t)\right)\right] \end{aligned} \tag{33}$$

Since the width of this GWP only depends on the curvature of the potential surface, the only way to decrease the error generated during the GWP propagation is to approximate the potential energy surface by a quadratic potential energy one. In this case, the remaining terms should be very small. This approach gives an insight in order to design the related algorithm [49]:

1. Given a molecular geometry system, $\mathbf{x}_0$, compute $V_t(\mathbf{x}_0)$, $\mathbf{g}_0$ and $\mathbf{B}_0$ and transform these vectors and matrix to the normal coordinate representation.
2. Solve the classical dynamic equations. This is done by expanding each $Q_k(t)$ coordinate by a Taylor series until fourth order:

$$Q_k(t_0 + \Delta t) = Q_k(t_0) + \sum_{j=1}^{4} \frac{1}{j!} \left(\frac{\partial^j Q_k(t)}{\partial t^j} \right)_{t=t_0} \Delta t^j \tag{34}$$

and find the increment of the variable t such that

$$[\mathbf{Q}(t_0 + \Delta t)]^T [\mathbf{Q}(t_0 + \Delta t)] = R^2 \tag{35}$$

where R is the so called *trust radius*.

3. Compute the GWP and the quantum dynamic properties associated to this wave function.
4. Go to step 1 to compute a new time step.

4.2 Eigenfunctions and Quantum probability density

We can extract an energy eigenfunction from the GWP by means of a Fourier transformation of $\Psi(\mathbf{Q}(t)|\mathbf{A}(t),\mathbf{P}(t),c(t))$:

$$\Psi(\mathbf{x}|E_c) = \lim_{\substack{t_0 \to -\infty \\ t_f \to \infty}} \frac{1}{h^{1/2}} \int_{t_0}^{t_f} \Psi(\mathbf{Q}(t)|\mathbf{A}(t), \mathbf{P}(t), c(t)) \exp\left(\frac{iE_c t}{\hbar} \right) dt \tag{36}$$

As the integration of the Schrödinger equation is carried out in a small time step, Δt, where the Hamiltonian H remains constant with respect to the time, then we obtain a dense set of GWP corresponding to a dense set of times [52]:

$$\begin{array}{l}\{\Psi_1(\mathbf{Q}(t)|\mathbf{A}(t),\mathbf{P}(t),c(t)),\ \Psi_2(\mathbf{Q}(t)|\mathbf{A}(t),\mathbf{P}(t),c(t)),\ \ldots\}\\ \{\Delta t_1,\Delta t_2,\ldots\}\end{array} \tag{37}$$

Based in this equation the eigenfunction is

$$\Psi(\mathbf{x}|E_c) \cong \frac{1}{h^{1/2}} \sum_{I=1}^{\text{number of steps}} \int_0^{\Delta t_I} \Psi_I(\mathbf{Q}(t)|\mathbf{A}(t),\mathbf{P}(t),c(t)) \exp\left(\frac{iE_c t}{\hbar}\right) dt \tag{38}$$

The quantum probability density is given by

$$\operatorname{Re}|\Psi(\mathbf{x}|E_c)|^2 \cong \frac{2}{h} \sum_{I=1}^{\text{number of steps}} \left(\frac{N^{(I)}}{b^{(I)}}\right)^2 \left[1 - \cos\left(b^{(I)}\Delta t_I\right)\right] \tag{39}$$

where

$$\begin{cases} N^{(I)} = \left(\prod_{k=1}^{3N-6} \frac{D_k^{(I)}}{\pi\hbar}\right)^{1/4} \exp\left(-\frac{1}{2\hbar}\sum_{k=1}^{3N-6} D_k^{(I)}\left(Q_k^{(I)}\right)^2\right) \\ b^{(I)} = \frac{1}{\hbar}\sum_{k=1}^{3N-6}\left(G_0^{(I)}\right)_k Q_k^{(I)} + \frac{1}{2}\sum_{k=1}^{3N-6} D_k^{(I)} \end{cases} \tag{40}$$

5. CONCLUSION

The proposed Bell-Evans-Polasny principle model provides an inexpensive and reasonable approximation to the potential energy surface for a variety of elementary reactions involving medium sized molecules. The model affords the contribution weights of reactants and products to the approximate transition states. Also, practical algorithms related to the Newton-Raphson procedures and the Powell formula were presented in a global context. The Bell-Evans-Polasny principle principle was reformulated in order to be applied to high dimension problems. Finally, a insight towards the gaussian wave packet application to medium sized molecules was also described.

ACKNOWLEDGMENTS

One of us (R.C.) thanks the CIRIT (Generalitat de Catalunya, Spain) for financial support. This research was supported by the Spanish DGICYT grants PB95-0278-C02-01 and PB95-0278-C02-02, the CICYT Research Project SAF 96-0158 and sponsored by the European Comission contract ENV4-CT97-0508.

REFERENCES

1. J. M. Anglada, E. Besalú, J. M. Bofill, and R. Crehuet, *J. Comput. Chem.* **20**, 1112 (1999).
2. L. Salem, *Electrons in Chemical Reactions: First Principles*, Wiley, New York (1982).
3. M. G. Evans and E. Warhurst, *Trans. Faraday Soc.* **34**, 614 (1938).
4. M. G. Evans, *Trans. Faraday Soc.* **35**, 824 (1939).
5. F. Bernardi, M. A. Robb, H. B. Schlegel, and G. Tonachini, *J. Am. Chem. Soc.* **106**, 1198 (1984).
6. F. Bernardi, M. Olivucci, M. A. Robb, and G. Tonachini, *J. Am. Chem. Soc.* **108**, 1408 (1986).
7. F. Bernardi, M. Olivucci, J. J. W. McDouall, and M. A. Robb, *J. Am. Chem. Soc.* **109**, 544 (1987).
8. F. Bernardi and M. A. Robb, *Adv. Chem. Phys.* **67**, 155 (1987).
9. J. J. W. McDouall, M. A. Robb, and F. Bernardi, *Chem. Phys. Lett.* **129**, 595 (1986).
10. F. Bernardi, J. J. W. McDouall, and M. A. Robb, *J. Comput. Chem.* **8**, 296 (1987).
11. F. Jensen, *J. Am. Chem. Soc.* **114**, 1596 (1992).
12. F. Jensen, *J. Comput. Chem.* **15**, 1199 (1994).
13. K. Ruedenberg and J.-Q. Sun, *J. Chem. Phys.* **101**, 2168 (1994).
A. Pross and S. S. Shaik, *Tetrahedron Lett.* **23**, 5467 (1982).
A. Pross and S. S. Shaik, *Acc. Chem. Res.* **16**, 363 (1983).
14. S. S. Shaik, *Prog. Phys. Org. Chem.* **15**, 197 (1985).
A. Pross, *Adv. Org. Chem.* **21**, 99 (1985).
15. S. S. Shaik, *Pure Appl. Chem.* **63**, 193 (1991).
16. S. S. Shaik, H. B. Schlegel, and S. Wolfe, *Theoretical Aspects of Physical Organic Chemistry. The S_N2 Mechanism*, Wiley, New York (1992).
A. Warshel and R. M. Weiss, *J. Am. Chem. Soc.* **102**, 6218 (1980).
A. Warshel, *Biochemistry* **20**, 3167 (1981).
A. Warshel, *Acc. Chem. Res.* **14**, 284 (1981).
17. J.-K. Hwang, G. King, S. Creighton, and A. Warshel, *J. Am. Chem. Soc.* **110**, 5297 (1988).
18. J. Aqvist and A. Warshel, *Biochemistry* **28**, 4680 (1989).
A. Warshel, *Computer Modeling of Chemical Reactions in Enzymes and Solutions*, John Wiley, New York (1991).
19. J. Aqvist and A. Warshel, *Chem. Rev.* **93**, 2523 (1993).
20. H. J. Kim and J. T. Hynes, *J. Am. Chem. Soc.* **114**, 10508 (1992).
21. P.-O. Löwdin, *J. Math. Phys.* **3**, 969 (1962).
22. P.-O. Löwdin, *J. Math. Phys.* **3**, 1171 (1962).
23. Y.-T. Chang and W. H. Miller, *J. Phys. Chem.* **94**, 5884 (1990).

24. J. M. Anglada, E. Besalú, J. M. Bofill, and R. Crehuet, *J. Comput. Chem.* **20**, 1130 (1999).
B. J. Cerjan and W. H. Miller, *J. Chem. Phys.* **75**, 2800 (1981).
25. J. Simons, P. Jørgensen, H. Taylor, and J. Ozment, *J. Phys. Chem.* **87**, 2745 (1983).
C. O'Neal, H. Taylor, and J. Simons, *J. Phys. Chem.* **88**, 1510 (1984).
A. Banerjee, N. Adams, J. Simons, and R. Shepard, *J. Phys. Chem.* **89**, 52 (1985).
26. H. Taylor and J. Simons, *J. Phys. Chem.* **89**, 684 (1985).
27. J. Baker, *J. Comput. Chem.* **7**, 385 (1986).
28. J. Nichols, H. Taylor, P. Schmidt, and J. Simons, *J. Chem. Phys.* **92**, 340 (1990).
29. T. Helgaker, *Chem. Phys. Lett.* **182**, 503 (1991).
30. P. Culot, G. Dive, V. H. Nguyen, and J. M. Ghuysen, *Theor. Chim. Acta* **82**, 189 (1992).
31. J. M. Anglada and J. M. Bofill, *Int. J. Quantum Chem.* **62**, 153 (1997).
D. Besalú and J. M. Bofill, *Theor. Chem. Acc.* **100**, 265 (1998).
32. H. B. Schlegel, *Adv. Chem. Phys.* **67**, 249 (1987).
33. H. B. Schlegel, in *Modern Electronic Structure Theory*, D. R. Yarkony (ed.), World Scientific, Singapore (1995).
34. J. M. Anglada, E. Besalú, J. M. Bofill, and J. Rubio, *J. Math. Chem.* **25**, 85 (1999).
E. Besalú and R. Carbó-Dorca, *J. Math. Chem.* **21**, 395 (1997).
F. Besalú and J. M. Bofill, *J. Comput. Chem.* **19**, 1777 (1998).
35. J. M. Anglada, E. Besalú, and J. M. Bofill, *Theor. Chem. Acc.* **103**, 163 (1999).
36. M. A. Robb, personal comunication.
37. P. A. M. Dirac, *The Principles of Quantum Mechanics*, Claredon Press, Oxford (1958).
38. E. J. Heller, *J. Chem. Phys.* **62**, 1544 (1975).
39. D. Neuhauser, *J. Chem. Phys.* **93**, 2611 (1990).

Chapter 10

Partitioning of Free Energies of Solvation into Fragment Contributions: Applications in Drug Design

J. Muñoz, X. Barril, F. J. Luque*
Departament de Fisicoquímica, Facultat de Farmàcia, Avgda Diagonal s/n, Barcelona 08028, Spain

J. L. Gelpí, M. Orozco*
Departament de Bioquímica i Biologia Molecular, Facultat de Química, Universitat de Barcelona, Martí i Franquès 1, Barcelona 08028, Spain
** Send correspondence to F. J. Luque or M. Orozco*

1. INTRODUCTION

It is recognised that solvation influences the activity of therapeutic agents at two different levels [1-6]: i) the binding with the receptor, and ii) the bioavailability of the drug in the organism. These aspects are complementary in drug discovery in the sense that lead optimisation should ideally concentrate on compounds having tight binding affinity and favorable ADME (absorption, distribution, metabolism, excretion) properties.

The binding affinity of a drug is determined not only by the interactions formed between functional groups of the drug and specific residues of the biological target, but also by the desolvation cost associated to the formation of the receptor-drug (R-D) complex (Figure 1). Polar groups in the drug generally establish contacts with polar sites in the receptor binding site or are pointing towards the bulk solvent. However, transfer of the drug from bulk solvent to cavities inside the receptor will be favored by apolar groups. Accordingly, there must be a balance between hydrophilic and hydrophobic units, which explains the amphipatic nature of most drugs. Thus, the 3-D

Fundamentals of Molecular Similarity, Edited by Carbó-Dorca *et al.*
Kluwer Academic/Plenum Publishers, New York 2001

distribution of hydrophilic/hydrophobic regions is expected to be essential for the pharmacologic activity of the drug.

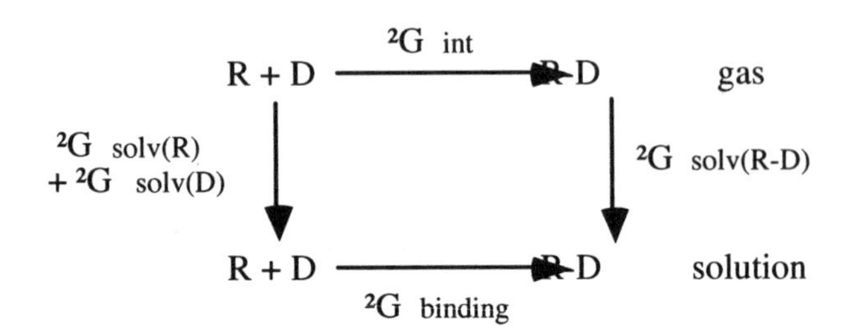

$$\Delta G\text{binding} = \Delta G\text{int} + \Delta G\text{solv(R-D)} - \Delta G\text{solv(R)} - \Delta G\text{solv(D)}$$
$$\Delta G\text{binding} = \Delta G\text{int} + \Delta\Delta G\text{solv}$$

Figure 1. Thermodynamic binding cycle between drug and receptor.

An optimal balance between polar and apolar groups is also required for the bioavailability of the drug. This is because the drug must be soluble in polar environments like blood and other physiological fluids, but it should be apolar enough to cross biological membranes that mediate drug absorption and transport. The most widespread physicochemical parameter utilised to estimate those properties is the octanol/water partition coefficient ($\log P_{oct}$), which measures the global lipophilicity of the molecule. Owing to the importance of this parameter, a variety of theoretical methods have been devised for the theoretical prediction of $\log P_{oct}$, and in most cases the partition coefficient is determined from the addition of empirical fragment contributions [3,7-24].

The advent of combinatorial chemistry [25-27] has stressed the importance of weeding out compounds with poor ADME properties in drug discovery. Combinatorial chemistry permits to synthesise large numbers of compounds in relatively short periods of time. For complex cases, however, it can be impossible to synthesise all potential compounds. For instance, for a core with 10 substitution sites and considering 50 possible substituents, a total of 50^{10} compounds could be made. Thus, the synthesis of 1 μg of each compound would represent an unaffordable synthetic task. These considerations evidence the need of developing techniques to manage efficiently such a big number of compounds, and more importantly to establish prescreening methods able to identify a priori compounds with undesirable properties [28-30], thus saving time and effort. In this context, concepts as similarity or diversity are now widely used to select few representative compounds of the whole data set [31-34]. Furthermore, consideration of pharmaceutical properties like bioavailability are also being considered in prescreening steps to eliminate compounds with poor pharmacologic and pharmacokinetic profiles.

The preceding discussion emphasises that drug discovery requires detailed knowledge of both the global solvation properties of drugs and the 3-D distribution of hydrophilic/hydrophobic regions. As noted above, knowledge of this latter information can be gained from fractional models, which decompose the free energy of solvation or the transfer free energy into fragment contributions. Those fractional models are generally based on empirical parameters, which often are combined with descriptors derived from quantum mechanical calculations. However, the use of such fractional contributions has been criticised [18,19,21,23,35-37], because it relies on the assumption that the fragmental contribution is largely independent of the molecular environment, which can limit the use of these techniques in the analysis of compounds having very different molecular structures.

Theoretical methods provide an alternative way to determine free energies of solvation, as well as the contribution to solvation exerted by specific groups. In these methods the free energy of solvation is determined as the difference between the reversible works needed to create the molecule in the gas phase and in solution. These calculations can be performed using a variety of methods, which combine a classical or quantum mechanical description of the compound (solute) and a discrete or macroscopic representation of the solvent. A thorough discussion of the strengths and weaknesses of the different methods can be found in several recent reviews [38-43].

We [43] and others [44-46] have recently proposed rigorous and simple schemes to partition the free energy of solvation into fragment contributions within the framework of self-consistent reaction field (SCRF) calculations. Particularly, our partition scheme has been developed within the polarizable continuum model developed by Miertus, Scrocco, and Tomasi (MST-SCRF; also denoted as polarizable continuum model) [47,48], which has been parameterised in the context of semiempirical MNDO [49], AM1 [50] and PM3 [51] methods and at the ab initio HF/6-31G(d) [52] level [53-59]. Our procedure divides the free energy of solvation into contributions assigned to the surface elements defining the solute/solvent interface. These surface contributions can be subsequently integrated to derive atomic or group contributions. The partitioning scheme provides a complete 3-D picture of the intrinsic solvation properties of molecules, which are valuable to describe the pattern of hydrophilic/hydrophobic properties. Here we outline the main features of the MST continuum model, and of the partitioning scheme. To illustrate the application of the fractional contributions in drug design studies, the absorption properties of a series of β-adrenoreceptor blocking agents are examined.

2. THEORETICAL DEVELOPMENT

2.1 The MST-SCRF method

In the MST-SCRF continuum method, the free energy of solvation, ΔG_{sol}, is determined as the reversible work necessary to transfer the solute from gas phase to bulk solvent (a 1M standard state is used for the two environments). Such a work is expressed by the addition of three contributions [41]: cavitation, van der Waals, and electrostatic (Eq. 1). The free energy of cavitation, ΔG_{cav}, accounts for the work necessary to generate in the bulk solvent a cavity large enough as to accommodate the solute. The van der Waals component, ΔG_{VW}, represents the contribution arising from dispersion-repulsion interactions between solute and solvent molecules. Finally, the electrostatic term, ΔG_{ele}, accounts for the work spent in charging up the solute in the bulk solvent.

$$\Delta G_{sol} = \Delta G_{cav} + \Delta G_{vw} + \Delta G_{ele} \tag{1}$$

The electrostatic contribution is determined assuming that the solvent is a continuum polarizable medium, which reacts against the solute charge distribution generating a reaction field. This reaction field is introduced as a perturbation operator, V_R, into the Schrödinger equation (Eqs. 2 and 3). In Eq. 3, M stands for the total number of surface elements (j) in which the solute/solvent boundary is divided, S_j is the area of the surface element j, σ_j is the charge density in this surface element, and q_j is the total charge representing the solvent response at the surface element j.

$$(\hat{H}^0 + \hat{V}_R)\Psi = E\Psi \tag{2}$$

$$\hat{V}_R = \sum_{j=1}^{M} \frac{\sigma_j S_j}{|r_j - r|} = \sum_{j=1}^{M} \frac{q_j}{|r_j - r|} \tag{3}$$

The mutual dependence between the solute wavefunction and the perturbation operator comes from the fact the solvent charge density depends on the electrostatic field created by the solute charge distribution through the

Laplace equation (Eq. 4), which is solved with suitable boundary conditions. In Eq. 4, V_T is the total electrostatic potential, which includes both solute and solvent contributions, n is the unit vector normal to the element surface j, and ε is the solvent dielectric permittivity. Owing to the mutual dependence between the solute wavefunction and the solvent reaction field, Eqs. 2-4 must be solved using an iterative procedure until consistency is achieved. Alternate solutions based on closure methods or matrix inversion procedures have also been recently proposed [60, 61].

$$\sigma_j = -\frac{\varepsilon-1}{4\pi\varepsilon}\left(\frac{\partial V_T}{\partial n}\right)_i \tag{4}$$

The electrostatic contribution is determined using Eq. 5, where the index "sol" means that the perturbation operator is adapted to the fully relaxed charge distribution of the solute in solution, and the index "0" stands for the gas phase environment.

$$\Delta G_{ele} = \langle \Psi^{sol} | \hat{H}^0 + \frac{1}{2} V^{sol} | \Psi^{sol} \rangle - \langle \Psi^0 | \hat{H}^0 | \Psi^0 \rangle \tag{5}$$

Regarding the nonelectrostatic terms, the cavitation component is computed following Pierotti´s scaled particle theory [62] adapted to molecular shaped cavities according to the procedure proposed by Claverie [63]. In this procedure the free energy of cavitation of a specific atom i is determined weighting the contribution of the isolated atom by the ratio between the solvent-exposed surface of such an atom and the total surface of the molecule, as noted in Eq. 6. In this expression, $\Delta G_{P,i}$ is the cavitation free energy of atom i in the Pierotti's formalism, S_i is the surface of atom i, S_T is the total molecular solvent-exposed surface, N is the number of atoms, and $\Delta G_{C\text{-}P,i}$ is the Claverie-Pierotti contribution of atom i.

$$\Delta G_{cav} = \sum_{i=1}^{N} {}^{2}G_{C-P,i} = \sum_{i=1}^{N} \frac{S_i}{S_T}\, {}^{2}G_{P,i} \tag{6}$$

Finally, the van der Waals term is determined using a linear relationship to the solvent-exposed surface of each atom [57-59]. In Eq. 7 $\Delta G_{VW,i}$ is the van der Waals free energy of atom i, and λ_i is the atomic surface tension, which was determined by fitting to the experimental free energy of solvation.

$$\Delta G_{vW} = \sum_{i=1}^{N} {}^{2} G_{vW,i} = \sum_{i=1}^{N} \lambda_i S_i \tag{7}$$

Eqs. 6 and 7 show that partitioning of the nonelectrostatic (cavitation + van der Waals) components of the free energy of solvation into contributions related to atoms is straightforward. These terms directly depend on the exposure of atoms to the solvent. Regarding the electrostatic term (Eq. 5), such a partition is not so simple. This difficulty is, nevertheless, solved by using a first order perturbation treatment of the solvation free energy, as reported by several groups [64-68].

2.2 Perturbation treatment of the free energy of solvation

In the charging up process of the solute in solution, the initial solute charge distribution, ρ^0, and the solvent response, σ^0 (represented by the solvent charge density spread over the cavity surface), are changed owing to the mutual dependence (Eqs. 2-4) until the final relaxed values in solution, ρ^{sol} and σ^{sol}, are reached. The polarization process can be followed (Figure 2) by means of an imaginary variable, ζ, which varies between 0 (gas phase) and 1 (solution). Note that when ζ=1 the solvent is fully adapted to the polarized solute, whereas in the state ζ=0 the solvent charge distribution is adapted to the non-polarized (gas phase) charge distribution of the solute.

Eq. 5 gives the electrostatic contribution to the free energy of solvation at the end of the polarization process (ζ=1). The electrostatic free energy corresponding to the transfer of the solute from gas phase to bulk solvent, while keeping the gas phase charge distribution ($\Delta G_{ele}{}^0$), can be expressed as written in Eq. 8, where V^0 stands for the solvent response created by the nonpolarized solute charge distribution. Accordingly, the polarization contribution to the free energy of solvation can be determined from Eq. 9, which gives the difference between the expressions shown in Eqs. 5 and 8.

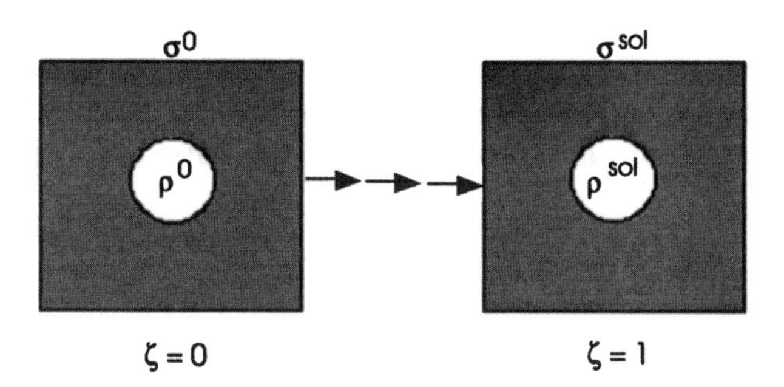

Figure 2. Representation of the polarization contribution upon solvation.

$$\Delta G_{ele}^{0} = <\Psi^{0} | \frac{1}{2} V^{0} | \Psi^{0}> \tag{8}$$

$$\Delta G_{pol} = <\Psi^{sol} | \hat{H}^{0} + \frac{1}{2} V^{sol} | \Psi^{sol}> - <\Psi^{0} | \hat{H}^{0} + \frac{1}{2} V^{0} | \Psi^{0}> \tag{9}$$

If we consider the final state as a perturbation of the initial state, both the energy and the wavefunction of the solute at ζ=1 can be evaluated from a series expansion at ζ=0 (see ref. 66 for detailed explanations). If we truncate such an expansion at the first-order term, the polarization energy can be expressed as noted in Eq. 10. Note that in this latter equation, in contrast to the exact expression given by Eq. 9, the wavefunction of the solute remains fixed at the gas phase value, while the solvent reaction field changes in the two terms of the right-hand side. Indeed, the polarization contribution to the free energy of solvation is described in terms of solute-solvent interactions.

$$\Delta G_{pol} = <\Psi^{0} | \frac{1}{2} V^{sol} | \Psi^{0}> - <\Psi^{0} | \frac{1}{2} V^{0} | \Psi^{0}> \tag{10}$$

Finally, addition of Eq. 10 to Eq. 8 allows us to derive a simple expression of the total electrostatic free energy, Eq. 11, where the electrostatic contribution is half the interaction energy between the solute having the nonpolarized charge distribution and the fully polarized solvent reaction field.

$$\Delta G_{ele} = \langle \Psi^0 | \frac{1}{2} V^{sol} | \Psi^0 \rangle \quad (11)$$

Eq. 11 has been shown to accurately reproduce the values of the electrostatic free energy computed rigorously from Eq. 5 [66]. However, Eq. 11 no longer depends on the solute Hamiltonian, which eliminates the intrinsic difficulty of Eq. 5 regarding the partition of the electrostatic component between atom contributions, as discussed in the next section.

2.3 Partition of the free energy of solvation into atomic contributions

The perturbational electrostatic contribution to the free energy of solvation provides a simple way to determine those contributions related to atom-exposed surfaces. Thus, when the perturbation operator is expanded as a set of "imaginary" point charges associated to surface elements j (see Eq. 3), Eq. 11 can be expressed as noted in Eq. 12, which permits to compute the electrostatic free energy from the addition of contributions associated to every surface element, ΔG^j_{ele} (Eq. 13).

$$\Delta G_{ele} = \langle \Psi^0 | \frac{1}{2} \sum_{j=1}^{M} \frac{q_j^{sol}}{| r_j - r |} | \Psi^0 \rangle \quad (12)$$

$$\Delta G_{ele} = \sum_{j=1}^{M} \Delta G^j_{ele} = \sum_{j=1}^{M} \langle \Psi^0 | \frac{1}{2} \frac{q_j^{sol}}{| r_j - r |} | \Psi^0 \rangle \quad (13)$$

Since the M surface elements can be assigned to atoms, the electrostatic term can be expressed in terms of contributions (ΔG^i_{ele}) associated to the N atoms in the molecule (Eq. 14). It is worth noting that such an "atomic contribution" should not be interpreted as the contribution arising from the charge distribution of a particular atom, but rather corresponds to the contribution exerted by the whole solute associated to the solvent-exposed surface of a particular atom.

$$\Delta G_{ele} = \sum_{i=1}^{N} \Delta G_{ele,i} = \sum_{i=1}^{N} \sum_{j=1}^{M \in N} <\Psi^0 | \frac{1}{2} \frac{q_j^{sol}}{| r_j - r |} | \Psi^0> \quad (14)$$

Finally, let us note that Eq. 14 can be easily rewritten in a classical framework, as shown in Eq. 15, where N_q stands for the number of point atomic charges representing the solute charge distribution. At this point, let us recall that N_q must not necessarily coincide with the number of atoms.

$$\Delta G_{ele} = \sum_{i=1}^{N} \Delta G_{ele,i} = \frac{1}{2} \sum_{i=1}^{N} \sum_{j=1}^{M \in N} \sum_{k=1}^{N_q} \frac{Q_k^{vac} q_j^{sol}}{| r_i - r_j |} \quad (15)$$

The free energy of solvation can then be obtained by adding the van der Waals and cavitation contributions (Eqs. 6 and 7) to the electrostatic ones (Eqs. 14 or 15).

$$\Delta G_{sol} = \sum^{N} \Delta G_{sol,i} = \sum^{N} (\Delta G_{vW,i} + \Delta G_{C-P,i} + \Delta G_{ele,i}) \quad (16)$$

2.4 Partition of the free energy of transfer into atomic contributions

Following the preceding discussion, the fractional contribution to the transfer free energy between two solvents (for instance, water and an organic solvent) can be obtained from the fractional contributions to the solvation free energy of each atom in the two solvents (Eq. 17, where X stands for each of the three contributions to the solvation free energy).

$$\Delta G_{w\to o} = \sum_{i=1}^{N} \Delta G_{w\to o,i} = \sum_{i=1}^{N} (\Delta\Delta G_{vW,i} + \Delta\Delta G_{C-P,i} + \Delta\Delta G_{ele,i}) \quad (17)$$

where $\Delta\Delta G_{X,i} = \Delta G_{X,i}(\text{organic}) - \Delta G_{X,i}(\text{water})$

The fractional contributions to the free energy of transfer can be valuable as molecular descriptors in quantitative structure-activity relationships (QSAR) studies, as well to provide a 3-D picture of the hydrophilic/hydrophobic properties of molecules.

3. FRACTIONAL CONTRIBUTIONS AND DRUG ABSORPTION

One of the pharmaceutical properties which has recently attracted most attention is drug absorption [69-74]. Information about drug absorption in vivo can be gained from experiments of drug permeability in cell cultures or intestinal tissues, such as monolayers of Caco-2 human intestinal cell line [75,76]. Cell culture models generally are better predictors of drug absorption than simple physicochemical parameters, but they demand significant quantities of the drugs and are labor intensive, which limits their applicability. In this context, the availability of theoretical methods for predicting reliably absorption properties would be extremely valuable.

Different molecular properties, like size, atomic charges, dipole moments and polar/apolar surfaces, have been examined for prediction of lipophilicity. However, the most important property has been and still is the calculation of partition coefficients from empirical fragment contributions, generally referred to partitioning between water and octanol, even though other systems like water/cyclohexane or water/chloroform have also been utilized. Indeed, theoretical methods can provide a 3-D picture of the distribution of hydrophilic/hydrophobic properties in the molecule, which can be used to modulate the bioavailability without altering other relevant pharmaceutical properties, like potency or receptor selectivity.

Here we focus our attention on the absorption properties of a series of six β-adrenoreceptor blocking agents (see Figure 3), whose permeability in Caco-2 cell line monolayers has been investigated. Our goal is to compute the fractional contributions of different structural fragments using the methodology described above, and to discuss the absorption properties in light of the results determined from our computational approach.

The β-adrenoreceptor compounds have similar molecular weights and a common structural core (Figure 3). The chemical differences concern both the nature of substituents and their position in the benzene ring. Indeed, there is information available about the major conformational families of these compounds [70], which permits to examine the influence of conformation on the partition properties. Finally, since these molecules have been used as a test system by other authors, our results can be compared with those determined using more empirical approaches.

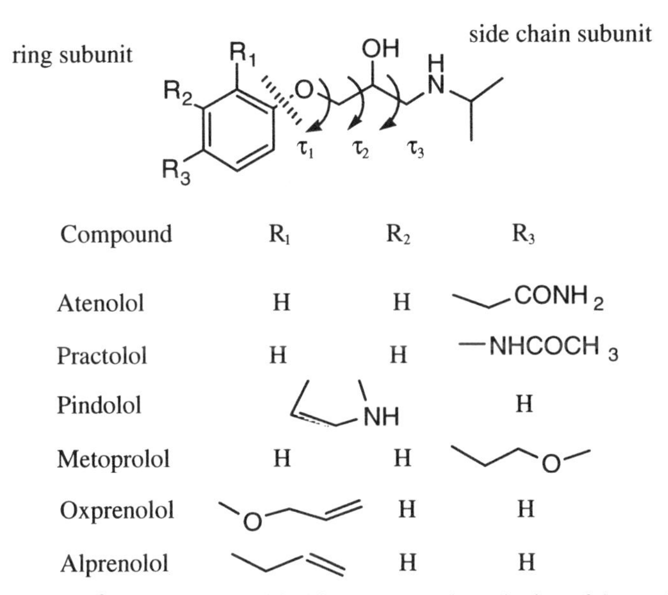

Figure 3. Structure of β-adrenoreceptor blocking agents and numbering of the torsional angles used to define the conformational preferences. The dashed line divides the molecule into side chain and ring subunits.

Table 1 reports selected physicochemical properties of the compounds and their permeability coefficients in Caco-2 monolayers. The partition coefficient between octanol and buffer at pH 7.4 for the neutral species, log P_{oct}, was calculated using the method developed by Hansch and co-workers [2,3,6]. The cellular permeability coefficient, Pc (cm/s x 10^6), was determined from apparent permeabilities at two different stirring rates (see ref. 70 for detailed explanations).

Transfer from water (pH 7.4) to an organic phase is expected to be unfavored owing to the predominance of the cationic species under these conditions in aqueous solution. The apparent octanol/buffer(pH 7.4) partition coefficient, log $D_{oct,7.4}$, is related to the partition coefficient of the neutral species, log P_{oct}, by log $D_{oct,7.4}$ = log P_{oct} - log($1+10^{(pKa-7.4)}$). Since the pK_a values are between 9.5 and 9.7, the correction term log($1+10^{(pKa-7.4)}$) shifts the log P_{oct} values by a factor of -2.2 units. For comparison purposes, we shall limit our attention to the partitioning properties of the neutral species.

Table 1. Selected physicochemical properties and permeability data of the six β-adrenoreceptor blocking agents.

Compound	pK_a	log Poct	Caco-2 P_c
Atenolol	9.6	-0.11	1.02±0.10
Practolol	9.5	+0.78	3.27±0.04
Pindolol	9.7	+1.65	50.9±3.2
Metoprolol	9.7	+1.20	91.9±4.0
Oxprenolol	9.5	+1.62	129±6
Alprenolol	9.6	+2.54	242±14

Figure 4 shows the variation in the logarithm of the permeability coefficient, log P_c, in front of the empirical log P_{oct} values. There is a reasonably good linear relationship between drug absorption and this traditional parameter, as noted in the value of $r^2 = 0.83$ corresponding to the regression equation log $P_c = -5.81 + 0.97$ log P_{oct}. Nevertheless, some fine details are clearly not well predicted, particularly the permeability of metoprolol.

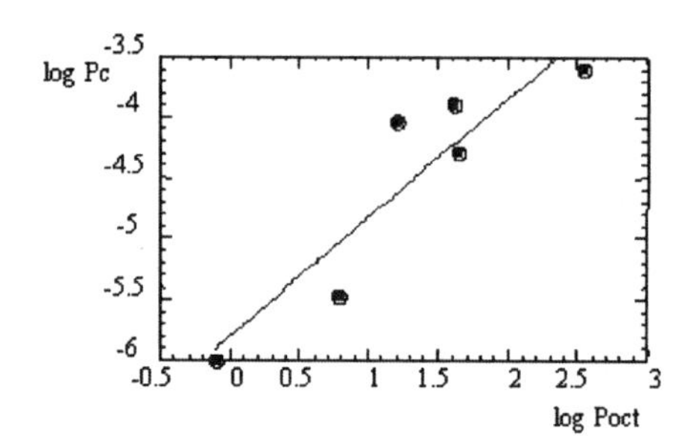

Figure 4. Variation of the logarithm of the cellular permeability in Caco-2 monolayers and the empirical log P_{oct} values.

To gain insight into the relationships between drug transport and hydrophilic/lipophilic properties of molecules, we have determined the partitioning properties of these compounds using the theoretical approach presented in the preceding section. To this end, we have determined the chloroform/water partition coefficient, as log P_{chl}, from the free energies of solvation in water and in chloroform (Eq. 18).

$$\log P_{chl} = -\frac{\Delta\Delta G_{wat\to chl}}{2.303\,R\,T} = -\frac{\Delta G_{chl} - \Delta G_{wat}}{2.303\,R\,T} \quad (18)$$

Calculations were performed with the AM1 version of the MST model, which was parameterised for both water and chloroform [57, 59] using a locally modified version of MOPAC [77]. Computations were carried out for a set of structures that represent the major conformational families in the gas phase [70]. The dihedral angles for the different families (see Figure 3) and the relative population determined from a Boltzmann distribution at 37°C [70] are given in Tables 2 and 3. With exception of oxprenolol, the preferred conformations correspond to families A and B, which encompass around 71% of the conformational space. All the compounds were built fixing the dihedral angles corresponding to each of the conformational families, and the rest of geometrical parameters were fully optimized at the AM1 level. The optimized structures were subsequently kept frozen in MST calculations, since geometry relaxation is expected to have little influence in the relative free energy of solvation for related structures [78].

Table 2. Values of the dihedral angles (degrees) that define the six major conformational families (A-F) of the β-adrenoreceptor blocking agents.

Dihedral Angle	A	B	C	D	E	F
$\tau 1$	-176	-176	177	175	74	74
$\tau 2$	179	180	63	59	178	178
$\tau 3$	-177	-63	180	-70	-176	-63

Table 3. Relative populations of the six major conformational families (A-F) of the β-adrereceptor compounds in the gas phase at 37°C.

Compound	A	B	C	D	E	F
Atenolol	46	26	9	7	6	3
Practolol	46	26	8	7	6	3
Pindolol	41	23	9	15	6	2
Metoprolol	46	25	8	7	6	3
Oxprenolol	14	9	3	32	25	9
Alprenolol	33	19	16	23	4	1

Table 4 gives the computed free energies of solvation in chloroform for the representative optimized structures families. There are generally little differences (around 1 kcal/mol) in the free energies of solvation for different conformations and even for different compounds, since the results lie in a relatively low range varying from -14.6 to -18.7 kcal/mol.

Table 4. Free energy of solvation (kcal/mol) in chloroform for the different conformational structures of β-adrenoreceptor blocking agents.

Compound	A	B	C	D	E	F
Atenolol	-18.2	-18.7	-18.1	-17.5	-17.8	-18.3
Practolol	-17.6	-18.3	-18.3	-17.0	-17.5	-18.0

Compound	A	B	C	D	E	F
Pindolol	-16.7	-16.2	-16.8	-15.7	-16.3	-16.0
Metoprolol	-17.0	-15.8	-16.7	-16.4	-16.8	-16.5
Oxprenolol	-16.6	-16.2	-17.0	-15.7	-16.1	-16.0
Alprenolol	-16.3	-15.7	-15.7	-14.6	-16.0	-15.4

The fractional contributions to the free energy of solvation in chloroform due to the side chain and ring subunits are given in Table 5. The side chain roughly makes a constant contribution ranging from -7 to -8 kcal/mol for all conformations and compounds. Likewise, the fractional contribution due to the ring subunit is also little affected by conformation. However, its contribution to the free energy of solvation is more sensitive to the nature of the substituents, since the fractional contributions vary from around -11 kcal/mol in atenolol to near –8 kcal/mol for alprenolol, which explains the range of variation in the ΔG_{chl} values reported in Table 4.

Table 5. Fractional contributions to the free energy of solvation in chloroform (kcal/mol) due to the ring (plain text) and side chain (italics) subunits of β-adrenoreceptor compounds.

Compound	A	B	C	D	E	F
Atenolol	-11.0	-10.7	-10.7	-10.3	-10.1	-10.2
	-7.1	*-7.9*	*-7.4*	*-7.2*	*-7.7*	*-8.1*
Practolol	-10.3	-10.3	-10.4	-10.1	-10.0	-9.9
	-7.3	*-8.0*	*-7.9*	*-6.9*	*-7.6*	*-8.1*
Pindolol	-9.0	-9.1	-9.0	-8.8	-8.5	-8.7
	-7.7	*-7.1*	*-7.8*	*-7.0*	*-7.8*	*-7.2*
Metoprolol	-8.3	-8.5	-7.8	-8.4	-8.1	-8.1
	-8.8	*-7.3*	*-8.8*	*-8.0*	*-8.6*	*-8.4*
Oxprenolol	-9.0	-8.9	-8.4	-8.3	-8.3	-8.6
	-7.6	*-7.3*	*-8.6*	*-7.4*	*-7.9*	*-7.4*
Alprenolol	-8.1	-8.0	-7.9	-7.6	-7.7	-7.8
	-8.2	*-7.7*	*-7.8*	*-7.0*	*-8.3*	*-7.6*

The behavior in chloroform is not completely unexpected considering that 1) all the compounds have similar molecular weights and sizes, and 2) the dominant contribution to solvation in chloroform comes from nonelectrostatic components [59], which are less sensitive than the electrostatic term to the nature of functional groups. Accordingly, it is not surprising to find similar contributions due to the side chain for all compounds and conformations, and that the main differences between compounds arise from the ring subunit, thus reflecting the different chemical nature of the functional groups in the benzene ring.

A fine picture of these trends is shown in Figure 5, which displays the atomic contributions to the free energy of solvation in chloroform. The larger contributions to solvation correspond to those surfaces enclosing polar groups, particularly the amido groups in atenolol and practolol, but also the nitrogen and oxygen atoms in the side chain. Indeed, the differences between the compounds are mostly restricted to the ring subunit, whereas little differences are observed between the surfaces for the side chains.

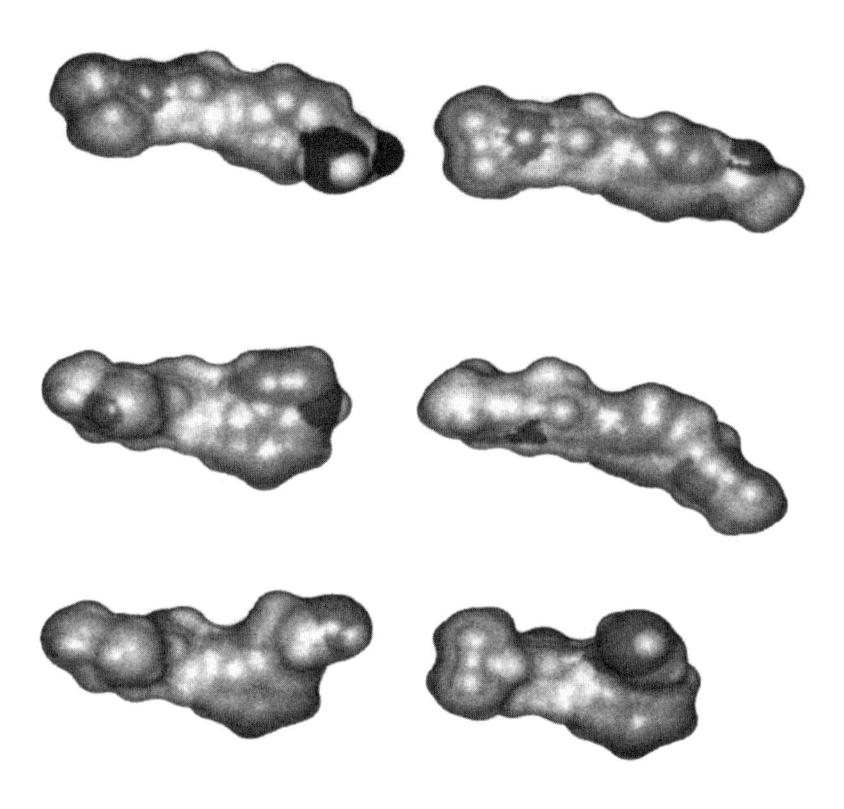

Figure 5. Fractional contributions to the free energy of solvation in chloroform for β-adrenoreceptor compounds. Top: atenolol (left), practolol (right). Middle: pindolol (left), metoprolol (right). Bottom: oxprenolol (left), alprenolol (right). The geometry of the compounds corresponds conformation A in Table 2. Colour code ranges from -3.0 (black) to 0.0 (light gray) kcal/mol.

The free energy of hydration shows larger variations than those found for solvation in chloroform, since the ΔG_{wat} values vary from -8.2 to -18.7 kcal/mol (Table 6), which reflects the greater importance of electrostatic effects in water. Indeed, the free energy of hydration is more sensible to conformation than the free energy of solvation in chloroform, since ΔG_{wat} values vary up to 3 kcal/mol between conformations for a given compound.

Table 6. Free energy of solvation (kcal/mol) in water for the different conformational structures of the β-adrenoreceptor blocking agents.

Compound	A	B	C	D	E	F
Atenolol	-16.9	-18.7	-16.9	-15.8	-16.8	-18.6
Practolol	-14.7	-16.9	-16.7	-14.0	-14.4	-16.7
Pindolol	-12.4	-11.4	-13.4	-12.5	-12.6	-12.6
Metoprolol	-12.9	-9.4	-11.5	-11.9	-12.5	-12.6
Oxprenolol	-10.9	-10.1	-11.4	-10.3	-10.7	-9.9
Alprenolol	-9.8	-8.6	-8.4	-8.2	-9.8	-8.7

The results in Table 7 permit us to analyse the origin of such variations in ΔG_{wat}. The dependence of ΔG_{wat} with conformation can be related to the fractional contribution of the side chain, which vary up to 3.7 kcal/mol with conformation. The ring contribution is little affected by conformation, since the largest variation amounts only to near 1 kcal/mol. The side chain contribution varies on average between –6 and –7 kcal/mol. The lowest value occurs for conformation D, which is the most compact structure.

Table 7. Fractional contributions to the free energy of solvation in water (kcal/mol) due to the ring (plain text) and side chain (italics) subunits for the β-adrenoreceptor blocking agents.

Compound	A	B	C	D	E	F
Atenolol	-12.2	-11.7	-11.7	-11.5	-11.8	-11.6
	-4.7	*-7.0*	*-5.2*	*-4.3*	*-5.0*	*-7.0*
Practolol	-9.1	-9.2	-9.6	-8.9	-8.9	-9.0
	-5.6	*-7.7*	*-7.1*	*-5.1*	*-5.6*	*-7.7*
Pindolol	-6.1	-6.2	-6.1	-6.3	-6.0	-6.2
	-6.3	*-5.2*	*-7.3*	*-6.2*	*-6.6*	*-5.7*
Metoprolol	-3.7	-3.9	-3.6	-4.1	-3.9	-3.9
	-9.2	*-5.5*	*-7.9*	*-7.7*	*-8.6*	*-8.7*
Oxprenolol	-4.0	-4.0	-3.1	-3.5	-3.6	-3.8
	-7.0	*-6.2*	*-8.3*	*-6.8*	*-7.0*	*-6.1*
Alprenolol	-1.9	-1.9	-2.0	-1.8	-1.8	-1.8
	-7.9	*-6.7*	*-6.5*	*-6.3*	*-7.9*	*-6.8*

The ring contribution determines the range of ΔG_{wat} values between compounds (Table 7), since it varies from -12 to -2 kcal/mol, reflecting the polarity of the substituents attached to the benzene ring. Thus, the contribution of $-CH_2CONH_2$ and $-NHCOCH_3$ in atenolol and practolol is around -10 and -8 kcal/mol, respectively, while it is around –3 kcal/mol for the ether chain in metoprolol and oxprenolol, and around –0.4 kcal/mol for the $-CH_2CHCH_2$ chain in alprenolol. This explains why this latter compound has the less favorable hydration. These differences can be appreciated in Figure 6, which shows the fractional contributions to the free energy of hydration for the compounds. The largest contributions to ΔG_{wat} are due to surfaces enclosing polar groups and the differences between compounds are mostly due to the ring substituents.

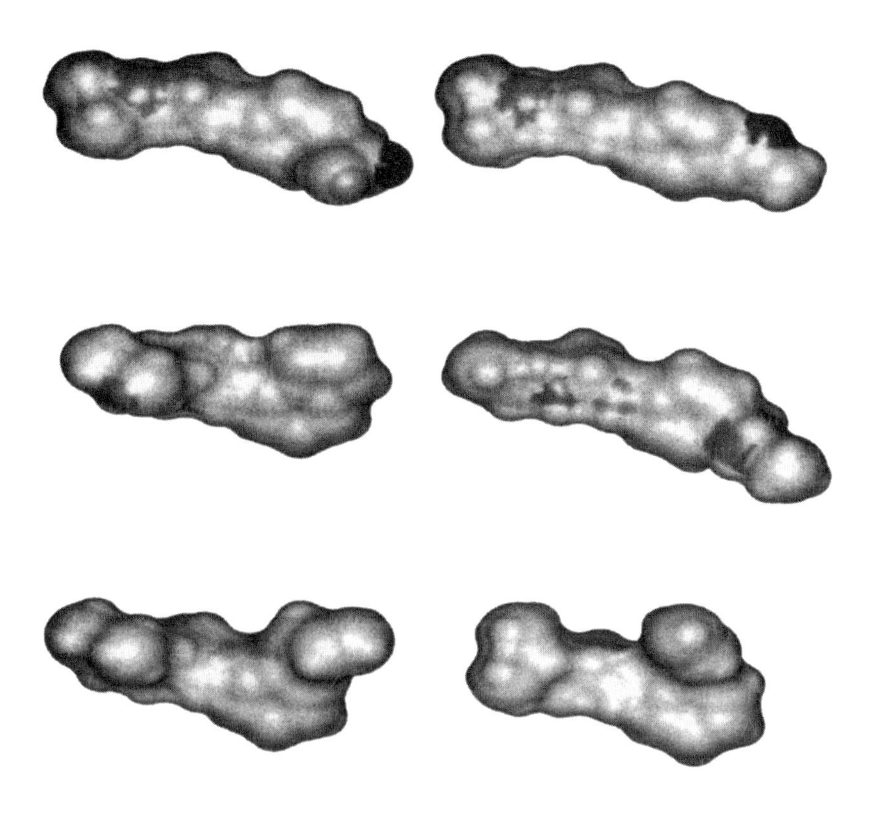

Figure 6. Fractional contributions to the hydration free energy for β-adrenoreceptor compounds (see legend in Figure 5). Colour code ranges from -8.0 (black) to 0.2 (light gray) kcal/mol.

The most favorable free energies of solvation both in chloroform and water are found for atenolol and practolol, i.e., the compounds bearing an amido group attached to the benzene ring. Conversely, metoprolol, oxprenolol, and particularly alprenolol possess the less favorable free energies of solvation, which agrees with the low polarity of their substituents (aliphatic ethers or unsaturated chains). As already mentioned, the difference between the ΔG_{sol} values for these two subsets of compounds is sensibly larger for water than for chloroform. Accordingly, one can expect the transfer from water to chloroform to be little favored for atenolol and practolol and more preferred for the other compounds, especially alprenolol. This can be verified from the data in Table 8, which reports the estimated partition coefficients determined using Eq. 18 and the data in Tables 4 and 6 for the series of compounds.

Table 8. Chloroform/water partition coefficients (log P_{chl}) for the different conformational structures of the β-adrenoreceptor blocking agents.

Compound	A	B	C	D	E	F
Atenolol	0.9	0.0	0.9	1.3	0.8	-0.2
Practolol	2.1	1.1	1.2	2.2	2.2	1.0
Pindolol	3.2	3.5	2.5	2.4	2.7	3.1
Metoprolol	3.0	4.7	3.7	3.3	3.1	2.8
Oxprenolol	4.2	4.5	4.2	3.9	4.1	4.4
Alprenolol	4.8	5.2	5.3	4.7	4.6	5.0

The log P_{chl} values can vary 0.6-1.5 units for a given compound, which mostly reflects the conformational dependence of the hydration free energy. With the only exception of conformers B and F of atenolol, there is marked preference for the chloroform phase. This trend increases following the ordering discussed above for the differences in solvation free energies of these compounds. Thus, atenolol and alprenolol are the compounds with the lowest and highest tendency to transfer from water to chloroform.

The log P_{chl} values can also be decomposed into fragment contributions, as stated in Eq. 19, where the $\Delta\Delta G_{wat\text{->}chl}$ values in Eq. 18 are written as a summation of differences between contributions to solvation in chloroform and water of the N fragments that constitute the molecule (see also Eq. 17).

$$\log P_{chl} = -\frac{\sum_{i=1}^{N} (\Delta G_{chl,i} - \Delta G_{wat,i})}{2.303\ R\ T} = -\sum_{i=1}^{N} \log P_{chl,i} \qquad (19)$$

Table 9 gives the fractional contributions of the ring and side chain to the log P_{chl} values. The side chain makes generally a positive contribution varying between 0.0 and 2.0 units. The largest contribution is found for atenolol, and the lowest values occur for metoprolol, oxprenolol and alprenolol. This effect mainly arises from the differences in fractional contributions to the hydration free energy of the side chain. However, the most important modulating factor of the log P_{chl} values is the contribution of the ring, which varies from around -1.0 in atenolol to around 4.4 in alprenolol. Again, this variation mainly reflects the differences in hydration mentioned above.

Table 9. Fractional contributions to the water/chloroform partition coefficient due to the ring (plain text) and side chain (italics) subunits for the β-adrenoreceptor blocking agents.

Compound	A	B	C	D	E	F
Atenolol	-0.8	-0.7	-0.8	-0.9	-1.2	-1.1
	1.8	*0.7*	*1.7*	*2.1*	*2.0*	*0.9*
Practolol	0.9	0.8	0.6	0.8	0.9	0.6
	1.2	*0.2*	*0.6*	*1.4*	*1.4*	*0.4*
Pindolol	2.2	2.1	2.1	1.8	1.9	1.9
	1.0	*1.4*	*0.4*	*0.5*	*0.9*	*1.1*
Metoprolol	3.3	3.4	3.1	3.1	3.1	3.1
	-0.3	*1.4*	*0.7*	*0.2*	*0.0*	*-0.3*
Oxprenolol	3.7	3.7	3.9	3.5	3.4	3.5
	0.5	*0.8*	*0.2*	*0.4*	*0.6*	*0.9*
Alprenolol	4.6	4.5	4.3	4.2	4.4	4.4
	0.2	*0.7*	*1.0*	*0.5*	*0.3*	*0.5*

Figure 7 shows the fractional contributions to the chloroform/water transfer free energy, expressed as ΔG_{wat}-ΔG_{chl}, of these compounds. The differences discussed above between hydrophilic/hydrophobic properties of the ring substituents, and between contributions from side chain and ring subunits, are clearly represented. This illustrates the potential capabilities of the method to obtain 3-D representations of the lipophilicity in molecules.

We now examine the relationships between theoretical log P_{chl} values and the Caco-2 absorption data. Table 10 gives the Boltzmann-averaged partition coefficients determined using the population weights given in Table 3 or correcting them for solvation effects in water and chloroform. The two sets of log P_{chl} values in Table 10 are very similar, suggesting that solvation has little influence on the conformational preference of these compounds. This justifies why good linear correlations have been obtained between Caco-2 permeability data and empirical parameters derived from the most populated conformations in the gas phase [70].

Figure 8 compares the computed log P_{chl} values with the empirical water/octanol partition coefficients, log P_{oct} (Table 1). The theoretical values correlate very well with the empirical ones (log P_{oct} = -0.07 + 0.49 log P_{chl}; r^2= 0.90), in agreement with previous results [44]. The computed log P_{chl} values tend to be twice the value of the log P_{oct}, which reflects the difference in dielectric response between octanol and chloroform [79]. The theoretical log P_{chl} values also correlate very well with the logarithm of the Caco-2 permeability (Figure 8; $\log P_c = -6.03 + 0.54 \log P_{chl}$; $r^2 = 0.95$), suggesting that the $\log P_{chl}$ values can be used as predictors of drug absorption in Caco-2 cell line monolayers models.

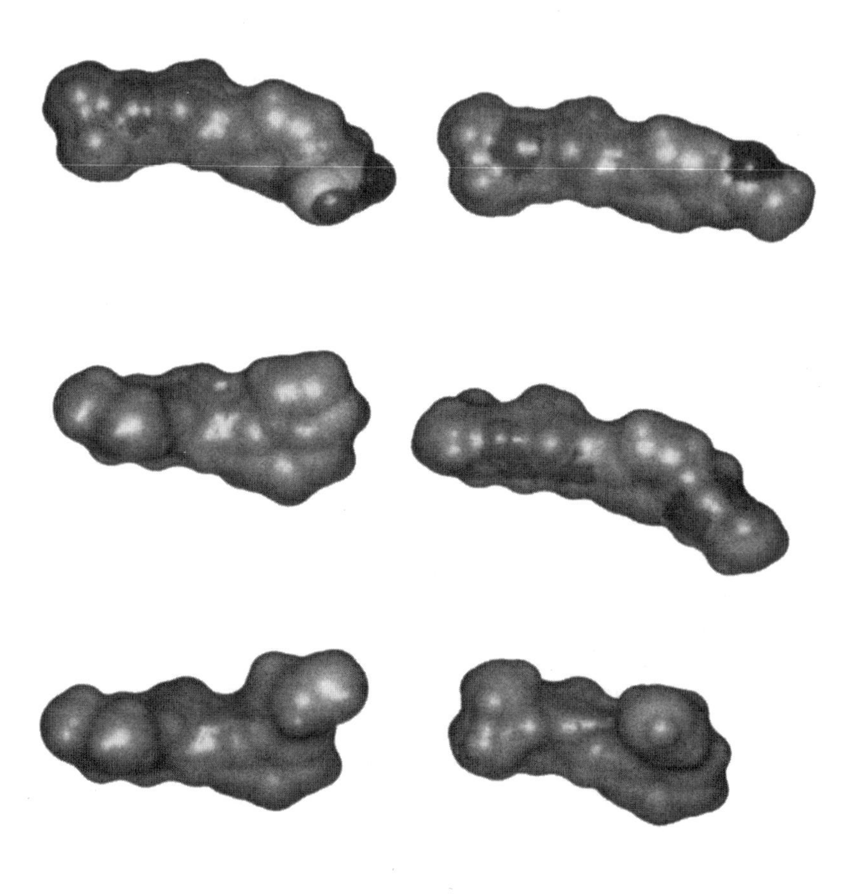

Figure 7. Fractional contributions to the transfer free energy (ΔG_{wat}-ΔG_{chl}) for the β-adrenoreceptor compounds (see legend in Figure 5). Colour code ranges from -5.0 (black) to 2.0 (light gray) kcal/mol.

Table 10. Theoretical estimates of the water/chloroform partition coefficients [a].

Compound	$\log P_{chl}$ (gas)	$\log P_{chl}$ (sol)
Atenolol	0.6	0.3
Practolol	1.7	1.2
Pindolol	3.0	2.8

Compound	log P_{chl} (gas)	log P_{chl} (sol)
Metoprolol	3.5	3.0
Oxprenolol	4.1	4.1
Alprenolol	4.9	4.9

[a] Determined using the gas phase population factors in Table 3 or correcting them for the solvent effect in water and chloroform.

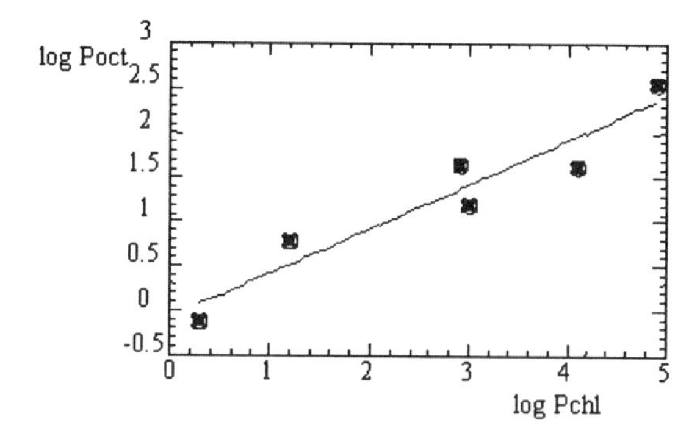

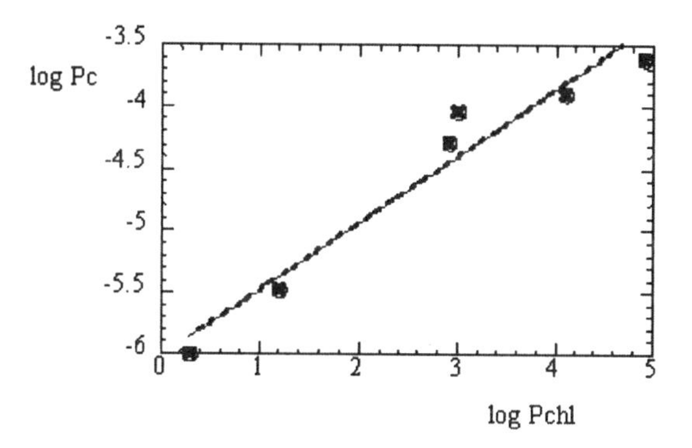

Figure 8. Comparison of empirical water/octanol (top) partition coefficients and the logarithm of Caco-2 permeability (bottom) versus theoretical water/chloroform partition coefficients.

Comparison of Figure 8 with the plot given in Figure 4 shows that the theoretical MST log P_{chl} values give the correct rank of drug permeability, in contrast with the results obtained with the log P_{oct} values, which failed in reproducing the ordering of metoprolol.

We have also examined the dependence of the log P_c values with the fractional contributions due to the ring (log $P_{chl,ring}$) and side chain (log $P_{chl,side\ chain}$) subunits. Figure 9 shows a good correlation between the log P_c data and the ring contribution (log P_c = -5.62 + 0.48; r^2= 0.96), which is in contrast to the poor correlation found between log P_c and the side chain contribution (r^2= 0.68; data not shown). Comparison with the regression equation reported above for log P_c versus log P_{chl} reveals that the side chain makes a roughly constant contribution to the permeability. This is not surprising considering that the main factor determining the differences in the water/chloroform partition coefficient arises precisely from the ring subunit.

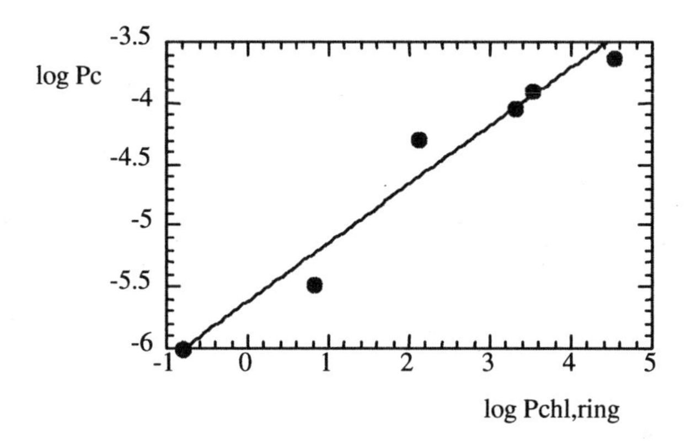

Figure 9. Dependence between the logarithm of the Caco-2 monolayer permeability and the contribution to the water/chloroform partition coefficients due to the ring (log $P_{chl,ring}$) subunit.

Because the variation in the ring contribution to log P_{chl} is mainly determined by the differences to the hydration free energy (see Tables 5 and 7) between compounds, one can expect that the Caco-2 permeability will be largely determined by the ring contribution to $\Delta G_{wat.}$ This is evident in Figure 10, which shows the existence of a close correlation between the log P_c values and the ring contributions to the free energy of hydration (r^2= 0.97). As expected from chemical intuition, the regression equation, log Pc = -3.04 + 0.25 $\Delta G_{wat,ring}$, clearly shows that the permeability decreases as the ring subunit becomes better hydrated, i.e., as the polarity of the substituents attached to the benzene ring is enlarged.

The dissection of the relationships between Caco-2 cell permeability and the fractional contributions to the free energy of solvation and partition coefficients illustrates the potential application of this methodology to drug design studies. Particularly, the preceding findings provides a theoretical justification to previous studies that have reported empirical relations between drug transport through biological membranes and the contribution of polar groups to the solvent accessible surface [69-72], the so called dynamic polar surface. This concept denotes the solvent accessible surface areas associated to nitrogen, oxygen and hydrogen bonded to these heteroatoms averaged for the most populated conformations. Likewise, it also agrees with the results of recent studies indicating a close relationship between the permeability data of these compounds and the surface areas corresponding to suitably chosen isocontour values of the interaction energy maps with a water molecule. [80].

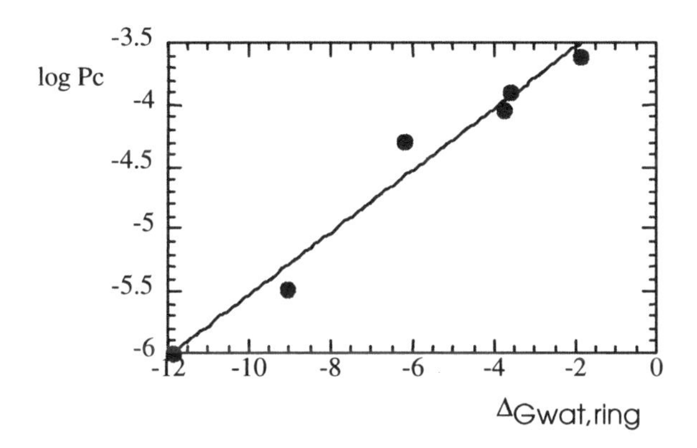

Figure 10. Dependence between the logarithm of the Caco-2 permeability and the ring contribution to the hydration free energy ($\Delta G_{wat,ring}$; kcal/mol).

The MST-based partitioning outlined here provides a rigorous basis to determine the contributions to the free energy of solvation or the transfer free energy of atoms or group of atoms. Even though the calculations reported here have been carried out at the semiempirical AM1 level, there is no limitation concerning either the use of ab initio quantum mechanical methods or even a quasi-classical treatment of the solute [64, 66-68]. It can then be applied to the study of the hydrophilic/hydrophobic properties of molecules of even medium or large size.

A valuable feature of the fractional partitioning scheme is that it gives a 3-D representation of the distribution pattern of hydrophilic/hydrophobic characteristics in molecules. This information can be used in molecular modeling studies to modulate the chemical and pharmaceutical properties of drugs. Thus, the method can be helpful for examining the effect of diverse substituents on the hydrophilicity/lipophilicity of specific regions in the molecule, and for designing chemical modifications appropriate to modulate the bioavailability of the drug without affecting other properties, like binding affinity to the target receptor or the selectivity between different receptors.

ACKNOWLEDGMENTS

We thank Drs. V. Segarra and M. López for making their results available. We also thank the DGICYT (grants PB97-0908 and PB96-1005) for financial assistance and the Centre de Supercomputació de Catalunya for computational facilities. Fellowship from the Ministerio de Educación y Cultura to X. B. is acknowledged.

REFERENCES

1. Hansch, P. P. Maloney, T. Fujita and R. Mui, *Nature* **194**, 178 (1963).
2. Hansch, *Acc. Chem. Res.* **2**, 232 (1969).
3. Leo, C. Hansch and D. Elkins, *Chem Rev.* **71**, 525 (1971).
4. Y. C. Martin, *Quantitative Drug Design: A Critical Introduction*, Marcel Decker, NY (1978).
5. L. Horvath, *Molecular Desing*. Elsevier, Amsterdam (1992).
6. Hansch and A. Leo, *Exploring QSAR: Fundaments and Applications in Chemistry and Biology*, ACS, Washington DC (1995).
7. Hansch and A. Leo, *Substituent Constants for Correlation Analysis in Chemistry and Biology*, John Wiley, New York (1979).
8. G. Nys and R. F. Rekker, *Chim. Therap.* **8**, 521 (1973).
9. Hopfinger and R. D. Battershell, *J. Med. Chem.* **19**, 569 (1976).
10. R. F. Rekker, *The Hydrophobic Fragmental Constant,* Elsevier, Amsterdam (1977).
11. Klopman and L. Iroff, *J. Comp. Chem.* **2**, 157 (1981).
12. K. Iwase, K. Komatsu, S. Hirono, S. Nakagawa and I. Moriguchi, *Chem. Pharm. Bull.* **33**, 2114 (1985).
13. Y. Suzuki and Y. Kudo, *J. Comput-Aided Mol. Des.* **4**, 155 (1990).
14. N. Bodor, Z. Gabany and C. K. Wong, *J. Am. Chem. Soc.* **111**, 3783 (1989).
15. K. Ghose and G. M. Crippen, *J. Comput. Chem.* **7**, 565 (1986).
16. V. N. Viswanadhan, A.K. Ghose, G. R. Revenkar and R. Robins, *J. Chem. Inf. Comput. Sci.* **29**, 163 (1989).
17. G. Klopman and S. Wang, *J. Comput. Chem.* **12**, 1025 (1991).
18. W. M. Meylan and P. H. Howard, *J. Pharm. Sci.* **84**, 83 (1995).
19. R. Mannhold and K. Dross, *Quant. Struct.-Act. Relat.* **15**, 403 (1996).
20. R. Wang, Y. Fu and L. Lai, *J. Chem. Inf. Comput. Sci.* **37**, 615 (1997).
21. V. Gombar and K. Enslein, *J. Chem. Inf. Comput. Sci.* **36**, 1127 (1996).
22. S. C. Basak, B. D. Gute and G. D. Grunwald, *J. Chem. Inf. Comput. Sci.* **36**, 1054 (1996).
23. van de Waterbeemd and R. Mannhold, *Quant. Struct.-Act. Relat.* **15**, 410 (1996).
24. T. Convard, J. P. Dubost, H. Le Solleu and E. Kummer, *Quant. Struct.-Act. Relat.* **13**, 34 (1994).
25. M. A. Gallop, et al., *J. Med. Chem.* **37**, 1233 (1994).
26. M. Gordon, et al., *J. Med. Chem.* **37**, 1385 (1994).
27. J. Martin, J. M. Blaney, M. A. Siani, D. C. Spellmeyer, A. K. Wong and W. H. Moos, *J. Med. Chem.* **38**, 1431 (1995).
28. W. P. W. Ajay and M. A. Murcko, *J. Med. Chem.* **41**, 3314 (1998).
29. Sadowski and H. Kubinyi, *J. Med. Chem.* **41**, 3325 (1998).
30. P. W. Finn, *Drug. Disc. Today* **1**, 363 (1996).

31. R. Carbó, L. Leyda and M. Arnau, *Int. J. Quantum Chem.* **17**, 1185 (1980).
32. R. Carbó (ed.), *Molecular Similarity and Reactivity: From Quantum Chemical to Phenomenological Approaches*, Kluwer, Amsterdam (1995).
33. R. Carbó-Dorca and P. G. Mezey (eds.), *Advances in Molecular Similarity*, Vol. 1, JAI Press, Greenwich, CT (1996).
34. Gorse, A. Rees, M. Kaczorek and R. Lahana, *Drug Disc. Today* **4**, 257 (1999).
35. M. Ter Laak, R. S. Tsai, G. M. Donné-Op den Kelder, P. A. Carrupt, B. Testa and H. Timmerman, *Eur. J. Pharm. Sci.* **2**, 373 (1994).
36. R. Mannhold, K. P. Dross and R. F. Rekker, *Quant. Struct.-Act. Relat.* **9**, 21 (1990).
37. P. Artursson and J. Karlsson, *Biochem. Biophys. Res. Commun.* **175**, 880 (1991).
38. M. Orozco, C. Alhambra, X. Barril, J. M. López, M. A. Busquets and F. J. Luque, *J. Mol. Model.* **2**, 1 (1996).
39. W. L. Jorgensen, in *Computer Simulation of Biomolecular Systems: Theoretical and Experimental Applications*, W.F van Gunsteren and P.K. Weiner (eds.), ESCOM, Leiden (1989), p. 60.
40. L. Rivail and D. Rinaldi, in *Computational Chemistry, Review of Current Trends*, J. Leszcynski (ed.), Word Scientific Publishing, Singapore (1996), p. 139.
41. J. Tomasi and M. Persico, *Chem. Rev.* **94**, 2027 (1994).
42. J. Cramer and D. G. Truhlar, in *Solvent Effects and Chemical Reactivity*, O. Tapia and J. Bertrán (eds.), Kluwer Academic Publishers, Dordrecht (1996), p. 1.
43. J. Cramer and D. G. Truhlar, *Chem. Rev.* **99**, 2161 (1999).
44. J. Luque, X. Barril and M. Orozco, *J. Comput.-Aided Mol. Design* **13**, 139 (1999).
45. J. Giesen, C. C. Chambers, C. J. Cramer and D. G. Truhlar, *J. Phys. Chem. B* **101**, 5084 (1997).
46. D. Hawkins, C. J. Cramer and D. G. Truhlar, *J. Phys. Chem. B* **102**, 3257 (1998).
47. S. Miertus, E. Scrocco and J. Tomasi, *Chem.Phys.* **55**, 117 (1981).
48. S. Miertus and J. Tomasi, *Chem.Phys.* **65**, 239 (1982).
49. J. S. Dewar and W. Thiel, *J. Am. Chem. Soc.* **99**, 4899 (1977).
50. J. S. Dewar, E. G. Zoebisch, E. F. Horsley and J. J. P. Stewart, *J. Am. Chem. Soc.* **107**, 3902 (1985).
51. J. J. P. Stewart, *J. Comput. Chem.* **10**, 209 (1989).
52. P. C. Hariharan and J. A. Pople, *Theor. Chim. Acta* **28**, 213 (1973).
53. J. Luque, M. J. Negre and M. Orozco, *J. Phys. Chem.* **97**, 4386 (1993).
54. Orozco and F. J. Luque, *Chem. Phys.* **182**, 237 (1994).
55. M. Bachs, F. J. Luque and M. Orozco, *J. Comput. Chem.* **15**, 446 (1994).
56. J. Luque, M. Bachs and M. Orozco, *J. Comput. Chem.* **15**, 847 (1994).
57. M. Orozco, M. Bachs and F. J. Luque, *J. Comput. Chem.* **16**, 563 (1995).
58. J. Luque, M. Bachs, C. Alemán and M. Orozco, *J. Comput. Chem.* **17**, 806 (1996).
59. J. Luque, Y. Zhang, C. Alemán, M. Bachs, J. Gao and M. Orozco, *J. Phys. Chem.* **100**, 4269 (1996).
60. L. Coitiño, J. Tomasi and R. Cammi, *J. Comput. Chem.* **16**, 20 (1995).
61. R. Cammi and J. Tomasi, *J. Comput. Chem.* **16**, 1449 (1995).
62. R. A. Pierotti, *Chem. Rev.* **76**, 717 (1976).
63. Claverie, in *Intermolecular Interactions: From Diatomics to Biomolecules*, B. Pullman (ed.), Wiley, Chichester (1978).
64. J. Luque, J. M. Bofill and M. Orozco, *J. Chem. Phys.* **103**, 10183 (1995).
65. J. G. Ángyán, *J. Chem. Phys.* **107**, 1291 (1997).
66. F. J. Luque, J. M. Bofill and M. Orozco, *J. Chem. Phys.* **107**, 1293 (1997).
67. F. J. Luque and M. Orozco, *J. Phys. Chem.* **101**, 5573 (1997).

68. M. Orozco, R. Roca, C. Alemán, M. A. Busquets, J. M. López and F. J. Luque, *J. Mol. Struct. (THEOCHEM)* **371**, 269 (1996).
69. L. H. Krarup, I. T. Christensen, L. Hovgaard and S. Frokjaer, *Pharm. Res.* **15**, 972 (1998).
70. K. Palm, K. Luthman, A.-L. Ungell, G. Strandlund and P. Artursson, *J. Pharm. Sci.* **85**, 32 (1996).
71. K. Palm, K. Luthman, A.-L. Ungell, G. Strandlund, F. Beigi, P. Lundahl and P. Artursson, *J. Med. Chem.* **41**, 5382 (1998).
72. D. E. Clark, *J. Pharm. Sci.* **88**, 815 (1999).
73. S. C. Basak, B. D. Gute and L. R. Drewes, *Pharm. Res.* **13**, 775 (1996).
74. M. H. Abraham, K. Takacs-Novak and R. C. Mitchell, *J. Pharm. Sci.* **86**, 310 (1997).
75. Artursson and J. Karlsson, *Biochem. Biophys. Res. Commun.* **175**, 880 (1991).
76. L. Amidon, P. J. Sinko and D. Fleisher, D., *Pharm. Res.* **5**, 651 (1988).
77. J. P. Stewart, MOPAC-93 Rev. 2. Fujitsu Limited, 1993. Modified by F. J. Luque and M. Orozco, University of Barcelona, Barcelona (1997).
78. F. J. Luque, J. M. López-Bes, J. Cemeli, M. Aroztegui and M. Orozco, *Theor. Chem. Acc.* **96**, 105 (1997).
79. D. R. Lide (ed.), *Handbook of Organic Solvents*, CRC Press, Boca Raton, FL (1995).
80. V. Segarra, M. Lopez, H. Ryder and J. M. Palacios, *Quant. Struct.-Act. Relat.* **18**, 474 (1999).

Chapter 11

Confronting modern valence bond theory with momentum-space quantum similarity and with pair density analysis

David L. Cooper
Department of Chemistry, University of Liverpool, P.O. Box 147, Liverpool L69 7ZD, UK

Neil L. Allan
School of Chemistry, University of Bristol, Cantocks Close, Bristol BS8 1TS, UK

Peter B. Karadakov
Department of Chemistry, University of Surrey, Guildford, Surrey GU2 5XH, UK

1. INTRODUCTION

The ever increasing level of sophistication of modern quantum chemical computations tends to make it more and more difficult to find direct links with the various classical models still employed by most chemists to visualize and to interpret molecular electronic structure. This has led to the development of a wide variety of schemes for the direct analysis of total wavefunctions and, especially, of total electron densities. Running against the general trend is the renaissance of valence bond (VB) theory, which aims to provide numerical accuracy with models that are relatively simple to interpret directly. This is especially true of modern developments such as spin-coupled theory [1-3], which combine useful accuracy with conceptually simple descriptions of the behaviour of correlated electrons. In the present work, we confront descriptions inferred directly from spin-coupled calculations with momentum-space quantum similarity indices for electron densities and with pair density analyses of total wavefunctions. By way of examples, we consider two gas-phase pericyclic reactions, namely the parent

Fundamentals of Molecular Similarity, Edited by Carbó-Dorca *et al.*
Kluwer Academic/Plenum Publishers, New York 2001

Diels-Alder process and the disrotatory ring-opening of cyclohexadiene. Additionally, we use the PF_4CH_3 molecule as an example to examine the nature of the bonding to hypercoordinate main group atoms.

2. SPIN-COUPLED WAVEFUNCTIONS

The original VB approach of Heitler and London for the ground state of H_2 is based on the wavefunction [4]

$$\Psi_{covalent}=[1s_A(1)\,1s_B(2)+1s_A(2)\,1s_B(1)]\sqrt{½}[\alpha(1)\beta(2)-\alpha(2)\beta(1)]\,, \quad (1)$$

in which 1,2 label electrons, and A,B label nuclei. This simple model provides a fairly realistic description of the binding, with the strength of the covalent bond linked to the degree of overlap between the two 1s orbitals, and of course it describes correctly the dissociation into neutral atoms.

A better description requires the admixture of a small component of "ionic" character (H^-H^+ and H^+H^-), via

$$\Psi_{ionic}=[1s_A(1)\,1s_A(2)+1s_B(1)\,1s_B(2)]\sqrt{½}[\alpha(1)\beta(2)-\alpha(2)\beta(1)]\,. \quad (2)$$

One physical interpretation of the role of Ψ_{ionic} is that it accounts for the distortion of the electron clouds around individual atoms as the two fragments are brought together.

Coulson and Fischer abandoned the strictly localized atomic orbitals and rewrote the Heitler-London covalent VB wavefunction in the form [5]:

$$\Psi_{CF}=[\phi_A(1)\,\phi_B(2)+\phi_A(2)\,\phi_B(1)]\sqrt{½}[\alpha(1)\beta(2)-\alpha(2)\beta(1)]\,, \quad (3)$$

in which $\phi_A=(1s_A+\lambda\,1s_B)$ and $\phi_B=(1s_B+\lambda\,1s_A)$. Notice that λ=0 corresponds to the covalent-only Heitler-London model, whereas λ=1 corresponds exactly to the basic molecular orbital theory description. Instead, Coulson and Fischer re-optimized λ variationally for each geometry. The optimal λ values correspond to relatively small distortions of the atomic orbitals, which thereby enhance their overlap. By abandoning strictly localized orbitals (leaving aside exactly what that might mean when using an extended basis set), Ψ_{CF} combines the conceptual simplicity of the Heitler-London model with enhanced accuracy.

The spin-coupled approach for many-electron systems builds directly on the simple Coulson-Fischer idea, with a single product of singly-occupied nonorthogonal orbitals ψ_μ expanded in an atom-centred basis set [1-3,6,7]:

$$\psi_\mu = \sum_p c_{\mu p} \chi_p \ . \tag{4}$$

In practice, of course, it is neither convenient nor appropriate to treat all of the electrons at the same level of theory. As in many other approaches, we start by partitioning the total system into inactive and active spaces, so that the total wavefunction takes the form

$$\Psi_{\text{SC}} = \mathcal{A}\left(\varphi_1^2 \varphi_2^2 \ldots \varphi_n^2 \, \Theta_{\text{pp}}^{2n} \, \psi_1 \psi_2 \ldots \psi_N \, \Theta_{SM}^N\right) , \tag{5}$$

in which the doubly occupied orbitals φ_i, which accommodate the inactive or "core" orbitals, are expanded in the same atom-centred basis set as the active orbitals:

$$\varphi_i = \sum_p c_{ip} \chi_p \ . \tag{6}$$

Without loss of generality, the φ_i may be taken to be orthonormal amongst themselves and to be orthogonal to the ψ_μ. The spin function [8] Θ_{pp}^{2n} for the inactive space is the perfect pairing function

$$\begin{aligned}\Theta_{\text{pp}}^{2n} = \sqrt{\tfrac{1}{2}}[\alpha(1)\beta(2) - \alpha(2)\beta(1)] \times \sqrt{\tfrac{1}{2}}[\alpha(3)\beta(4) - \alpha(4)\beta(3)] \\ \ldots \times \sqrt{\tfrac{1}{2}}[\alpha(2n\text{-}1)\beta(2n) - \alpha(2n)\beta(2n\text{-}1)] \ . \end{aligned} \tag{7}$$

The spin function for the active space, Θ_{SM}^N, is an optimal N-electron spin eigenfunction with quantum numbers S and M for the total spin and its z component, respectively; it is expanded in the full spin space of f_S^N linearly independent modes of spin coupling:

$$\Theta_{SM}^N = \sum_{k=1}^{f_S^N} c_{Sk} \, \Theta_{SM;k}^N \ . \tag{8}$$

In general, we optimize simultaneously the inactive orbitals φ_i, the "spin-coupled orbitals" ψ_μ and the "spin-coupling coefficients" c_{Sk} *without* any constraints on the overlaps between the ψ_μ. The outcome of the variational procedure is unique, in the sense that Ψ_{SC} is not in general invariant to linear transformations of the orbitals. The N-electron spin-coupled wavefunction is often an excellent approximation to the analogous many-configuration "N electrons in N orbitals" CASSCF wavefunction, but it is of course very much simpler to interpret. Indeed, the spin-coupled configuration dominates an "N

in N' full CI based on the spin-coupled orbitals, so that we may say that the additional configurations do not alter the basic qualitative picture.

We may also use various multiconfiguration variants of spin-coupled theory [9], but it is more common to obtain further refinements and to describe excited states by means of nonorthogonal CI calculations [1-7], involving excitations into suitable virtual orbitals. The resulting wavefunctions are accurate, but still compact and easy to interpret, not least because we invariably find that the final wavefunction for the lowest state is dominated by the spin-coupled configuration whereas those for excited states are dominated by a handful of excited configurations. Efficient computational algorithms, often relying on group theory and on graphical indexing techniques, as well as the extraordinary advances in affordable computing power, have led to the re-emergence of (modern) valence bond theory as a serious tool for computational chemistry. Applications of the spin-coupled valence bond approach, which provides compact, highly visual representations of the correlated behaviour of electrons in molecules, whilst also producing results of very high accuracy, now span all of the main branches of chemistry [1-3].

3. PAIR DENSITY ANALYSIS

The normalization condition for a general spinless two-particle density matrix may be written

$$\binom{N}{2} = \sum_{abcd} D(ab|cd)\, \langle ab|cd \rangle \tag{9}$$

It is constructive[i] to insert the identities

$$\tfrac{1}{2}(1+P_{12}^{r})+\tfrac{1}{2}(1-P_{12}^{r}) \tag{10}$$

or

$$\tfrac{1}{2}(1+P_{12}^{\sigma})+\tfrac{1}{2}(1-P_{12}^{\sigma}) \,, \tag{11}$$

in which P_{12}^{r} permutes the spatial coordinates of two electrons and P_{12}^{σ} permutes the corresponding spin coordinates. This leads to [10]

$$\Lambda^{\pm} = \sum_{abcd} D(ab|cd)\, \langle ab|\tfrac{1}{2}(1 \pm P^{\mathrm{r}}_{12})|cd\rangle$$

$$= \tfrac{1}{2} \sum_{abcd} D(ab|cd)\, (\langle a|c\rangle\langle b|d\rangle \pm \langle a|d\rangle\langle b|c\rangle)$$

$$= \tfrac{1}{4} N(N-1) \mp \tfrac{1}{2}(\tfrac{1}{4}N^2 - N + S(S+1))\,. \tag{12}$$

from which we may define the numbers of "effective pairs" Λ^{eff} according to [10,11]

$$\Lambda^{\mathrm{eff}} = \Lambda^{+} - \tfrac{1}{3}\Lambda^{-}\,. \tag{13}$$

Useful information is extracted by restricting the summations over *abcd*. For example, in the particular case of a spin-coupled wavefunction, we may choose *abcd* to label spin-coupled orbitals, ψ_μ, and thus write

$$\Lambda^{\pm}_{\mu\nu} = \tfrac{1}{2} \sum_{\sigma\tau} D(\mu\nu|\sigma\tau)\, (\langle\mu|\sigma\rangle\langle\nu|\tau\rangle \pm \langle\mu|\tau\rangle\langle\nu|\sigma\rangle)\,, \tag{14}$$

from which we may define values of $\Lambda^{\mathrm{eff}}_{\mu\nu}$. Alternatively, a link with "classical VB" is achieved by identifying *abcd* with basis functions centred on particular atoms A,B [10]:

$$\Pi^{\pm}_{\mathrm{AB}} = \tfrac{1}{2} \sum_{p\in \mathrm{A}} \sum_{q\in \mathrm{B}} \sum_{rs} D(pq|rs)\, (\langle p|r\rangle\langle q|s\rangle \pm \langle p|s\rangle\langle q|r\rangle)$$

$$+ \tfrac{1}{2}(1-\delta_{\mathrm{AB}}) \sum_{p\in \mathrm{B}} \sum_{q\in \mathrm{A}} \sum_{rs} D(pq|rs)\, (\langle p|r\rangle\langle q|s\rangle \pm \langle p|s\rangle\langle q|r\rangle)\,. \tag{15}$$

Of course, the resulting values of $\Pi^{\mathrm{eff}}_{\mathrm{AB}}$ rely on the assumption that it is possible to associate each basis function with one particular atomic centre. In this context, it is worth recalling that an s-type Gaussian function on oxygen with exponent 0.15 has a maximum in its radial distribution at 97 pm, i.e. at the position of a hydrogen atom in H_2O. As is the case with Mulliken charges, enlarging a small or medium sized basis set with diffuse functions may lead to unpredictable changes in values of Π_{AB}.

4. MOMENTUM-SPACE QUANTUM SIMILARITY

Position, r, and momentum, p, are conjugate variables in quantum mechanics and so we may choose to work with $\Psi(\mathbf{p})$ instead of $\Psi(\mathbf{r})$, and with the corresponding electron density $\rho(\mathbf{p})$ instead of $\rho(\mathbf{r})$. For molecular systems, it is difficult to obtain directly approximate p-space solutions to the Schrödinger equation. Instead, it is much more convenient first to obtain $\Psi(\mathbf{r})$ and then to transform it analytically to momentum space. The appropriate Fourier transform,

$$\Psi(\mathbf{p})=(2\pi)^{-3/2}\int\Psi(\mathbf{r})\,e^{-i\mathbf{p}.\mathbf{r}}\,d\mathbf{r}\,, \tag{16}$$

preserves the *form* of wavefunction, in the sense that the links between density, wavefunction, orbitals and (transformed) basis functions are exactly the same in p space as in r space. Additionally, it preserves direction, so that we can identify components parallel or perpendicular to a bond or plane, for example. In momentum space, all *explicit* information about the nuclear coordinates occurs in the $\exp(-i\mathbf{p}.\mathbf{R}_A)$ terms; in the case of individual orbitals, these "phase factors" can give rise to diffraction features, but the total densities are rather less exotic. This makes it easy to compare electron densities based on different nuclear frameworks. We note in passing that it is also somewhat easier in momentum space to distinguish sp^x hybrids than it is in position space [12-13] and that changes in individual p-space orbitals can illustrate rather well the process of bond formation [14].

The total electron density $\rho(\mathbf{p})$ falls off with a high power of p, so that it is dominated by the low values of p which correspond in position space to the long-range slowly-varying valence electron density. This is to be contrasted with an X-ray diffraction map, which emphazises heavy atoms.

A central quantity in momentum-space quantum similarity is the integral

$$I_{XY}(n)=\int p^n\,\rho_X(\mathbf{p})\,\rho_Y(\mathbf{p})\,d\mathbf{p} \tag{17}$$

in which n is usually chosen from the range -2 to $+2$. Values of $I_{XY}(n)$, which may be computed for total densities or for those corresponding to individual orbitals or particular fragments, depend on the relative orientation of X and Y, but not on the translation vector between them. We have used $I_{XY}(n)$ to define a variety of similarity and dissimilarity indices and have published a number of applications [15], mostly in the context of rationalizing and predicting drug activity. In the present work, we use the "Tanimoto index":

$$T_{\mathrm{XY}}(n) = 100 \times \frac{I_{\mathrm{XY}}(n)}{I_{\mathrm{XX}}(n)+I_{\mathrm{YY}}(n)-I_{\mathrm{XY}}(n)} \,. \tag{18}$$

5. PERICYCLIC REACTIONS

In a number of recent studies, we have investigated what changes in electronic structure accompany the variations in energy and geometry of a system on its way from reactants to products [16-18]. A convenient way to follow reactions is in terms of the minimum energy path or intrinsic reaction coordinate (IRC) which consists of steepest-descent paths (in mass-weighted coordinates) leading from the transition states towards reactants or products. Our preferred strategy is to locate the transition states and several points along the minimum energy paths using high-level calculations and then to check that an appropriate "*N* in *N*" CASSCF description is qualitatively correct. We may then perform spin-coupled calculations at each geometry. In the present work, we reexamine two gas phase pericyclic reactions, namely the Diels-Alder reaction of *cis*-butadiene and ethene to give cyclohexene [16]:

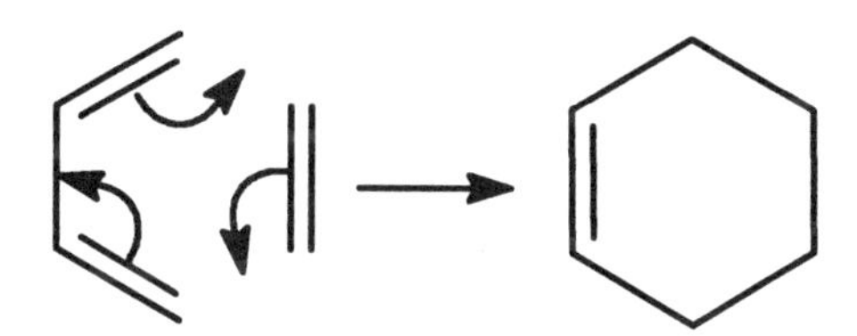

and the disrotatory electrocyclic ring-opening of cyclohexadiene [18]:

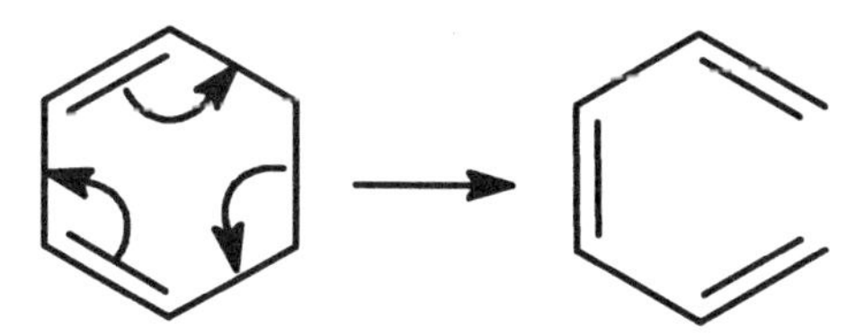

The choice of basis sets and the generation of geometries along the IRC are described in detail in our previous work, together with the corresponding energies and a full analysis of the evolution of electronic structure revealed directly by the spin-coupled wavefunctions [16,18].

Symmetry-unique spin-coupled orbitals for the Diels-Alder process are shown in Figure 1 for IRC values (in amu$^{1/2}$bohr) of 0.6 (towards reactants), zero (transition state) and –0.6 (towards products). Each orbital remains associated with the same carbon atom throughout the reaction, with the main changes being in the degree of sp^x character and in the amount and direction of the deformations of the orbitals. An attractive alternative to plotting a large number of sets of orbitals is to evaluate the numerical values of the momentum-space quantum similarity index $T_{XY}(-1)$ between spin-coupled orbitals at neighboring points along the IRC. Incidentally, such a procedure provides a very simple scheme for putting the spin-coupled orbitals into a consistent order, given that the "diagonal" similarity indices are all very high (typically greater than 99%) whereas the others are somewhat smaller. Our values of $T_{XY}(-1)$ suggest that the orbitals change rather slowly along the IRC, but that the most rapid changes occur close to the transition state geometry. This is entirely consistent with the pattern of overlap integrals shown in Figure 2a. The simultaneous evolution of the active-space spin coupling pattern along the IRC is illustrated in Figure 2b. The pair density quantities $\Lambda_{\mu\nu}^{\text{eff}}$ (see Figure 3a) also show rather clearly that the most dramatic changes occur in the vicinity of the transition state.

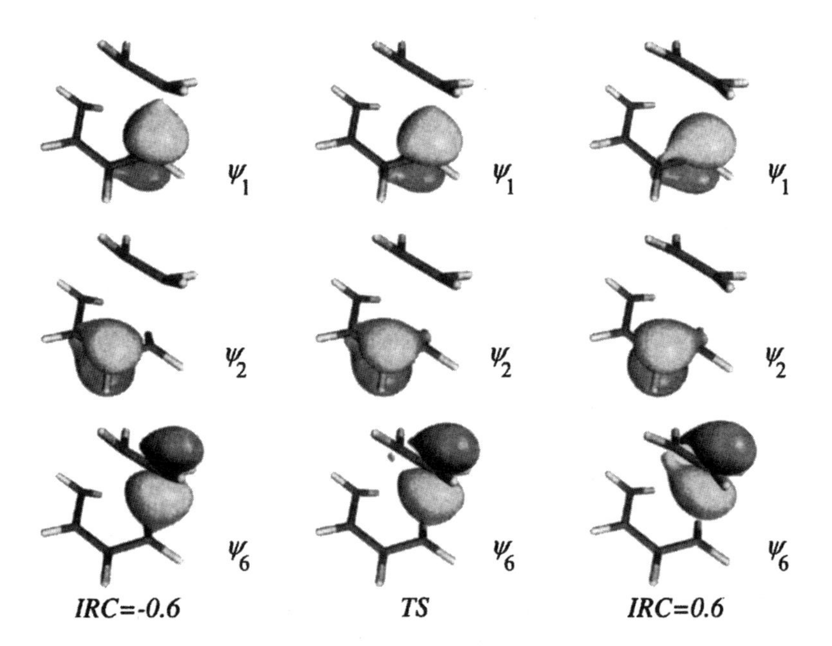

Figure 1. Symmetry-unique spin-coupled orbitals for the Diels-Alder reaction.

(a)

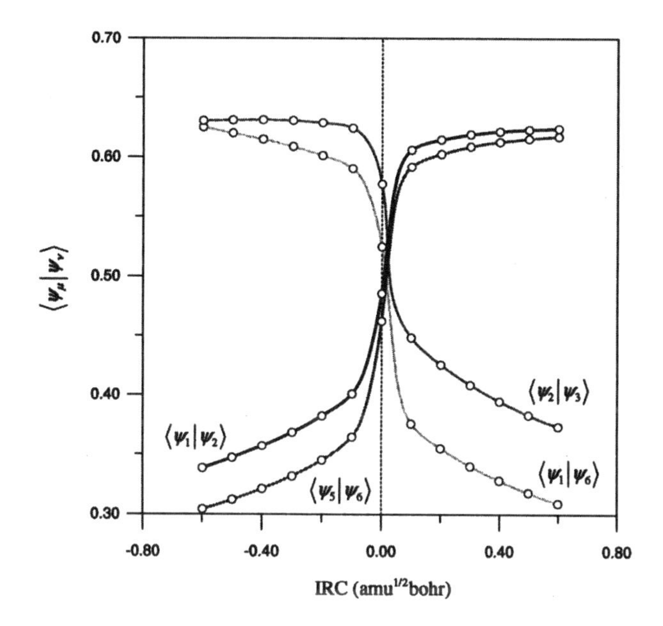

(b)

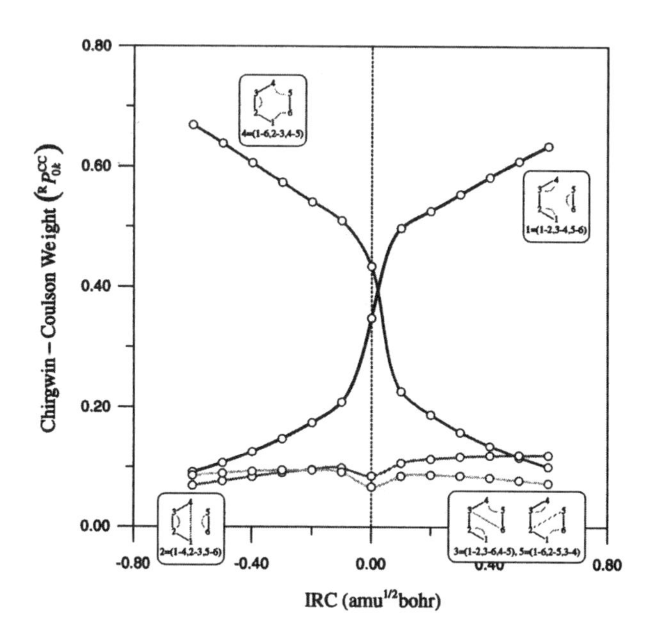

Figure 2. (a) Overlap integrals and (b) spin-coupling weights for the Diels-Alder reaction.

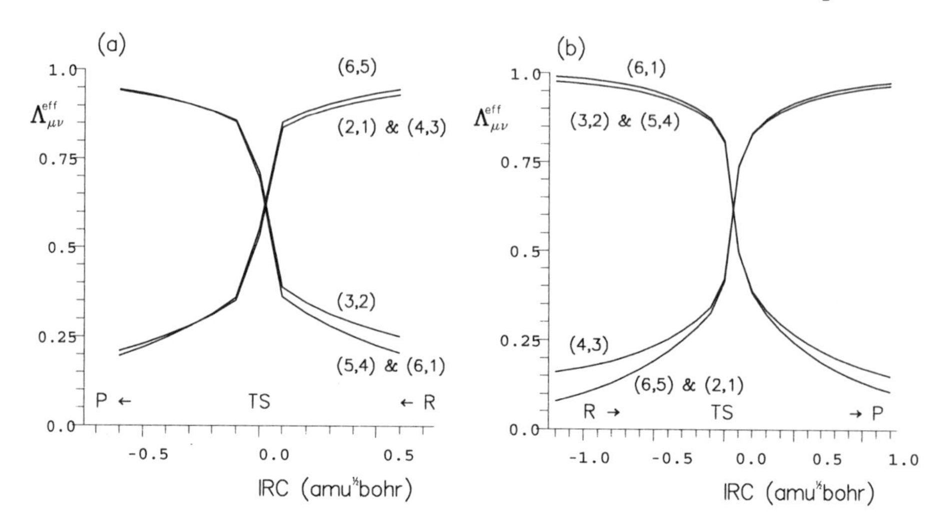

Figure 3. Pair density quantities $\Lambda^{eff}_{\mu\nu}$: (a) Diels-Alder reaction; (b) disrotatory ring-opening of cyclohexadiene.

For the disrotatory ring-opening of cyclohexadiene, the most rapid changes in bond lengths, orbital overlaps and the mode of spin coupling occur for IRC values a little after the transition state [18]. We find that the numerical values of $T_{XY}(-1)$ again suggest rather slow evolution of the orbitals along the IRC, with the most rapid changes occurring fairly close to the transition state geometry. An entirely consistent story emerges from the pair density quantities $\Lambda^{eff}_{\mu\nu}$ illustrated in Figure 3b.

At the transition state of the Diels-Alder reaction [16], the orbitals and orbital overlaps, the mode of spin coupling, the estimated "resonance energy" and the nature of the first excited singlet state are all strongly reminiscent of the spin-coupled description of benzene. It is tempting to argue in favour of an "aromatic" transition state. Many organic chemistry textbooks use the (heterolytic) representation:

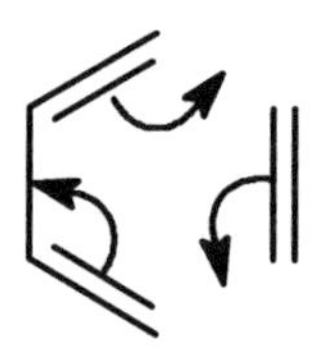

but often with remarks that such a scheme is not meant to illustrate the real mechanism. If one must use a simplistic representation, then it seems that the following ("homolytic") scheme is much more appropriate:

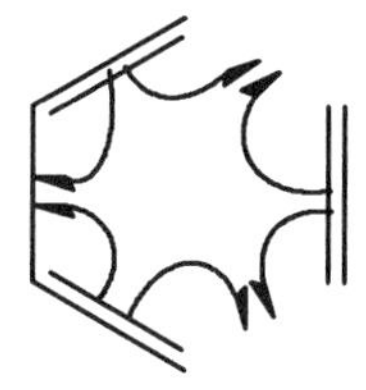

We reached analogous conclusions for the disrotatory ring-opening of cyclohexadiene [18], but not for some of the other systems we have studied, such as the 1,3-dipolar cycloaddition of fulminic acid to ethyne [17].

6. HYPERCOORDINATE MAIN GROUP ATOMS

It is clear that the central atoms in PF_5 and SF_6, for example, are hypercoordinate, meaning that they appear to be involved in more bonds than one can expect from a single Lewis structure that satisfies the octet rule. In spite of all the theoretical evidence to the contrary [19-22], some chemists still choose to rationalize hypercoordinate main group chemistry in terms of the extensive use of d orbitals, such as in d^2sp^3 hybrids for SF_6. In the dsp^3 model of PF_5, the three equatorial bonds, based on $P(sp^2)$, are considered to be very different from the two axial bonds, based on P(pd). However, the two sets of bond lengths are rather similar, differing by no more than might be anticipated for different steric repulsions. Furthermore, the two sets (at low temperature) of ^{19}F chemical shifts are very similar.

If the dsp^3 model of PF_5 were appropriate, then one should expect significant qualitative differences between the spin-coupled orbitals for the two sets of bonds, but this turns out not to be the case [20]. Moreover, there is no clear distinction in the response to the removal of d functions from the central atom. We find that the phosphorus atom contributes five equivalent, nonorthogonal sp^x-like hybrids, each of which delocalizes onto a fluorine atom. Each two-centre hybrid overlaps with a distorted F(2p) function, with little very differentiation between axial and equatorial bonds. The perfect pairing spin function dominates. Our results show that there is no need to invoke d orbitals to rationalize the bonding. Expressed in the language of classical VB, the bonding would be best described in terms of resonance between the plethora of "ionic" structures. The spin-coupled description is of course rather more compact.

Spin-coupled calculations have been now carried out for many molecules featuring hypercoordinate main group atoms [20-22], and a consistent picture emerges. We find that d basis functions play much the same qualitative role in hypercoordinate and normal molecules, acting essentially as polarization functions. We found no evidence for sp^nd^m hybridization and no obvious demarcations in the energy penalty per bond of excluding d functions. Polar bonds which shift density from the central atom appear to be favoured, particularly if the formal number of bonds is very high.

We have suggested that the octet rule should not be taken so seriously and have promoted a "democracy principle": *It is the democratic right of every valence electron to take part in bonding if provided with a sufficient energetic incentive* [20]. Key factors in the formation of hypercoordinate compounds of main group elements include steric factors (NF_5 is not plausible but we believe that CH_2N_2 and NF_3O do feature a hypercoordinate N atom [21,22]) and electronegativity differences (P and Cl differ sufficiently for PCl_5 to exist, but S and Cl are too similar for SCl_4 to be a stable covalent compound; H is not sufficiently electronegative in either case).

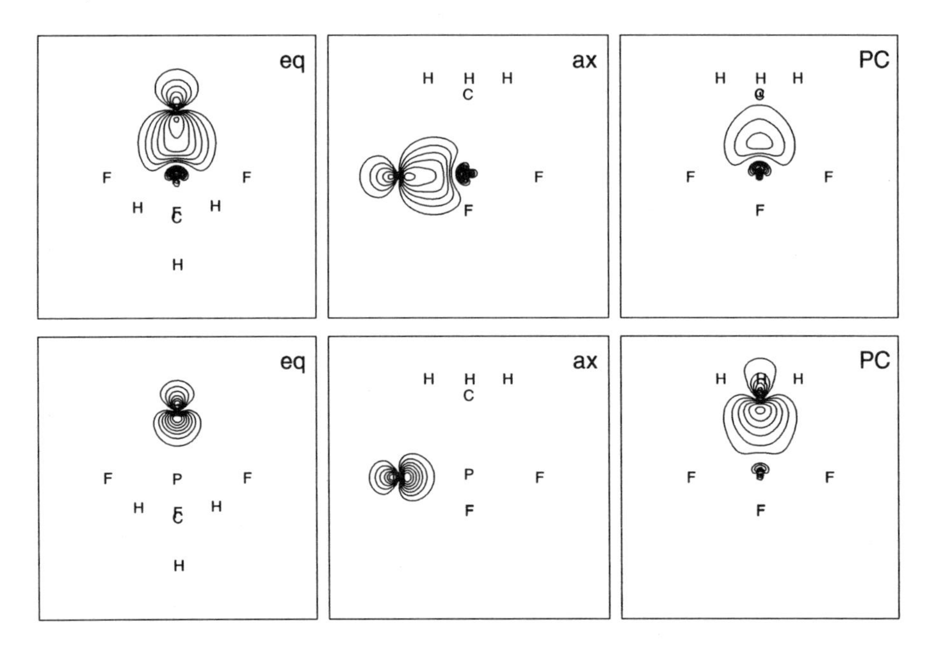

Figure 4. Spin-coupled orbitals for $PF_4(CH_3)$.

For the purposes of the present work, we optimized the geometry of the fluorophosphorane $PF_4(CH_3)$ at the restricted Hartree-Fock level using a modest 3-21G basis set augmented with a (Cartesian) polarization function with exponent d_P=0.465. We located a structure based on a trigonal bipyramid, with the methyl group occupying an equatorial site, in harmony with experiment. The valence localized molecular orbitals, obtained according to the population or overlap criterion introduced by Pipek and Mezey [23], could easily be identified with particular two-centre bonds or lone pairs. It was thus straightforward to carry out 10-electron spin-coupled calculations, concentrating on the P–F and P–C bonds. We found that the optimized spin-coupled orbitals consist of pairs, each clearly associated with a particular two-centre bond, and with predominantly singlet coupling of the electron spins. From the spin-coupled orbitals shown in Figure 4 as contour plots in appropriate planes, it is immediately apparent that the corresponding axial and equatorial orbitals are very similar to one another and, indeed, to those for the parent PF_5 molecule [20]. As we would have anticipated, the P–C bonds appear to be much less polar than are the P–F bonds. The closer match of the electronegativities of P and C results in a phosphorus-based sp^x-like hybrid that exhibits less delocalization onto the methyl group, and some delocalization of the methyl-based sp^x-like hybrid back onto phosphorus. Nonetheless, the overlaps $\langle\psi_\mu|\psi_\nu\rangle$ between the two orbitals making up each bond (see Table 1) attain essentially the same value. There are also small, but not negligible, overlaps between the various P-based hybrids.

Table 9. Values of $\langle\psi_\mu|\psi_\nu\rangle$ for PF_4CH_3.

	PF_{ax} ψ_1	F_{ax} ψ_2	PF_{ax} ψ_3	F_{ax} ψ_4	PF_{eq} ψ_5	F_{eq} ψ_6	PF_{eq} ψ_7	F_{eq} ψ_8	P_C ψ_9	C ψ_{10}
ψ_1	1	0.83	0.02	0.00	0.22	0.07	0.23	0.07	0.28	0.11
ψ_2	0.83	1	0.00	0.01	0.06	0.00	0.07	0.00	0.08	0.00
ψ_3	0:02	0.00	1	0.83	0.22	0.07	0.23	0.07	0.28	0.11
ψ_4	0.00	0.01	0.83	1	0.06	0.00	0.07	0.00	0.08	0.00
ψ_5	0.22	0.06	0.22	0.06	1	0.83	0.13	0.03	0.20	0.06
ψ_6	0.07	0.00	0.07	0.00	0.83	1	0.03	–0.01	0.05	–0.01
ψ_7	0.23	0.07	0.23	0.07	0.13	0.03	1	0.83	0.19	0.06
ψ_8	0.07	0.00	0.07	0.00	0.03	–0.01	0.83	1	0.05	–0.01
ψ_9	0.28	0.08	0.28	0.08	0.20	0.05	0.19	0.05	1	0.83
ψ_{10}	0.11	0.00	0.11	0.00	0.06	–0.01	0.06	–0.01	0.83	1

Table 10. Values of $\langle\hat{s}_\mu\cdot\hat{s}_\nu\rangle$ (lower triangle) and $\Lambda_{\mu\nu}^{eff}$ (upper triangle) for PF_4CH_3.

	1	2	3	4	5	6	7	8	9	10
1		1.00	0.00	0.00	0.01	–0.01	0.01	–0.01	0.01	–0.01
2	–0.75		0.00	0.00	–0.01	0.01	–0.01	0.01	–0.01	0.01
3	0.00	0.00		1.00	0.01	–0.01	0.01	–0.01	0.01	–0.01
4	0.00	0.00	–0.75		–0.01	0.01	–0.01	0.01	–0.01	0.01
5	0.00	0.01	0.00	0.00		1.00	0.00	0.00	0.00	0.00
6	0.00	0.00	0.00	0.00	–0.75		0.00	0.00	0.00	0.00
7	0.00	0.00	0.00	0.00	0.00	0.00		1.00	0.00	0.00
8	0.00	0.00	0.00	0.00	0.00	0.00	–0.75		0.00	0.00
9	–0.01	0.01	–0.01	0.01	0.00	0.00	0.00	0.00		1.00
10	0.00	–0.01	0.00	–0.01	0.00	0.00	0.00	0.00	–0.75	

The dominance of the perfect pairing mode of spin coupling is apparent from the spin matrix elements reported in the lower triangle of Table 2. Special values of such matrix elements are $\langle \hat{s}_\mu \cdot \hat{s}_\nu \rangle = -\frac{3}{4}$ for purely singlet coupling of the spins associated with orbitals ψ_μ and ψ_ν, zero for uncoupled spins, and $+\frac{1}{4}$ for pure triplet coupling [24]. The orbitals plots and the patterns of numerical values of $\langle \psi_\mu | \psi_\nu \rangle$ and of $\langle \hat{s}_\mu \cdot \hat{s}_\nu \rangle$ certainly suggest five polar two-centre bonds around the central phosphorus atom in $PF_4(CH_3)$.

Turning now to pair density analysis of the spin-coupled wavefunction, it is clear that the values of $\Lambda_{\mu\nu}^{\text{eff}}$ reported in the upper triangle of Table 2 reflect the dominance of the perfect pairing mode of spin coupling, with values close to unity (i.e. one pair) for each of the five bonds. All other values are close to zero. In order to understand the numerical values of Π_{AB}^{eff}, it is useful first to consider the following rather idealized values for a two-electron XY unit:

	Π_{XX}^{eff}	Π_{XY}^{eff}	Π_{YY}^{eff}
X–Y	0	1	0
$X^+ Y^-$	0	0	1
$X^- Y^+$	1	0	0

Looking first at an axial PF unit, we find (Table 3) $\Pi_{FF}^{\text{eff}}=0.57$ and $\Pi_{PF}^{\text{eff}}=0.39$. In terms of classical VB theory, analysis based on the idealized XY values suggests that this bond is 57% ionic in the sense P^+F^- and 39% purely covalent. The corresponding values for the equatorial PF units are very similar to these, whereas (in terms of classical VB) the PC unit appears to be 29% ionic in the sense P^+C^- and 63% covalent. Of course, when considering these various numbers, it is important to bear in mind that the "covalent" bond in H_2 has a non-negligible ionic character in classical VB theory. As described in Section 2, this last is subsumed in the Coulson-Fischer and spin-coupled models into deformations of the orbitals. Nonetheless, the values of Π_{AB}^{eff} do indicate rather well the polarity of the various bonds, and the high similarity between axial and equatorial PF units.

Table 11. Values of Π_{AB}^{eff} for PF_4CH_3.

	P	F_{ax}	F_{ax}	F_{eq}	F_{eq}	C
P	0.23	0.39	0.39	0.42	0.42	0.63
F_{ax}	0.39	0.57	0.00	0.01	0.01	0.01
F_{ax}	0.39	0.00	0.57	0.01	0.01	0.01
F_{eq}	0.42	0.01	0.01	0.54	0.00	0.00
F_{eq}	0.42	0.01	0.01	0.00	0.54	0.00
C	0.63	0.01	0.01	0.00	0.00	0.29

7. CONCLUSIONS

We have re-examined the spin-coupled descriptions of two gas phase pericyclic reactions, namely the parent Diels-Alder process and the disrotatory ring-opening of cyclohexadiene. Information provided from momentum-space quantum similarity indices $T_{XY}(-1)$, evaluated for pairs of orbitals, was found to be entirely consistent with that inferred directly from the spin-coupled calculations. In particular, we observe a rather slow pre- and post-transition state evolution of the orbitals in these two reactions, with the most rapid changes occurring in the vicinity of the transition state. The pair density quantities $\Lambda_{\mu\nu}^{eff}$ also show rather clearly that the most dramatic changes in the mode of spin coupling occur in the vicinity of the transition state. In the case of the fluorophosphorane PF_4CH_3, pair density analysis supports the spin-coupled interpretation of the nature of the bonding to hypercoordinate main group elements, but also provides additional useful information.

All in all, this confrontation of modern valence bond theory, in its spin-coupled form, with analysis based on momentum-space quantum similarity indices and on quantities derived from pair densities has turned out to be a rather peaceful one. This last observation does provide some additional confidence in the use of such tools for wavefunctions or densities that are too complex to analyze directly.

REFERENCES

1. D. L. Cooper, J. Gerratt and M. Raimondi, *Adv. in Chem. Phys.* **69**, 319 (1987); D. L. Cooper, J. Gerratt and M. Raimondi, *Int. Rev. Phys. Chem.* **7**, 59 (1988); J. Gerratt, D. L. Cooper and M. Raimondi, The spin-coupled valence bond theory of molecular electronic structure, in D. J. Klein and N. Trinajstić (eds) *Valence Bond Theory and Chemical Structure* (Elsevier, Amsterdam, 1990).
2. D. L. Cooper, J. Gerratt and M. Raimondi, *Chem. Rev.* **91**, 929 (1991).
3. J. Gerratt, D. L. Cooper, P. B. Karadakov and M. Raimondi, *Chem. Soc. Rev.* **26**, 87 (1997); J. Gerratt, D. L. Cooper, P. B. Karadakov and M. Raimondi, in *Encyclopaedia of Computational Chemistry* (John Wiley, 1998); pp 2672-88.
4. W. Heitler and F. London, *Z. Phys.* **44**, 455 (1927).

5. A. Coulson and I. Fischer, *Phil. Mag.* **40**, 386 (1949).
6. J. Gerratt, *Adv. Atom. Mol. Phys.* **7**, 141 (1971).
7. J. Gerratt and M. Raimondi, *Proc. Roy. Soc. (Lond.) Ser. A* **371**, 525 (1980).
8. R. Pauncz, Spin Eigenfunctions (Plenum, New York, 1979).
9. M. Raimondi, M. Sironi, J. Gerratt and D. L. Cooper, *Int. J. Quant. Chem.* **60**, 225 (1996); F. Penotti, *Int. J. Quant. Chem.* **59**, 349 (1996); D. L. Cooper, T. Thorsteinsson and J. Gerratt, *Int. J. Quant. Chem.* **65**, 439 (1997); N. J. Clarke, M. Raimondi, M. Sironi, J. Gerratt, D. L. Cooper, *Theor. Chem. Acc.* **99**, 8 (1998).
10. L. Cooper, R, Ponec, T. Thorsteinsson and G. Raos, *Int. J. Quant. Chem.* **57**, 501 (1996).
11. R. Ponec and M. Strnad, *Int. J. Quant. Chem.* **50**, 43 (1994).
12. L. Cooper, S. D. Loades and N. L. Allan, *J. Mol. Struct. (THEOCHEM)* **229**, 189 (1991).
13. J. Clark, H. L. Schmider and V. H. Smith, On hybrids in momentum space, in Z. B. Maksić and W. J. Orville-Thomas (eds) *Pauling's Legacy – Modern Modelling of the Chemical Bond* (Elsevier, Amsterdam, 1999).
14. L. Cooper and N. L. Allan, *J. Chem. Soc. Faraday Trans. 2* **83**, 449 (1987).
15. L. Cooper and N. L. Allan, *J. Comput.-Aided Mol. Design* **3**, 253 (1989); D. L. Cooper and N. L. Allan, *J. Am. Chem. Soc.* **114**, 4773 (1992); N. L. Allan and D. L. Cooper, *J. Chem. Inf. and Comp. Sci.* **32**, 587 (1992); D. L. Cooper, K. A. Mort, N. L. Allan, D. Kinchington and C. McGuigan, *J. Am. Chem. Soc.* **115**, 12615 (1993); N. L. Allan and D. L. Cooper, *Topics in Curr. Chem.* **173**, 85 (1995); D. L. Cooper and N. L. Allan, Molecular similarity and momentum space, in R. Carbó (ed.) *Molecular Similarity and Reactivity: From Quantum Chemical to Phenomenological Approaches* (Kluwer, Dordrecht, 1995); P. T. Measures, K. A. Mort, N. L. Allan and D. L. Cooper, *J. Comput.-Aided Mol. Design* **9**, 331 (1995); P. T. Measures, N. L. Allan and D. L. Cooper, *Adv. in Molec. Similarity* **1**, 61 (1996); P. T. Measures, K. A. Mort, D. L. Cooper and N. L. Allan, *J. Mol. Struct. (THEOCHEM)* **423**, 113 (1998); N. L. Allan and D. L. Cooper, *J. Math. Chem.* **23**, 51 (1998).
16. P. B. Karadakov, D. L. Cooper and J. Gerratt, *J. Am. Chem. Soc.* **120**, 3975 (1998).
17. P. B. Karadakov, D. L. Cooper and J. Gerratt, *Theor. Chem. Acc.* **100**, 222 (1998).
18. P. B. Karadakov, D. L. Cooper, T. Thorsteinsson and J. Gerratt, in A. Hernández-Laguna, J. Maruani, R. McWeeny and S. Wilson (eds) *Quantum Systems in Chemistry and Physics. Volume 1: Basic problems and models systems.* (Kluwer, Dordrecht, 2000).
19. See, for example: P. J. Hay, *J. Am. Chem. Soc.* **99,** 1003 (1977); H. Wallmeier and W. Kutzelnigg, *J. Am. Chem. Soc.* **101**, 2804 (1979); W. Kutzelnigg, *Angew. Chem., Int. Ed. Engl.* **23**, 272 (1984); E. Magnusson and H. F. Schaefer, *J. Chem. Phys.* **83**, 5721 (1985); A. E. Reed and F. Weinhold, *J. Am. Chem. Soc.* **108**, 3586 (1986); C. H. Patterson and R. P. Messmer, *J. Am. Chem. Soc.* **111**, 8059 (1989); C. H. Patterson and R. P. Messmer, *J. Am. Chem. Soc.* **112**, 4138 (1990); A. E. Reed and P. v. R. Schleyer, *J. Am. Chem. Soc.* **112**, 1434 (1990); E. Magnusson, *J. Am. Chem. Soc.* **112**, 7940 (1990); E. Magnusson, *J. Am. Chem. Soc.* **115**, 1051 (1993); R. P. Messmer, *J. Am. Chem. Soc.* **113**, 433 (1991); M. Häser, *J. Am. Chem. Soc.* **118**, 7311 (1996).
20. L. Cooper, T. P. Cunningham, J. Gerratt, P. B. Karadakov and M, Raimondi, *J. Am. Chem. Soc.* **116**, 4414 (1994).
21. T. P. Cunningham, D. L. Cooper, J. Gerratt, P. B. Karadakov and M. Raimondi, *Int. J. Quant. Chem.* **60**, 399 (1996).
22. L. Cooper, J. Gerratt, M. Raimondi and S. C. Wright, *Chem. Phys. Lett.* **138**, 296 (1987); D. L. Cooper, J. Gerratt and M. Raimondi, *J. Chem. Soc. Perkin Trans. 2*, 1187 (1989); T. Thorsteinsson, D. L. Cooper, J. Gerratt, P. B. Karadakov and M. Raimondi, *Theor. Chim. Acta* **93**, 343 (1996); T. P. Cunningham, D. L. Cooper, J. Gerratt, P. B. Karadakov

and M. Raimondi, *J. Chem. Soc. Faraday Trans.* **93**, 2247 (1997); D. L. Cooper, J. Gerratt and M. Raimondi, Hypercoordinate bonding to main group elements: the spin-coupled point of view, in Z. B. Maksić and W. J. Orville-Thomas (eds) *Pauling's Legacy: Modern Modelling of the Chemical Bond* (Elsevier, Amsterdam, 1999); D. L. Cooper, to be published.

23. J. Pipek and P. G. Mezey, *J. Chem. Phys.* **90**, 4916 (1989).
24. Raos, S. J. McNicholas, J. Gerratt, D. L. Cooper and P. B. Karadakov, *J. Phys. Chem. B* **101**, 6688 (1997); B. Friis-Jensen, D. L. Cooper, S. Rettrup and P. B. Karadakov, *J. Chem. Soc. Faraday Trans.* **94**, 3301 (1998).

Chapter 12

Quantum Molecular Similarity: Theory and Applications to the Evaluation of Molecular Properties, Biological Activities and Toxicity

Ramon Carbó-Dorca, Lluís Amat, Emili Besalú, Xavier Gironés and David Robert
Institut de Química Computacional, Universitat de Girona, Girona 17071, Catalonia (Spain)

1. INTRODUCTION

1.1 Historical Development

In this chapter we present an updated revision of the mathematical interpretation leading to further development of the theory and practice associated with quantum similarity measures (QSM) [1-40]. The role of QSM can be resumed on their ability to be the vehicle producing discrete n-dimensional mathematical representations of molecular structures. This property transforms QSM into a general source of unbiased molecular descriptors. QSM descriptors are general indeed, because of their quantum chemical origin: They can be computed, in principle, for any quantum system or any molecular structure possessing arbitrary geometrical conformation or state. Moreover, QSM descriptors are unbiased, because their values are not chosen by a priori designs: They are built up as a consequence of the theoretical quantum framework results and only depend on the nature and topological characteristics of the studied molecular set.

Quantum object (QO) definition and analysis, as provided in early work [41-44] and used in present times [45-78], quite often constitute the main argument of the present study. QO definition requires a fundamental

Fundamentals of Molecular Similarity, Edited by Carbó-Dorca *et al.*
Kluwer Academic/Plenum Publishers, New York 2001

ancillary structure, which may be related to a generalisation of fuzzy set theory [79,80], but can be independently redefined as a new collection of mathematical devices: tagged sets [81]. Within this point of view, QO concept appears inseparably connected to density functions (DF).

Mainly due to practical purposes, first order DF become good candidates to be used in this kind of molecular QO discrete description and comparison [1]. Although higher order DF can be employed as well [48,50,52], and other kinds of QO can be studied in the same way as molecular structures are [53,54]. Even other probability distributions of no quantum origin can be used for the same purpose [74]. Several new possibilities become apparent along the path of this theoretical development, among others: kinetic energy, electrostatic potential distributions, and DF transformations [82]. A naïve mathematical discussion permits to uncover the way some general DF can be structured. At the same time, a natural fashion to describe the particular kind of operations and representations involved in a general formulation of DF emerges easily from theory development. Thus, tagged sets DF and QO open the way to an easy definition of QSM and their generalisation [41-44,81].

Once established the appropriate theoretical framework as a whole, the possibility of using QSM appears as a sound tool to construct n-dimensional QO descriptions. This circumstance produces, in turn, the way to study several aspects related to quantitative structure-properties relationships (QSPR). Among other possibilities: a general theoretical foundation, and thus a better understanding of these techniques, the possible computation of a large collection of new QO descriptors, a general discussion on molecular topology...

1.2 Initial Considerations

The evolution, relative to the relationship between quantum chemistry and QSPR-QSAR computational formalisms, can be understood by examining the ideas associated to the original effort to clarify the theory frame as much as possible.

Bell's [82] presentation of various proposals exposed can shed light into the remaining ambiguous theoretical aspects of quantum mechanics. One of Bell's suggestions to start with this, perhaps never ending and arduous, task was to produce a set of clear background definitions, where the theory can find out secure logical foundations as starting points to easily develop new ideas afterwards. Following the Bell's spirit of reinterpretation, in the field of QSM, this chapter presents a set of definitions, which can be employed to cover the whole area, starting from the basic aspects and ending at the practical applications.

Thus, this chapter is organised in the following way: Tagged sets and vector semispaces are defined. Density functions and their extension and generalisation are analysed next. Quantum objects and convex sets are discussed afterwards. Quantum similarity measures follow, and the related subjects of discretization and similarity matrices presented. A discussion on both old and novel aspects of molecular superposition will end the theoretical part and serve as the initial step towards the application of QSM into QSAR. In this way, this previous collection of ideas moves effortlessly into a study of the theoretical background of QSPR and illuminates several varied aspects of the problem. Finally, an introduction is given to the possible definition of alternative topological quantum similarity matrices as a source of new molecular descriptors. Numerical or graphical application examples, as they relate to the pure theoretical development, are provided.

2. TAGGED SETS

Consider an arbitrary collection of objects, forming a set, and a collection of mathematical elements: Boolean strings, vectors, matrices, functions, etc. forming another set, which can be chosen independently from the initial one. Both sets can be related by means of a new composition rule, according to the following:

Definition 1: Tagged sets

Let us suppose a given set, the object set, S, and another set, made of mathematical elements, which will be hereafter called *tags*, forming a tag set, T. A *tagged set*, Z, can be constructed using the ordered product: Z=S×T:

$$\mathrm{Z}=\left\{\forall\theta\in\mathrm{Z}\,\middle|\,\exists s\in\mathrm{S}\wedge\exists t\in\mathrm{T}\rightarrow\theta=(s,t)\right\}.\square$$

Tagged sets constitute a mathematical structure, frequently present in many areas, but particularly within the chemical information universe. Atomic or molecular parametric description may be made and studied inside this general, simple tagged set construct. An unnoticed tagged set structure began when chemistry stepped out of alchemy. Such a situation in contemporary chemistry can be put into evidence by setting a tagged set formal building up rule, where molecules become elements of the object set and their ordered attributes can be attached to the tag set. The atomic periodic table is a trivial paradigm of a tagged set. More definitions, examples and extensions of tagged sets can be found in Appendix A.

2.1 Vector Semispaces

Plausible tag set candidates: Boolean strings, positive definite (PD) n-dimensional vectors, and functions permit a great flexibility when a particular molecular tagged set has to be defined. It is important to consider which kind of tag sets can be chosen as candidates to represent molecules in a discrete n-dimensional framework.

2.1.1 Definition of VSS

A natural choice of the mentioned tag set options may be constituted by n-dimensional vector spaces with some appropriate restrictions. The next auxiliary definition can be used for this purpose:

Definition 2: Vector semispaces.

A vector semispace (VSS) over the PD real field $\mathbf{R}^+$, is a vector space (VS) with a structure of abelian *semigroup* associated to the vector addition.

By an additive semigroup [84], it is understood an additive group, whithout reciprocal elements. All VSS elements can be considered directed towards the region of the whole set of positive axis directions. Throughout this contribution, if nothing is added, it will be assumed that *null elements* are both included in the scalar field as well as in the VSS structure. Such a VSS will be noted as: $V(\mathbf{R}^+)$. However, there could be defined another specific VSS structure, lacking of null elements completely. In this particular case, the adjective *Strict* is written before the VSS name (SVSS) and the notation $V(\mathbf{R}_0^+)$, could be used. SVSS, besides the restrictions of the additive semigroup, do not possess the neutral element, becoming a set with just a closed additive abelian composition.

VSS linear combinations to be considered are made with positive coefficients, and so, vector coordinates are always positive or, in some special cases, null. This last possibility is ruled out in SVSS. Metric VSS may be constructed in the same manner as the usual metric VS, taking into account that scalar products will also become PD in VSS. Any VSS could be taken as generated by some VS, if some rule is defined, describing the transformation of the VS elements into those of the VSS. Some relevant VSS generating rules are discussed in the next sections.

From the tagged set side, nothing opposes to the fact that, using a computational point of view, any VSS tag set part can be constituted by normalised vectors, whose elements will then be (rational) numbers belonging to the unit segment [0,1]. Thus, in this way VSS tag sets could be naturally associated to Boolean tag sets as those described in Appendix A.

2.1.2 An example of SVSS

A very interesting SVSS example, referred to molecular structure may be constructed in the following manner. Suppose a subset **X** made of p elements taken from some n-dimensional metric VS:

$$\mathbf{X} = \{\mathbf{x}_1, \mathbf{x}_2, \ldots, \mathbf{x}_p\} \subseteq \boldsymbol{V}_n(\mathbf{R})$$

Suppose a distance is defined in the usual way in **X**, then a vector constructed collecting all nondiagonal and nonredundant distances over the elements of the subset **X**:

$$|\boldsymbol{D}\rangle = \{d(\mathbf{x}_i;\mathbf{x}_j); \forall \mathrm{i} < \mathrm{j}\} \in \boldsymbol{V}_{\frac{1}{2}(p(p-1))}(\mathbf{R}_0^+).$$

The elements of the distances vector belong to a SVSS of dimension ½(p(p-1)). One can think of these SVSS as the collection of the nonredundant PD distance vectors associated to all the possible conformations of a given molecular structure possessing p noncoincident atoms.

The elements of the vectors $|\boldsymbol{D}>$ are used as variables in some molecular functions, such as nuclear repulsion, where **X** is the set of molecular atomic coordinates defined in a 3-dimensional space.

2.1.3 SVSS and Topological matrices

Another interesting application can be found in the topological matrix definitions. A molecular topological matrix, corresponds to a vector like $|\boldsymbol{D}>$, but with the following differences: The space where the topological vector belongs is now: $\boldsymbol{V}_{\frac{1}{2}(p(p-1))}(\mathbf{R}^+)$. The elements of $|\boldsymbol{D}>$ are transformed into the bits {1;0} if the associated distance element corresponds to a pair of neighbour bonded atoms or not respectively. Also $|\boldsymbol{D}>$, when made according to these rules, can be observed as the vertex of some unit hypercube, with the appropriate dimensions.

Similar transformations can be made using the atomic self-similarity matrix associated to a given molecular structure. Then $|\mathbf{D}>$ is transformed into a new vector of a similar SVSS $|\boldsymbol{T}>$, by using some continuous PD function such that: $t: d(\mathbf{x}_i;\mathbf{x}_j) \rightarrow t(d(\mathbf{x}_i;\mathbf{x}_j)) \in \mathbf{R}_0^+$. The nature and use of this function will be discussed in section 6.

3. DENSITY FUNCTIONS (DF) AS OBJECT TAGS

This section discusses the background ideas to explain the fundamental relationship between quantum systems and tagged sets. It can be admitted that DF are, since the early times of quantum mechanics, an indispensable tool to precisely define all the properties of mechanical systems at the adequate microscopic scale, see for example [85-87]. Therefore, tag set parts made of quantum DF ought to be associated to quantum system's information. The quantum theoretical structure attached to DF, fits perfectly into the tagged set formalism and permits the definition of valuable new elements. Moreover, DF can be described in a very general and plausible way, as it is shown in sections 3.1 and 3.4 below and in Appendices B, C and F.

3.1 Density Functions

DF are originated within the main body of computational Quantum Mechanics, resumed in the following Algorithm:

Algorithm 1: Classical Quantum Mechanics

The main computational scheme of quantum mechanics can be described in three steps:

a) Construction of the Hamiltonian operator, H.

b) Computation of the state energy-wavefunction pairs, {E, Ψ}, by solving Schrödinger equation: H Ψ= E Ψ.

c) Evaluation of the state DF: $\rho = |\Psi|^2$.

It is well known [88,89] how DF can be variable reduced. Integrating the raw DF definition, which appears as the third step of Algorithm 1, over the entire system particle coordinates, except r of them, produces a r-th order DF. This kind of reduction has been studied in many ways [90,91] since the early days of quantum chemistry and will not be repeated here.

When practical implementation of QSM has been considered in this laboratory, a simplified manner to construct the first-order DF form [92-97] has also been proposed and named Atomic Shell Approximation (ASA) DF, see Appendix B for detailed information on this technique. A computational procedure has been recently described [96], containing the correct necessary conditions to obtain PD ASA DF, possessing appropriate probability distribution properties. The ASA DF development could be trivially related with the first-order DF form as expressed within MO theory, because the first-order MO DF structure can be defined as:

$$\rho(\mathbf{r}) = \sum_i \mathrm{w}_i |\varphi_i(\mathbf{r})|^2 . \qquad (1)$$

MO DF can be written generally as a double sum of products of function pairs, coupled with a set of matrix coefficients [98]. However, a simple matrix diagonalization, followed by a unitary MO basis set transformation, can revert DF to the formal expression in Equation (1), see for instance [88,89] or [99] for more technical details.

The coefficient set: $\mathbf{W}=\{w_i\}\subset\mathbf{R}^+$, usually interpreted as MO occupation indices, corresponds in any case to a collection of positive real numbers. The original MO function set: $\mathbf{f}=\{\varphi_i(\mathbf{r})\}\subset\mathbf{H}(\mathbf{C})$ belongs to a Hilbert VS, but appears when used in DF expressions by means of a squared modular form, that is: $F=\{|\varphi_i(\mathbf{r})|^2\}\subset H(\mathbf{C})$. This new basis is a set of PD functions belonging to an ∞-dimensional VSS. The result of the PD linear combination of PD functions is a PD first order DF, $\rho(\mathbf{r})$.

Here, a unit norm convention has been adopted:

$$\int|\varphi_i|^2 dr = 1; \forall i \;\Rightarrow\; \int\rho(\mathbf{r})d\mathbf{r} = \sum_i w_i \int|\varphi_i|^2 d\mathbf{r} = \sum_i w_i = 1, \tag{2}$$

and this result can be interpreted considering the coefficient set $\mathbf{W}=\{w_i\}$ as a discrete probability distribution.

$\mathbf{W}$ can be considered, in the most general terms, a PD discrete real, in computational practice rational, set and it can be organised as some n-dimensional vector, $\mathbf{w}=(w_1,w_2,\ldots,w_n)$. Moreover, either $\mathbf{W}$ or $\mathbf{w}$ can be generated using a complex coefficient set: $\mathbf{X}=\{x_i\}\subset\mathbf{C}$. A new coefficient set can always be obtained using the modules of the $\mathbf{X}$ set elements, as: $w_i = |x_i|^2; \forall i$. Supposing we defined a vector with the $\mathbf{X}$ elements: $\mathbf{x}=(x_1,x_2,\ldots,x_n)$, then the norm of such a vector can be forced to become unit: $N(\mathbf{x}) = \langle\mathbf{x}|\mathbf{x}\rangle = 1$, corresponding in this way to the last condition in Equation (2).

Together all of this defines a device in close parallelism to the more general quantum mechanical ∞-dimensional generating rule, presented in Equation (10) in Section 3.3.2 below. For this purpose a *discrete generating rule* can be described as follows:

$$R(\mathbf{x}\to\mathbf{w}) = \left\{\begin{array}{l}\forall\mathbf{x}\in V_n(\mathbf{C})\to\exists\mathbf{w}=\left\{w_i = x_i^* x_i = |x_i|^2\right\}\in V_n(\mathbf{R}^+)\\ \wedge\langle\mathbf{x}|\mathbf{x}\rangle = \sum_i x_i^* x_i = \sum_i |x_i|^2 = 1 \to \langle\mathbf{w}\rangle = \sum_i w_i = 1\end{array}\right\}. \tag{3}$$

If an equivalent set of conditions as those shown in Equation (2) hold for some r-th order DF basis functions, a discrete generating rule, such as the one described above, can be extended to DF of arbitrary r-th order too.

A question not really important in practice, but interesting from the purely theoretical point of view can be posed as whether in the space the definition of the scalar field, appearing in the previous generating rule and in other places in the following discussion, has to be $\mathbf{R}^+$ or $\mathbf{R}_0^+$. Although the first and usual notation seems more general, the logical coherence of the functions and vectors appearing in the tag set parts seem to favour the strict PD of the involved scalars. The definitions and properties of all next sections are made over $\mathbf{R}^+$, but they can easily be associated to $\mathbf{R}_0^+$, too.

3.2 Convex Conditions

Optimisation of the coefficient vector, **w**, in order to obtain an approximate function completely adapted to ab initio DF, must be restricted within the boundaries of some VSS: $V_n(\mathbf{R}^+)$ and the element sum, **<w>**, shall be unity. This feature could be cast into a unique symbol, which can be referred to as *convex conditions*, $K_n(\mathbf{w})$, applying over the n-dimensional vector **w** and written as:

$$K_n(\mathbf{w}) \equiv \left\{ \mathbf{w} \in V_n(R^+) \wedge \langle \mathbf{w} \rangle = \sum_i w_i = 1 \right\} \tag{4}$$

The symbol: **<w>**, is used as a particular case of the following definition, which was proposed in another context [100], some time ago:

Definition 3: Elements Sum of a (M×N) Matrix **A**
Known a (m×n) matrix A={a_{ij}} by the symbol **<A>** it is meant:

$$\langle \mathbf{A} \rangle = \sum_{i=1}^{m} \sum_{j=1}^{n} a_{ij} . \square$$

It must be noted that when $\mathbf{A}$=Diag(a_i), then **<A>**=Tr|**A**|. Also, the matrix operation **<A>** can be considered a linear transformation: $\forall \mathbf{A} \in M_{(m\times n)}(\mathbf{F})$:**<A>**→**F**, from the (m×n) matrix vector space defined over a field **F** into the same field **F**.

In a similar notation, as in Equation (4), the elements of the set **W,** or these of the vector, {w_i} can be used instead in the symbol defining the convex conditions, that is:

$$K_n(\{w_i\}) \equiv \left\{ \forall i : w_i \in R^+ \wedge \sum_i w_i = 1 \right\}. \tag{5}$$

Together, Equations (4) and (5) can be considered the discrete counterparts of continuous convex conditions, defining a convex DF in some ∞-dimensional VSS:

$$K_\infty(\rho) \equiv \left\{ \rho \in H(R^+) \wedge \int \rho(r)\, dr = 1 \right\}. \tag{6}$$

It is obvious that convexity may be in close relationship with the definition of QO. Also, vector coefficients may be easily transformed by means of norm conserving orthogonal transformations, such as Elementary Jacobi Rotations (EJR) [101]. EJR, or other unitary or orthogonal transformations, can be applied over a generating vector to obtain the desired optimal coefficients, while preserving convex conditions [42,78,96] (see also Appendix B).

As commented at the end of Section 3.1, the DF form shown in Equation (1) can be used to build up new DF elements, preserving $K_\infty(\rho)$. If **w** is taken as a vector, assuming the convex conditions $K_n(\mathbf{w})$, while $\mathrm{P} = \{\rho_i(\mathbf{r})\} \subseteq \mathrm{H}(\mathbf{R}^+)$, is used as a given set of homogeneous order DF, then the linear combination:

$$\rho(r) = \sum_i w_i \rho_i(r) \in \mathrm{H}(\mathbf{R}^+) \tag{7}$$

produces a new DF with the same order and characteristic properties like the elements in the set P. It can be said that convex conditions over vector coefficients, affecting DF linear combinations, are the way to allow the construction of new DF of the same nature, possessing the same properties.

General forms of density functions can be defined within other contexts (see Section G3.3 of Appendix G).

3.3 Quantum Objects

The idea of Quantum Object (QO) without a well-designed definition has been already used in the literature [45-49,52]. However, the background mathematical structure leading towards the recently published [41,42] definition of QO has to be considered as a consequence of the previous section. In order to obtain a convenient QO definition, some preliminary considerations are needed.

3.3.1 Expectation Values in Quantum Mechanics

These can begin considering the fact that customary quantum study of microscopic systems is essentially associated with Algorithm 1, defined above. According to quantum mechanics, known the state DF, all observable property values of the system, ω, can be formally extracted from it, as expectation values, $\langle\omega\rangle$ see for example [102], of the associated Hermitian operator, Ω, acting over the corresponding function, ρ. In the same way as in theoretical statistics it can be written:

$$\langle\omega\rangle = \int \Omega(\mathbf{r})\,\rho(\mathbf{r})\,d\mathbf{r}, \tag{8}$$

where $\mathbf{r}$ shall be considered a p-dimensional particle coordinate matrix. It must be also noted that Equation (8) can be interpreted as some scalar product or linear functional: $\langle\omega\rangle = \langle\Omega|\rho\rangle$, defined within the space where both the involved $\Omega(\mathbf{r})$ and $\rho(\mathbf{r})$ p-particle operators belong. The expectation value integral associated to Equation (8) will be referred hereafter as the *statistical form* of expectation values, to distinguish it from the quantum mechanical form, which could be written as:

$$\langle\omega\rangle = \int (\Psi(\mathbf{r}))^{*}\,\Omega(\mathbf{r})\,\Psi(\mathbf{r})\,d\mathbf{r}. \tag{9}$$

3.3.2 An Application Example of Statistical Form of Expectation Values: Generalised Hohenberg-Kohn Theorem

A generalisation of the well-known Hohenberg-Kohn theorem [106] can be related to the statistical form taken by expectation values. This section will give a short discussion, published in another context [107], as an example of how powerful the use of the expectation value statistical form may be. This discussion has been raised by the recent work of Mezey [108] who has reformulated the Hohenberg-Kohn theorem, using an information theoretical approach.

Consider a quantum system, with an attached Hamiltonian H_0, with nondegenerate ground state, characterised by the energy-wavefunction pair: $\{E_0,\Psi_0\}$, and another system with known Hamiltonian H, with a non-degenerate ground state described by the pair: $\{E,\Psi\}$. Assuming that both wavefunctions are normalised, then the following relationships will hold: $E_0 = \langle \Psi_0 | H_0 | \Psi_0 \rangle$ and $E = \langle \Psi | H | \Psi \rangle$. Also, approximate energies for both ground states can be defined, by exchanging the position of both wavefunctions, respectively, as $E_{a,0} = \langle \Psi | H_0 | \Psi \rangle$ and $E_a = \langle \Psi_0 | H | \Psi_0 \rangle$. In general, and independent of the system description, the inequalities: $\Delta_0 = E_0 - E_{a;0} \leq 0$ and $\Delta = E - E_a \leq 0$, will be valid.

Formally, from both wavefunctions, arbitrary order density functions can be made [88,89,102,104]. Here, the process of constructing density functions of arbitrary order, will be symbolised, for both ground states considered above, by the projectors: $\rho_0 = |\Psi_0\rangle\langle\Psi_0|$ and $\rho = |\Psi\rangle\langle\Psi|$, respectively. Employing this convention, another set of energy expressions can be used taking into account the statistical form of expectation values as in Equation (8), namely: $E_0 = \langle H_0 | \rho_0 \rangle$ and $E = \langle H | \rho \rangle$. This can be extended to approximate energy expressions as well, that is: $E_{a;0} = \langle H_0 | \rho \rangle$ and $E_a = \langle H | \rho_0 \rangle$.

The negative energy difference, previously defined for the first system, can be formally rewritten now as follows:

$$\Delta_0 = \langle H_0 | \rho_0 \rangle - \langle H_0 | \rho \rangle = \langle H_0 | \rho_0 - \rho \rangle = \langle H_0 | \Delta\rho \rangle,$$

and a similar sequence can be written for the second system:

$$\Delta = \langle H | \rho \rangle - \langle H | \rho_0 \rangle = -\langle H | \rho_0 - \rho \rangle = -\langle H | \Delta\rho \rangle.$$

Since both energy differences are non-PD, the following inequality holds rigorously: $\Delta_0 + \Delta \leq 0$. This will give the following sequence of differences:

$$\Delta_0 + \Delta = \langle H_0 | \Delta\rho \rangle - \langle H | \Delta\rho \rangle = \langle H_0 - H | \Delta\rho \rangle = \langle \Delta H | \Delta\rho \rangle \leq 0.$$

Essentially the same kind of information can be obtained if all the relationships are written in such a way that the last inequality becomes PD instead of non-PD. A sign reversal and a substitution of the symbol $\leq$ by $\geq$ will produce essentially the same information. This symmetrical characteristic has to be related to the VSS structure of the spaces associated to quantum mechanical Tag Sets. However, here, as well as in the original

Hohenberg-Kohn reasoning the usual quantum mechanical variational convention has been kept.

The last inequality above shows that the action of the Hamiltonian increment upon the density difference will be, in any case, a non-PD quantity, which can be symbolised by the expression: $<\Delta H|\Delta\rho>\leq 0$. Four cases can be considered, which arise out of the respective increments to be null or not. But only the diagonal situations, the ones with both increments simultaneously null or not null, will be relevant. Precisely, the situation: $\Delta H\neq 0$ and $\Delta\rho=0$, produces the reductio ad absurdum, and was used [106] to deduce the Hohenberg-Kohn Theorem, as the argument to accept the functional interdependence of density and potential. The similar and equally nonrealistic increment values: $\Delta H=0$ and $\Delta\rho\neq 0$, can be discarded. The couple of remaining possibilities provide the trivial equality: $<\Delta H|\Delta\rho>=0$, when both increments are null; and, finally, when: $\Delta H\neq 0$ and $\Delta\rho\neq 0$ the result: $<\Delta H|\Delta\rho><0$, will hold. This discussion completes a general form of Hohenberg-Kohn theorem.

The discussion above serves to find out that the only relevant situation is the one, where both Hamiltonian and density increments are not null. This is the same as to admit that both ΔH and $\Delta\rho$ must belong to some SVSS.

Finally, it must be stressed that the main argument throughout this section has been founded on the use of the statistical form of quantum mechanical expectation values. The applied statistical form makes the theorem completely general, as it is not necessary to consider any restricted structure associated a priori to ΔH or $\Delta\rho$. However, this statistical form has been applied here without proof that it can be used in all circumstances. Even in cases when the custom admits the use of the quantum chemical expectation value structure. This can be employed as an indirect proof that the general H-K theorem should be taken as a consequence of some previous theorem, dealing with the possible universal application of the statistical-like form of quantum mechanical expectation values. Section 3.4 below will prove this.

3.3.3 Quantum Object (QO) Definition and Generating Rules

Thus, after these preliminary considerations, the next definition can be made.

Definition 4: Quantum Object

A quantum object is an element of a tagged set, made by quantum systems in well-defined states taken as the object set part and the corresponding density functions constituting the tag set part. Such a tagged set will be named a quantum object Set (QOS). □

The interesting fact, which must be stressed here, is the DF leading role played in quantum mechanical systems description and, as a consequence, in the QO definition. The DF generation in varied wavefunction environments has been studied since the early times of quantum chemistry [88,89]. The most appealing aspect of this situation corresponds to the way DF, ρ, are constructed, starting from the original system's wavefunctions, Ψ. This formation process, already used in Section 3.1, has been called a *generating rule* [42], which can be shortened by using the symbol: $R(\Psi \rightarrow \rho)$. A continuous generating rule can easily be written, summarising the three steps of quantum mechanical Algorithm 1:

$$R(\Psi \rightarrow \rho) = \{\forall \Psi \in H(\mathbf{C}) \rightarrow \exists \rho = \Psi^* \Psi = |\Psi|^2 \in H(\mathbf{R}^+)\}. \quad (10)$$

In Equation (10) there are given explicitly the wavefunction Hilbert VS [109], $H(\mathbf{C})$, and the DF VSS, $H(\mathbf{R}^+)$, defined over the complex and the positive real fields, respectively.

3.4 Positive Definite Operators, Kinetic energy and Extended Hilbert Spaces

The relationship between DF and PD Operators will be studied in this section. The DF themselves may be considered as elements of a VSS or, alternatively, as members of a PD operator set, which can be collected in turn into another isomorphic VSS, whose elements may be considered PD operators.

The most relevant feature to be noted in the context of PD Operator VSS, as well as in the isomorphic VSS structure, is the *closed nature* of such VSS, when appropriate PD coefficient sets are known. That is: PD linear combinations of PD Operators remain PD Operators. Discrete matrix representations of such PD Operators are PD too, and PD linear combinations of PD matrices will remain PD in the same way. These properties can be expressed in a compact and elegant way, using convex conditions symbols, as previously discussed in Section 3.2: If $\{K_\infty(\rho_i);\forall i\}$ and $K_n(\mathbf{w})$ hold, then Equation (7) is a convex function fulfilling $K_\infty(\rho)$.

3.4.1 Differential Operators and Kinetic Energy

Another interesting question could be attached to the interpretation of the differential operators role, as connected with momentum within the framework of classical quantum mechanics when the position space point of view is chosen, which constitutes the usual, most frequent, computational chemistry option.

3.4.1.1 Statement of the Problem

There appears to be a formal puzzle when one tries to connect a second order differential operator, representing the QO kinetic energy (KE), using the expression of an expectation value. KE expectation values do not fulfil the usual statistical formalism represented by Equation (8), but they possess a kind of expression, according to the quantum mechanical expectation value form (9), which adequately transformed and, avoiding scalar factors, looks like a norm, when writing the equalities:

$$2\langle K\rangle = -\int \Psi^{*}\nabla^{2}\Psi \, dV = \int (\nabla\Psi)^{*}(\nabla\Psi)\, dV , \qquad (11)$$

where the change of sign can be attributed to Green's first identity [110].

3.4.1.2 Extended Hilbert Spaces

KE integrands, as in Equation (11), present in the form of a module, can formally be considered behaving as DF. Thus, such KE distributions can supposedly belong to a given VSS. Admitting a modification of the original Hilbert space, H(**C**), where wavefunctions belong, into another extended one, $H^{(\nabla)}(\mathbf{C})$, which also contains the wavefunction first derivatives, the quantum mechanical momentum representation, can be expressed by means of the implication:

$$\forall\Psi \in H(\mathbf{C}) \Rightarrow \Psi \in H^{(\nabla)}(\mathbf{C}) \wedge \exists\nabla(\Psi) \in H^{(\nabla)}(\mathbf{C})$$

Then, considering the attached DF VSS, $H(\mathbf{R}^{+})$, it can also be accepted that the next implication holds:

$$\forall\rho = |\Psi|^{2} \in H(\mathbf{R}^{+}) \Rightarrow \exists\kappa = |\nabla\Psi|^{2} \in H(\mathbf{R}^{+}),$$

so, to every DF there exists a momentum DF. In a better descriptive way, this kind of distributions can be named as: KE DF, assuming they belong to some Hilbert VSS. The KE DF when integrated yields the expectation value of the QO KE.

It can be concluded that KE can be considered related to the momentum norm, the QO wavefunction gradient. In molecules shall be expected a similar behaviour of both electronic and KE DF distributions at large distances, and also it can be easily deduced that they will possess very different forms at the nuclei neighbourhoods. Some visual examples can be found in Appendix D.

3.4.1.3 Generating Rules in Extended Wavefunctions

In order to obtain a coherent picture, with KE occupying a sound place, among other quantum mechanical structures, the extended Hilbert VS, $H^{(\nabla)}(\mathbf{C})$, could be defined containing not only QO wavefunctions but their first derivatives, too. This allows us to construct the associated DF VSS, $H^{(\nabla)}(\mathbf{R}^+)$, as containing not only DF but also KE DF. The elements of this peculiar Hilbert VS, where both wavefunctions and their gradients are contained, can be ordered in a first instance as column vectors, such as:

$$|\Phi\rangle = |\Psi ; \nabla\Psi\rangle \in H^{(\nabla)}(\mathbf{C}),$$

this can be considered a scalar to vector transformation. Using a vector operator, involving the gradient, such as:

$$|1;\nabla\rangle[\Psi] = |\Psi;\nabla\Psi\rangle = |\Phi\rangle .$$

Or by an isomorphic diagonal transformation, employing the same elements as before:

$$\mathrm{Diag}(1,\nabla)|\Psi\rangle = \mathrm{Diag}(|\Psi\rangle, \nabla|\Psi\rangle) = |\Phi\rangle . \tag{12}$$

The generating rule within the extended wavefunction domain, can be written as:

$$\begin{aligned} R(|\Phi\rangle \rightarrow |\rho;\kappa\rangle) = \{ & \forall |\Phi\rangle = |\Psi;\nabla\Psi\rangle \in H^{(\nabla)}(\mathbf{C}) \\ & \rightarrow \exists \rho = \Psi^*\Psi = |\Psi|^2 \\ & \wedge \exists \kappa = (\nabla\Psi)^*(\nabla\Psi) = |\nabla\Psi|^2 \\ & \Rightarrow |\rho;\kappa\rangle \in H^{(\nabla)}(\mathbf{R}^+) \}. \end{aligned} \tag{13}$$

The DF ρ can be considered normalised in general, according to an equivalent form of Equation (2). The KE DF κ, can be normalised too, the gradient density norms being in absolute value twice the expectation value of the kinetic energy, $\langle K \rangle$. This amounts the same as to consider the extended wavefunction, $|\Phi\rangle$, normalised: $\langle \Phi | \Phi \rangle = (1 + 2\langle K \rangle)$. This is an outcome of the characteristics of the spaces containing both, the wavefunction and their gradient. As a consequence, the extended Hilbert space elements can be considered square summable functions.

3.4.1.4 Extended wavefunction projectors

The projectors associated to extended quantum mechanical wavefunctions will possess a matrix structure like:

$$|\Phi\rangle\langle\Phi| = \begin{pmatrix} |\Psi|^2 & \Psi^*(\nabla\Psi) \\ (\nabla\Psi)^*\Psi & |\nabla\Psi|^2 \end{pmatrix} = P$$

then, using symmetrisation: $Q = 1/2\left(P^+ + P\right)$, the new projector could be written as the matrix:

$$Q = \begin{pmatrix} \rho & |j\rangle \\ \langle j| & \kappa \end{pmatrix}$$

and thus: $\mathrm{Tr}(Q) = \mathrm{Tr}(P) = \rho + \kappa$. Examples of other possible extended DF characteristics can be found in Appendix C, while visualisation of isodensity surfaces are located in Appendix D.

3.4.2 Diagonal Hamiltonian Operators

There is only a final detail to underline and thereafter to discuss: the calculation of energy expectation values, within the extended Hilbert space framework.

This can be done, considering Born-Oppenheimer approach, defining an electronic Hamiltonian operator with a diagonal matrix structure and, in addition, supposing the wavefunction, $|\Psi\rangle$, normalised:

$$\mathbf{H} = \mathrm{Diag}\left(V; \tfrac{1}{2}\mathbf{I}\right) \wedge \langle \Psi | \Psi \rangle = 1 \Rightarrow \boldsymbol{E} = \langle \Phi | \mathbf{H} | \Phi \rangle = \langle \Psi | V | \Psi \rangle + \tfrac{1}{2}\langle \nabla\Psi | \nabla\Psi \rangle \equiv \langle V \rangle + \langle K \rangle. \quad (14)$$

In the diagonal Hamiltonian operator definition, V is the potential part and **I**, a unit matrix with the appropriate dimension. This result shows that in the extended Hilbert space $\boldsymbol{H}^{(\nabla)}(\mathbf{C})$, it is possible to use the standard variational procedures.

3.4.2.1 Extended Hilbert space framework Energy as a trace

Not only can the Hamiltonian operator be written as a diagonal matrix, but also the elements of the extended Hilbert space too, as shown in Equation (12). The system energy expressed as in Equation (14) can be also written as a trace of a diagonal matrix:

$$E = \langle\Phi|\mathbf{H}|\Phi\rangle = \left\langle \mathrm{Diag}\left(\langle\Psi|V|\Psi\rangle; \tfrac{1}{2}\langle\nabla\Psi|\nabla\Psi\rangle\right)\right\rangle,$$

where the symbol <**A**> is used according the Definition 3.

3.4.2.2 Statistical form of the extended framework energy

It has not been shown yet that, using diagonal Hamiltonian operators and extended wavefunction forms, a general statistical expression of the energy can be deduced.

Within a diagonal matrix formalism for both extended operators and wavefunctions, the statistical form of expectation values is preserved in any circumstance, at least when system's energy is sought. Indeed, if the extended function diagonal form is chosen, then the following sequence of integral expressions can be easily written:

$$\begin{aligned}
E &= \langle\Phi|\mathbf{H}|\Phi\rangle = \int \left\langle \mathrm{Diag}\left(\Psi^*;(\nabla\Psi)^*\right)\mathrm{Diag}(V;\tfrac{1}{2}\mathbf{I})\mathrm{Diag}(\Psi;(\nabla\Psi))\right\rangle d\mathbf{r} \\
&= \int \left\langle \mathrm{Diag}(V;\tfrac{1}{2}\mathbf{I})\mathrm{Diag}\left(\Psi^*;(\nabla\Psi)^*\right)\mathrm{Diag}(\Psi;(\nabla\Psi))\right\rangle d\mathbf{r} \\
&= \int \left\langle \mathrm{Diag}(V;\tfrac{1}{2}\mathbf{I})\mathrm{Diag}\left(|\Psi|^2;|\nabla\Psi|^2\right)\right\rangle d\mathbf{r} = \int \left\langle \mathrm{Diag}(V;\tfrac{1}{2}\mathbf{I})\mathrm{Diag}(\rho;\kappa)\right\rangle d\mathbf{r} \\
&= \left\langle \int \mathbf{H}\tau\, d\mathbf{r} \right\rangle \equiv \int \left\langle \mathrm{Diag}(V\rho;\tfrac{1}{2}\kappa)\right\rangle d\mathbf{r} = \left\langle \mathrm{Diag}\left(\int V\rho\, d\mathbf{r}; \tfrac{1}{2}\int \kappa\, d\mathbf{r}\right)\right\rangle \\
&= \left\langle \mathrm{Diag}\left(\langle V\rangle; \tfrac{1}{2}\langle\kappa\rangle\right)\right\rangle = \langle V\rangle + \langle K\rangle .
\end{aligned}$$

where has been used the commutative property of the diagonal matrices product. An extended DF in diagonal form has been represented by the symbol $\tau = \mathrm{Diag}(\rho;\kappa)$.

Such a coherent statistical picture is possible in the context of extended Hilbert spaces. The Hamiltonian operator no longer has the gradient in it, but the corresponding momentum operator is embedded into the extended Hilbert space wavefunction.

It can be concluded that, if the DF role must be preserved, then, the statistical formalism of expectation values, represented by Equation (8), can be used in classical quantum mechanics for stationary states, in every circumstance; but appropriate diagonal operator forms must be sought. In order that the extended framework becomes practical, the following conditions must hold:

- A computational environment shall be chosen. That is: The Schrödinger equation has to be transformed into the corresponding variational Rayleigh quotient.
- A diagonal form of both observable operators and extended Hilbert space wavefunctions must be adopted.
- Generalisation of these possibilities to relativistic quantum mechanics [111] has been also proved elsewhere [82].

3.4.2.3 ASA Kinetic Energy DF

ASA DF formalism could be easily transported to KE DF formalism, discussed in the extended Hilbert space context. In this case, as ASA DF can be considered constructed by the general MO DF form, as in Equation (1). The MO KE DF may be written accordingly as:

$$\kappa(\mathbf{r})=\sum_i \omega_i |\nabla \varphi_i(\mathbf{r})|^2 . \tag{15}$$

Because in ASA the basis functions $\{\varphi_i(\mathbf{r})\}$ are assumed to be normalised 1s GTO functions, with centre at the position $\mathbf{R}_\mathrm{I}$, the gradient vector can be written as:

$$\nabla \varphi_i(\mathbf{r})=\nabla\left(N(\alpha_i)\exp\left(-\alpha_i|\mathbf{r}-\mathbf{R}_\mathrm{I}|^2\right)\right)=-2\alpha_i(\mathbf{r}-\mathbf{R}_\mathrm{I})\varphi_i(\mathbf{r}),$$

and after this, the ASA KE DF expression can be obtained:

$$\kappa(\mathbf{r})=4\sum_i \gamma_i |\mathbf{r}-\mathbf{R}_\mathrm{I}|^2 |\varphi_i(\mathbf{r})|^2 ,$$

using the scaled coefficients: $\gamma_i=\omega_i\alpha_i^2$. This result tells that ASA KE DF basis functions acquire a structure such as the 3s GTO functions.

See Appendix D for more details on ASA KE DF, as well as for other extended DF shape visualisation.

4. DISCRETE QO REPRESENTATIONS

Theoretical chemistry possesses, if studied from the computational side, a discrete numerical structure. Quantum chemical descriptions are based on a mixture of continuous functions and discrete coefficients, having the last ones great computational importance. This poses no great problem, as discontinuity in space shall not be a constraint, as pointed out by Dedekind [113] long time ago. The computational quantum description of molecular systems ought to be associated per force to a discrete environment. The origin of this situation must be found in the approximate nature of the Schrödinger equation solutions for atomic and molecular systems.

In this section the possible issues of this fact will be discussed. The mathematical definitions and properties given so far and the ones provided next are directly applicable to molecules, but also they are effortlessly generalised to any quantum system, nuclei for instance [53,54]. Hence, the repeated use of the QO general concept throughout this work.

4.1 Quantum Similarity Measures

Suppose a tagged set T, formed by QO, defined from an object set, M, made by microscopic systems and taking the tag set, P, as the collection of the systems DF in a given state and computed within a uniform order for every system, that is: $T = M \times P$. Then, choose a PD operator, Ω, provided with the appropriate homogeneous dependence on the tag set DF coordinates. The following definition can be used to describe QSM [41].

Definition 5: Quantum Similarity Measures

Suppose known a quantum system set M, and a chosen DF Tag Set P. A QSM, $Z(\Omega)$, weighted by a PD operator Ω, is an application of a quantum object tagged set, $T = M \times P$, direct product: $T \otimes T$, into the PD real field, $\mathbf{R}^+$, such as $Z(\Omega) : T \otimes T \to \mathbf{R}^+$.

4.1.1 A description of some particular QSM

Here, a short review on the possible definitions associated to the QSM computation will be given. First the most used usual formulae is presented, and other possibilities are discussed. A final generalisation will be provided at the end. The role of extended DF, as presented in the previous Section 3.4.2 and in Appendix C will be also discussed.

4.1.1.1 Main Practical QSM form

In practice, the previous definition can be immediately translated into an *integral measure* computation involving two QO $\{A,B\} \in T$:

$$z_{AB}(\Omega)=\int\int \rho_A(\mathbf{r}_1)\Omega(\mathbf{r}_1,\mathbf{r}_2)\rho_B(\mathbf{r}_2)d\mathbf{r}_1 d\mathbf{r}_2 \in \mathbf{R}^+ \quad (16)$$

where $\{\rho_A,\rho_B\}\in P$, are the respective tag set DF of the involved QO pair. The tag set DF can be taken here in a very broad sense, following the previous discussion on the possible extension of the DF concept. This form as presented in Equation (16) is the one currently quoted in the literature [45-69]. Other alternative possibilities have been discussed in previous work [48], but will not be given here.

4.1.1.2 Quantum Self-Similarity Measures

When both QO, {A,B}, in Equation (16) are the same, the QSM can be called a quantum self-similarity measure (QS-SM). This last form is nothing else than a weighted norm of the involved DF:

$$z_{AA}(\Omega)=\int\int \rho_A(\mathbf{r}_1)\Omega(\mathbf{r}_1,\mathbf{r}_2)\rho_A(\mathbf{r}_2)d\mathbf{r}_1 d\mathbf{r}_2 \in \mathbf{R}^+$$

4.1.1.3 Main Properties of QSM

The values of the integrals z_{AB} are always PD and real, being all the integrand elements PD functions or operators. In Equation (16), the QSM can be interpreted as a weighted scalar product between the DF: associated with the involved QO, the weight being the operator Ω.

Finally, the PD nature of all the involved integrands, provided within the structure of this kind of measures, can be alternatively interpreted as some kind of generalised molecular volume.

4.1.1.4 Overlap-like QSM

The usual choice for the PD weight operator Ω in Equation (16) has been Dirac delta function $\delta(\mathbf{r}_1-\mathbf{r}_2)$. This transforms the general QSM definition into the so-called overlap-like QSM:

$$z_{AB}=\int\int \rho_A(\mathbf{r}_1)\delta(\mathbf{r}_1-\mathbf{r}_2)\rho_B(\mathbf{r}_2)d\mathbf{r}_1 d\mathbf{r}_2=\int \rho_A(\mathbf{r})\rho_B(\mathbf{r})d\mathbf{r} \quad (17)$$

4.1.1.5 Triple Density QSM

A choice of a third DF tag as the PD weight operator: $\Omega(\mathbf{r}_1,\mathbf{r}_2)\equiv\rho_C(\mathbf{r})$, transforms the general definition (16) into a triple density QSM [114]:

$$z_{AB;C}=\int \rho_A(\mathbf{r})\rho_C(\mathbf{r})\rho_B(\mathbf{r})d\mathbf{r} \quad (18)$$

and in the same way multiple density QSM can be defined [48,50,52,73].

4.1.2 Coulomb energy as a QS-SM

In this section, various forms of the Coulomb interaction will be analysed from a QSM point of view.

4.1.2.1 Coulomb Energy as an expectation value

Several choices are open to the QSM definition, reverting at the end to a formal structure as the one appearing in Definition 5. This may be illustrated by the expectation value of the Coulomb energy for a p-particle system, which may be written employing Equation (8), as:

$$\mathbf{R}=(\mathbf{r}_1,\mathbf{r}_2,..\mathbf{r}_p) \wedge \Omega(\mathbf{R})=\sum_{i<j} r_{ij}^{-1} : \langle C\rangle = \int \Omega(\mathbf{R})\rho(\mathbf{R})\mathrm{d}\mathbf{R}>0, \tag{19}$$

where $\mathbf{R}$ is the collection of particle coordinates and the density matrix is evaluated using the generating rule (10) over the total particle coordinates. It is well known that this produces an expression, when taking into account electronic systems in the framework of MO theory closed shell monoconfigurational case, where Coulomb $\{\langle ii|jj\rangle\}$ and exchange $\{\langle ij|ij\rangle\}$ two-electron integrals play a leading role [115]:

$$\langle C\rangle=\sum_i\sum_j(2\langle ii|jj\rangle-\langle ij|ij\rangle)>0. \tag{20}$$

The two parts of the expression can be independently used as self-similarity measures over MOs. In fact, Coulomb and exchange integrals have been used long time ago to classify MO in a geometrical manner [40]. The PD nature of the molecular quantum Coulomb energy, as a whole, can be considered such a similarity measure too. The same can be said when observing the multi-configurational equivalent of Equation (20), see for example [116]. This simple reasoning opens the way to the potential use as a molecular descriptor of this self-similarity measure, which appears computed customarily in the available quantum chemical programs.

4.1.2.2 Electrostatic potentials

Coulomb operators can be used to compare a given first-order DF $\rho(\mathbf{r})$ with the electronic part of the electrostatic molecular potential (eEMP), V($\mathbf{R}$), used here as a PD distribution, defined as in Equation (69), see

Appendix C for more details. The appropriate QSM, can be easily associated to a QS-SM, weighted with a Coulomb operator. However, this QS-SM is also nothing else than a classical Coulomb energy. The proof is evident if the following QS-SM is defined, and afterwards studied their equivalent sequence of integral forms:

$$\begin{aligned}
Z\left(|\mathbf{r}-\mathbf{R}|^{-1}\right) &= \iint \rho(\mathbf{R})|\mathbf{r}-\mathbf{R}|^{-1}\rho(\mathbf{r})\,d\mathbf{r}\,d\mathbf{R} \\
&= \int \rho(\mathbf{R})\left[\int |\mathbf{r}-\mathbf{R}|^{-1}\rho(\mathbf{r})\,d\mathbf{r}\right]d\mathbf{R} \\
&= \int \rho(\mathbf{R})V(\mathbf{R})\,d\mathbf{R} = \int \rho(\mathbf{r})V(\mathbf{r})\,d\mathbf{r} \\
&= \int \rho(\mathbf{r})\left[\int |\mathbf{r}-\mathbf{R}|^{-1}\rho(\mathbf{R})\,d\mathbf{R}\right]d\mathbf{r} \\
&= \int \left[\int \rho(\mathbf{r})|\mathbf{r}-\mathbf{R}|^{-1}\,d\mathbf{r}\right]\rho(\mathbf{R})\,d\mathbf{R} \\
&= \int V(\mathbf{R})\,\rho(\mathbf{R})\,d\mathbf{R} = \int V(\mathbf{r})\rho(\mathbf{r})\,d\mathbf{r}.
\end{aligned} \tag{21}$$

The final integrals also connect Coulomb energy with Equation (8). From this point of view, Coulomb energy can be interpreted as the expectation value of eEMP, as it is a well known fact.

4.1.2.3 Gravitational-like QSM

Comparison of two eEMP produces a new breed of QSM: the so-called gravitational-like form of QSM [48,73]. It must be said here that gravitational QSM, do not produce more significant information than overlap QSM, as one can deduce from the following integral sequence:

$$\begin{aligned}
g_{AB} &= \iint \rho_A(\mathbf{r}_1)|\mathbf{r}_1-\mathbf{r}_2|^{-2}\rho_B(\mathbf{r}_2)\,d\mathbf{r}_1 d\mathbf{r}_2 \\
&\approx \iint V_A(\mathbf{r}_1)\delta(\mathbf{r}_1-\mathbf{r}_2)V_B(\mathbf{r}_2)\,d\mathbf{r}_1 d\mathbf{r}_2 \\
&= \int V_A(\mathbf{r})V_B(\mathbf{r})\,d\mathbf{r}
\end{aligned}$$

As can be seen from the above expression, the gravitational QSM can almost be associated to an overlap between two eEMP distributions.

4.1.2.4 Coulomb energy as a QS-SM and the definition of n-th order QS-SM

As it was pointed out above, Coulomb energy may also be seen as a QSM. To start, Equation (19) may be considered. After this, using Equation

(16) and admitting A=B, a QS-SM involving a square DF, may be rewritten in the form of Equation (8):

$$z_{AA}^{(2)}(\Omega)=\int \Omega(\mathbf{R})\rho_A^2(\mathbf{R})d\mathbf{R}\,. \tag{22}$$

Thus, nothing opposes that this last second-order DF integral can be generalised to a n-th order DF form:

$$z_{AA}^{(n)}(\Omega)=\int \Omega(\mathbf{R})\rho_A^n(\mathbf{R})d\mathbf{R}\,. \tag{23}$$

In this manner, a first order QS-SM form could be easily written as a particular case of Equation (23):

$$z_{AA}^{(1)}(\Omega)=\int \Omega(\mathbf{R})\rho_A(\mathbf{R})d\mathbf{R}\,. \tag{24}$$

Then, if the Ω operator structure as given in Equation (19), is used in Equation (16), it is easy to see that: $Z_{AA}^{(1)}(\Omega)=\langle C\rangle$.

Continuing the above discussion and definitions, in Section 4.1.4 below a generalised QSM structure will be presented.

4.1.3 Other possible QSM forms

The definition in Section 3.4.2 of the KE DF, leads to the possible comparison between this new collection of DF and electronic DF, eEMP and KE DF themselves. The following QSM can be taken as examples of the application of Definition 5. The new possible measures are described in the following integral, featuring the corresponding QSM within a general formalism. For this purpose it is only needed to be aware of the possibility of employing DF other than the electronic ones.

Thus, it is only necessary to rewrite Equation (16) in an alternate way:

$$\theta_{AB}(\Omega)=\int\int \lambda_A(\mathbf{r}_1)\Omega(\mathbf{r}_1,\mathbf{r}_2)\mu_B(\mathbf{r}_2)d\mathbf{r}_1 d\mathbf{r}_2 \in \mathbf{R}^+ \tag{25}$$

where the $\{\lambda;\mu\}$ pair may represent diverse kinds of DF. When $\lambda\equiv\mu\equiv\rho$, the symmetric integral (16) is obtained. Thus, hybrid similarity matrices could be easily computed. Also an obvious QS-SM example of this definition is Equation (21), where $\lambda_A = V(\mathbf{r}_1)$, $\mu_B = \rho(\mathbf{r}_2)$ and $\Omega(\mathbf{r}_1,\mathbf{r}_2) = \delta(\mathbf{r}_1 - \mathbf{r}_2)$.

It is indubitable that it will be worth trying this new way to obtain QO descriptors, being symmetric or hybrid, as a new and varied source of QO descriptors. Also they can be taken as the starting point of new molecular superposition devices, in the same way as QSM based on electronic DF have been used [77]. It must be finally remarked that they can suffer the same manipulations as these discussed previously.

Some questions unanswered remain: how will look a KE, eEMP or a Coulomb expression substituting DF by KE DF? Coulomb QSM based on KE DF will appear as Equation (15), for instance, but with the left-hand side DF substituted by the proper KE DF: $\kappa_A(\mathbf{r})$. The use of KE DF, will permit to order the elements of a QOS according their momentum distributions acting as tag set parts and, consequently, a possible different ordering pattern could emerge.

Here is the adequate place to speak about other probability density distributions such as Boltzmann functions, as candidates to support molecular comparisons. This kind of similarity measures have been proposed and studied elsewhere [74].

4.1.4 A general definition of QSM

The characteristic properties of Coulomb energy, as discussed in Section 4.1.2 above, connect Equation (8) with Definition 5 too. Considering that a QSM can be also computed, looking at Equation (8) as a scalar product, constructed within the VSS, containing the PD operators and DF, as pointed out before. From this point of view, Equation (17) appears as a particular case of Equation (8), where another DF has substituted the weight operator on the left. This precludes the possible generalisation of Definition 5 in a very easy manner:

Definition 6: General QSM

A General QSM, G(Ω), can be considered a PD multiple scalar product defined by a contracted ν-direct product of a QOS, *T*:

$$G(\Omega): \bigotimes_{K=1}^{\nu} T \rightarrow \mathbf{R}^{+}. \square$$

This allows to mix ν DF: $\{\rho_I(\mathbf{r}), I=1,\nu\}$ of the QOS with ω arbitrary operators, collected into a set: $\Omega=\{\Omega_K(\mathbf{r}), K=1,\omega\}$, for example:

$$G(\Omega)=\int\left[\prod_{K=1}^{\omega}\Omega_K(\mathbf{r})\right]\left[\prod_{I=1}^{\nu}\rho_I(\mathbf{r})\right]d\mathbf{r}, \qquad (26)$$

and if

$$\omega(\mathbf{r})=\prod_{k=1}^{\omega}\Omega_k(\mathbf{r})$$

then $\Omega(\mathbf{r})=\omega^{+}(\mathbf{r})\omega(\mathbf{r})$ in order to obtain a PD weighting operator. The coordinate vector, **r,** shall be taken here in a broad general sense, in order to make possible the calculation of the integral. Thus, Equations (8), (17),(19),(21) and in Appendix C, Equation (69), can be all of them considered as diverse forms of QSM. At the same time, the DF set $\{\rho_I\}\{\rho_I\}$ can be taken in a broad sense too and be supposed formed by a blend of various kinds of extended DF, depending on homogeneous coordinates.

Finally, one can note that from inspecting Equation (26), taking $\omega=\nu=1$, Equation (8) can be deduced and, thus, any quantum expectation value can be considered as a particular form of QSM. A general picture of QSM was already given and studied in reference [48], but the present formal structure has a much better adaptation to the tagged sets theoretical background.

4.1.5 Superposition of two molecular structures: A constrained topological algorithm

A question of great importance, when QOS are molecular structures, already described in the first paper on molecular quantum similarity [1], has been the fact that QSM will depend in this case of the relative positions of both molecules. That is, supposed frozen the molecular involved frames, as accepted within the Born-Oppenheimer approximation; then, the QSM integral z_{AB} as defined in Equation (16) will depend on six parameters describing the relative three-dimensional position of a molecular frame with respect to the other: three translation components: $\mathbf{t}=(t_x;t_y;t_z)$ and three rotation angles: $\Omega=(\alpha;\beta;\gamma)$. As any QSM is PD: $z_{AB}\geq 0$, then the most usual way to proceed is to somehow reach an optimal value of the QSM, which must be a maximum. This has been the basic idea, which has inspired the recent work of our laboratory on the subject of molecular superposition; see for example [77]. However, using this maximal QSM criterion the obtained superpositions, when heavy atoms, other than the second row ones, are present on both structures, appear connecting these heavy atoms, providing sometimes unclear relative positions, far away from the chemical intuition.

In order to obtain chemically intuitive superpositions in a general way, there is no need to maximize an overlap-like QSM, but align in acceptable way atomic subsets of both molecular structures. If this is done using a set of plausible rules, then any attached QSM will possess a local maximal attribute. If this is sufficient or not for application purposes can be easily tested: comparing results obtained by using a global QSM criterion and a constrained one. The QSAR or QSPR examples presented here are obtained under this constrained criterion, in accord with chemical intuition.

The molecular superposition new algorithm proposed is fast and efficient. The most computationally slow part, associated in former procedures to the evaluation of QSM, has been avoided within this new strategy, resembling geometrical strategies already described by other authors [117]. The present superposition algorithm is based upon the geometrical and topological features of both active molecules, and is structured according the following procedure:

Algorithm 2: Topo-Geometrical Superposition of two Molecules

The topo-geometrical superposition algorithm for molecular alignment can be schematised in a few steps:

a) Input of atomic coordinates of both molecules.
b) Definition of *atomic diads*: For each molecule, establish which atoms are bonded.
c) Comparison of diad distances of both molecules. Keep those similar within a given threshold.
d) Creation of *atomic triads*: Adding a new atom, bonded to a selected diad.
e) Comparison of the two added distances of the triads for both molecules: Store the similar ones within the threshold.
f) If no similar triads are present, increase threshold value and return to c).
g) Keep the pair of triads that superpose as near as possible the largest number of atoms of the same kind between both molecules.
h) If there exists more than one triad fulfilling conditions of step g), select the one that minimises the distances between the superposed atoms.
i) Optimal topo-geometrical alignment: Superpose the pair of selected triads. □

See Appendix E for more details on the topo-geometrical superposition algorithm.

4.2 Discrete Representation of Quantum Objects: Similarity matrices

In this section the discrete description of QO will be studied. Discrete numerical images of molecules have been used since the past century, but it is only within a quantum mechanical support that this task acquires a well-founded meaning.

4.2.1 Theoretical considerations

The possibility of obtaining multiple relationships between the appropriate QOS elements, via their DF tags, in terms of QSM, as discussed in the previous section, has other interesting consequences, besides the calculation of the possible relationship between QO. The most relevant one constitutes the potential representation of a QO as a discrete vector or matrix. The next definition will present an ordered way to describe this important process:

Definition 7: Similarity Matrices

Suppose a QOS: $T = M \times P$ of cardinality n. The symmetric (n×n) matrix: $\mathbf{Z}=\{Z_{IJ}\}$ whose elements are made using QSM between homogeneous pairs of QO in T, will be called a similarity matrix (SM). □

By construction, provided that all the involved QO are different, any SM could be considered a PD metric matrix, belonging to some matrix VSS: $\mathbf{Z} \in M_{(n\times n)}(\mathbf{R}^+)$ [52]. Such a matrix can also be interpreted as the representation of the PD operator, Ω, in the basis set defined by the QO DF. Considering the SM column vectors: $\mathbf{Z}=\{\mathbf{z}_I\}$, this set also belongs to some n-dimensional VSS: $\forall I:\mathbf{z}_I \in V_n(\mathbf{R}^+)$. It is not clear, as it has been commented several times before, if the final space is a VSS or a SVSS one. It is difficult to imagine that some QSM values can be zero. Experience tells that this is not the case, so although the current literature describes a QSM as in Definition 5, the correct numerical set where the corresponding QO discretization belongs is: $\mathbf{R}_0^+$.□

Moreover, every column, $\mathbf{z}_I$, of the SM can be considered as some *n*-dimensional discrete representation of the I-th QO in the Tagged QOS. The set of columns of the SM was also referred within earlier papers [45-49], in an obvious descriptive manner, as a molecular point cloud.

Generalisation to similarity hypermatrix structures by using the general form of Equation (25) is trivial. The only point to be taken into account is the hypersymmetry of the final arrays if all the employed DF possesses the same form.

A warning must be provided here. If both DF are of different nature, as in the integral of Equation (25), then the resultant hybrid QSM is no longer a symmetric matrix: $\mathbf{Z}(\lambda;\mu) \neq \mathbf{Z}(\mu;\lambda)$. Symmetrisation involving an average of both matrices may solve this problem in a trivial manner. This feature can be used to induce a partial order in the associated QOS [120]. Other kinds of symmetrisation can be envisaged: $\mathbf{A}=\mathbf{Z}^T\mathbf{Z}$ or $\mathbf{B}=\mathbf{Z}\mathbf{Z}^T$, for instance.

4.2.2 Similarity Matrices and Quantum Similarity Indices

A brief account of quantum similarity indices will be given in this section.

4.2.2.1 General Considerations

The overall relationships between the elements of a molecular data set can be expressed in matrix form, yielding the SM: $\mathbf{Z}=\{z_{ij}\}$ As discussed, any homogeneous DF SM is symmetric, indicating that the QSM between two molecules is identical independently of the order of the comparison of the QO. This is used to compress the information only in the upper triangle of the matrix. The order of magnitude of the different types of QSM is highly connected to the structural form of the molecule, and to the presence of heavy atoms. Due to the particular construction of the SM, the diagonal elements of SM bring out information on the size of the compound.

4.2.2.2 Quantum Similarity Indices

Several transformations of the SM can be performed, yielding the so-called molecular QS indices (MQSI), which are mainly used to scale or normalise the SM elements. In particular, a normalisation of the MQSM, known in the literature as Carbó index, [1] can be defined as:

$$C_{IJ} = Z_{IJ}\left(Z_{II}Z_{JJ}\right)^{-\frac{1}{2}}.$$

The overall set of Carbó indices can also be expressed in matrix form. Carbó index can be interpreted as the cosine of the angle subtended by both involved DF in ∞-dimensional space, and so it ranges from zero to one. The closer to one, more similar can be considered the compared QO. Therefore, for two identical QO, that is, the main diagonal Carbó index elements the value of one is found, irrespective of the nature of the analysed QO. See Appendix F for a possible extension of Carbó index involving an arbitrary number of QO. A discussion on the relationship between QSM and Indices can be also found in reference [75,76].

5. DISCRETE REPRESENTATIONS AND QSAR

Discrete QO Tagged Sets may be defined at the same time as the original ∞-dimensional ones, based on DF Tag Sets. Suppose known a QOS, $\mathrm{T} = \mathrm{M}\times\mathrm{P}$, and a Similarity Matrix, $\mathbf{Z}=\{\mathbf{z}_I\}$, considered as a hypermatrix, formed by column vectors as elements, evaluated using the Definition 7 procedure.

Definition 8: Discrete Quantum Object Sets

A discrete QOS can be constructed as a new tagged set : $Z = M \times \mathbf{Z} \wedge \mathbf{Z}=\{\mathbf{z}_I\}$, with the same object part as the original QOS tagged set, T, but with the tag part formed by the columns of the similarity matrix, **Z**. □

5.1 Discrete Expectation values

The point of view appearing in Definition 8 is the same as finding out the way to project a set of points, defined in ∞-dimensional VSS, into an *n*-dimensional vector. It can be also considered that any similarity matrix, collecting a similarity relationship between any studied QO and a set of parent QO structures, is a source of an *unbiased* QO representation in the form of *n*-dimensional discrete information. SM columns are to be considered molecular descriptors chosen in such a way, that the remaining arbitrariness of choice corresponds to the nature of the weight operator appearing in the MQSM calculation.

Obviously, from this point of view discrete QO representations will depend on the PD weight operator, $\mathbf{\Omega}$. Suppose that a collection of discrete QO tagged sets is formed using various PD operators, $\{\Omega_i\}$, producing a collection of PD SM: $\{\mathbf{Z}(\Omega_i)\}$. A new PD SM, can be obtained, choosing a set of PD scalars, $\{\alpha_i\}\in\mathbf{R}^+$, by forming the linear combination:

$$\mathbf{Z}=\sum_i \alpha_i \mathbf{Z}(\Omega_i).$$

The columns of the combined SM, **Z**, can be considered linear combinations with the same coefficients, as those employed to construct the new PD SM. That is:

$$\forall \mathbf{z}_I \in \mathbf{Z}: \mathbf{z}_I = \sum_i \alpha_i \mathbf{z}_I(\Omega_i).$$

This discrete kind of QO Tags can be considered as a source of descriptors, which can be of use in the field of QSAR or QSPR. It has been recently shown, how QSM can be used as the origin of multilinear QSAR [55-69]. This can be made associating a given QO property, π, to the expectation value of some unknown operator in the way of Equation (8). Taking into account that the DF and the unknown PD operator belong to the same ∞-dimensional VSS, and then both can possess an associated discrete representation in the appropriate n-dimensional VSS. In a discrete framework, Equation (8) adopts the appropriate form:

$$\pi = \langle \omega \rangle \approx \langle \mathbf{w} | \mathbf{z} \rangle = \mathbf{w}^{\mathrm{T}} \mathbf{z} = \sum_{K} w_K z_K \tag{27}$$

Where **w** is an n-dimensional vector attached to the unknown operator, to be determined, in a least-squares sense, and **z** is a n-dimensional discrete QO tag.

This seems not surprising if the origin of Equation (27) is taken into account, as given by the ∞-dimensional counterpart in Equation (8). Indeed, both equations represent scalar products in conveniently chosen VSS. This will be discussed and broadened in the forthcoming sections.

It must be said here that QSM, collected as a vector **z**, associated to a molecular structure, can be considered from a quantum mechanical optic, as an ultimate way to represent QO in a discrete manner. Empirical QSAR or QSPR parameters, whatever be their origin and number, shall be considered as more or less successful attempts to simulate such QSM vectorial description. At the light of Equation (27) and the following discussion in deep of the QSPR problem, the QSM vectors cannot be considered as another set of molecular descriptors, chosen in the usual arbitrary way. They are obtained as a result of analysing the geometrical structure of quantum mechanics applied to the description of QO, as atoms and molecules. It has been shown that even the origin of topological matrices can be traced up to similarity matrices and, thus, topological indices can be constructed from the similarity measures contained in them [78]. Some more discussion on this last issue will be presented in section 6.

It can be concluded that any attempt to describe a known QO by using an arbitrary number of parameters of any kind, other than those computed using QSM, shall be considered as a rough way to simulate the theoretically based QO descriptors, represented, in turn, by the discrete QSM vectors **z**. From this point of view, the usual QSPR techniques, employed in chemistry for more than a hundred years, ought to be accepted as an empirical procedure to obtain approximate expectation values within a discrete framework.

In some of the following examples, a technique associated to the principal components of the SM, **Z**, will be used. In this case, use is made of the SM spectral decomposition:

$$\mathbf{Z} = \mathbf{U} \Lambda \mathbf{U}^{+} = \sum_{i} \lambda_i \mathbf{u}_i \mathbf{u}_i^{+},$$

where $\mathbf{UU}^+ = \mathbf{U}^+\mathbf{U} = \mathbf{I}$, and the diagonal matrix of the eigenvalues is defined as: $\Lambda = \mathrm{Diag}(\lambda_i)$. Then, Equation (27) can be written as:

$$|\pi\rangle = \mathbf{w}^{\mathrm{T}}\mathbf{Z} = \sum_i \theta_i \mathbf{u}_i \tag{28}$$

with: $\theta_i = \lambda_i \omega_i \wedge \omega_i = \mathbf{w}^{\mathrm{T}}\mathbf{u}_i$. This means that one can write the following expression for each molecular property collected into the vector $|\pi\rangle$:

$$\pi_I = \sum_i^n \theta_i \mathrm{u}_{Ii} \approx \sum_i^m \theta_i \mathrm{u}_{Ii} \tag{29}$$

with m<n. Then, afterwards $\{\theta_i; i = 1, m\}$ can be obtained with a least-squares technique.

5.1.1 Simple linear QSPR model involving QS-SM

For a given QO in a studied discrete tagged QOS, Equation (27) can be written as:

$$\pi_I \approx \mathbf{w}^{\mathrm{T}}\mathbf{z}_I = \sum_K \mathrm{w}_K \mathrm{z}_{KI} . \tag{30}$$

Equation (30) can be rewritten isolating the self-similarity part of the SM elements $\{\mathrm{z}_{KI}\}$:

$$\pi_I \approx \mathrm{w}_I \mathrm{z}_{II} + \sum_{K \neq I} \mathrm{w}_K \mathrm{z}_{KI} . \tag{31}$$

The terms: $a = \mathrm{w}_J$, $b = \sum_{K \neq J} \mathrm{w}_K \mathrm{z}_{KJ}$, can be considered as constants, or at least varying slowly within a QO homogeneous series, made of molecular structures, for example. Then, it is trivial to see that a linear relationship may be present between some properties and self-similarity:

$$\pi_I \approx a\mathrm{z}_{II} + \mathrm{b} . \tag{32}$$

This kind of relationship has been used successfully to assess some QSAR and QSPR, see for example [62-64,68].

This simplified equation and the previous section 4.1.2 discussion, about Coulomb two-electron energy as a QS-SM can be used together. They permit to think that these expectation values, as well QS-SM themselves, constitute very good candidates to compete, under some favourable circumstances and within homogeneous molecular series, with empirical parameters, like Hammett σ constant or log P, the octanol-water partition coefficient.

Some application examples are given in section 5.5.

5.2 Full SM as QSPR descriptors

When one deals with practical QSPR problems, the discrete expectation value law must be slightly modified. Thus, such algebraic relation can be transformed into a reduced expression where the number of parameters is less than the number of equations, yielding a true structure-activity relationship. In this case, the equality turns out into an approximation:

$$\pi_I \approx \mathbf{w}^T f(\mathbf{z}_I).$$

Now $\mathbf{w}$ is an m-dimensional vector, being $m<n$, and $f(\mathbf{z}_I)$ is a transformation of the similarity vectors that reduces their dimensionality. The solution to this problem is found by a least squares technique.

There exist several possible transformations of the similarity matrices $\mathbf{Z}$ in order to reduce their dimensionality. Principal component analysis (PCA) [121], partial least squares (PLS) [122,123], classical scaling or more general multidimensional scaling (MDS) [124] are examples of this kind of techniques. It is commonly accepted that MDS techniques are the most appropriate ones for the treatment of similarity and dissimilarity data. The different methods for the dimensionality reduction of similarity matrices search a common final objective: the initial similarities $\{Z_{ij}\}$ are wanted to be mapped into distances $\{d_{ij}\}$ in a multidimensional Euclidean space. Molecular coordinates in this new space are collected within a coordinate matrix $\mathbf{X}$. The mapping between proximities and distances cannot be exactly fulfilled, and in the practical case, $\mathbf{X}$ configuration is searched in such a way that the application is satisfied as best as possible, i.e. that the interpoint distances match as close as possible the original proximities. The condition "as best as possible" has a mathematical translation through the aggregation of mapping errors for each individual. The way to define this measure, called *loss of information function*, yields to the different MDS techniques. As can be noted, these methods have a clearly defined geometrical origin.

Classical scaling was one of the first multivariate analysis tools for similarity data treatment, and it is the one used here to transform the quantum similarity matrices. The central idea of this approach is considering the similarities as Euclidean distances, and then finding coordinates to recover them. A detailed explanation of this technique can be found in reference [124]. The coordinates in the new space are known as *principal coordinates* (PCs), and they will be used as parameters in the construction of QSAR regressions.

Once the SM **Z** is transformed, some type of variable selection ought to be also performed. The most usual choice is made by selecting the variables, arranged following the order given by the explained variance [121]. However, this is not the only possible way to proceed. In particular, the so-called *most predictive variables method* (MPVM) [125] has been employed successfully to select optimal QSPR variables. It consists of defining an individual correlation coefficient by projecting each PC on the property vector, and then arranging the variables in relation to that correlation coefficient. This leads to an specific order for each studied property. A 1% variance threshold has been imposed in order to avoid undesirable background noise parameterisation. This filter rejects all those variables that explain a variance lower than the threshold value.

In order to quantify the quality of the model proposed, several conventional statistical parameters can be used. The most frequently employed are: A) multiple determination coefficient (r^2), a goodness-of-fit measure. B) predictive coefficient (q^2), found by cross-validation by the leave-one-out procedure [126], which indicates the predictive capacity of the model. Together with these indicators, internal validation is completed with the so-called randomization test [127]. It consists of generating a series of vectors formed by the components of the activity vector, but randomly permuted in their positions. These altered responses are used to construct new models, exactly as if they were real. In the case that statistically significant adjustments and predictions are found when using randomized data, the model is suspect to possess an excess of parameters, and it is then able to correlate any type of data. A possible variation of this technique consists of using randomized parameters to correlate real data. If the randomized SM, after the statistical treatment, yields satisfactory results, the procedure has too many degrees of freedom and must be revised.

For computational cost reasons, the molecular density functions used in practical QSAR calculations are not computed ab initio, but accurately fitted using ASA [92-97], as explained in Appendix B. The similarity measures are made employing either overlap or Coulomb operators, and the matrices are normalised by means of the Carbó index transformation. All the QSM computations have been carried out using MOLSIMIL97 software [128], and the statistical calculations using the TQSAR-SIM program [129].

5.2.1 Example of QSAR using SM: Protein SAR

The application of SM matrices to QSAR will be illustrated with a practical example. One of the most recent fields of application of the QSAR analysis techniques is the study of the function or stability variation in certain proteins when a single amino acid residue, relevant for activity, is substituted. Within protein engineering, these techniques are known as quantitative structure-function and structure-stability relationships (QSFR and QSSR). As an example of the application of SM to this kind of problem, the stability of 18 mutants of α-subunits tryptophan synthase from *Escherichia Coli* in position 49 is analyzed. Yutani and co-workers [130] determined that glutamic acid located in position 49 is essential for protein function, and they carried out controlled mutations on this position to elucidate their effect on the conformational stability of the protein [130-133]. The stability is measured by Gibbs free energy of unfolding ($\Delta_d G$) in water at 25°C and pH 7.0, expressed in kcal mol^{-1}. The Gibbs energy of unfolding in water was calculated from the equilibrium constant in the absence of denaturation ($K_{0,\mathrm{nd}}$) using the equation: $\Delta_d G = -RT \ln K_{0,\mathrm{nd}}$. *Table 1* shows the amino acid residues studied and their stability measures. Experimental data was taken from reference [134].

Table 1. Stability data of wild-type and mutant α-subunits tryptophan synthases

Amino Acid Residue	$\Delta_d G$ kcal mol^{-1}	Amino Acid Residue	$\Delta_d G$ kcal mol^{-1}
Ala	8.5	Lys	7.9
Asn	8.2	Met	13.3
Asp	8.5	Phe	11.2
Cys	11.0	Pro	8.2
Gln	6.3	Ser	7.4
Leu	15.0		
Glu	8.8	Thr	8.8
Gly	7.1	Trp	9.9
His	10.1	Tyr	8.8
Ile	16.8	Val	12.0

Following the protocol explained above, overlap QSM were computed from amino acid SPARTAN geometries [135], and transformed into Carbó indices. SM for the 19 amino acids involved was dealt with classical scaling. The PCs employed as parameters in the multilinear regression were chosen using the MPVM method. The statistical coefficients of the different QSAR models are given in *Table 2*. These results are satisfactory, since a value of $q^2 > 0.5$ is considered to be the minimal threshold bearing statistical sense. The optimal model is attained when using 4 PCs, yielding $r^2 = 0.803$ and $q^2 = 0.661$. A plot showing the experimental versus the cross-validated activities for the optimal model is given in *Figure 1*.

Table 2. QSAR results from the treatment of SM

Number of PCs	r^2	q^2
3	0.704	0.604
4	0.803	0.661
5	0.827	0.653

In order to assess possible chance correlations, the activity vector components were randomly permuted in their positions, and the new altered vector was used as an external data, exactly as it were real. This procedure was repeated hundred times. The results are shown in *Figure 2*. As can be observed, the maximal r^2 and q^2 values correspond to the correctly ordered activities, whereas all the random responses cannot be correlated satisfactorily. The major part of the random models yields q^2.

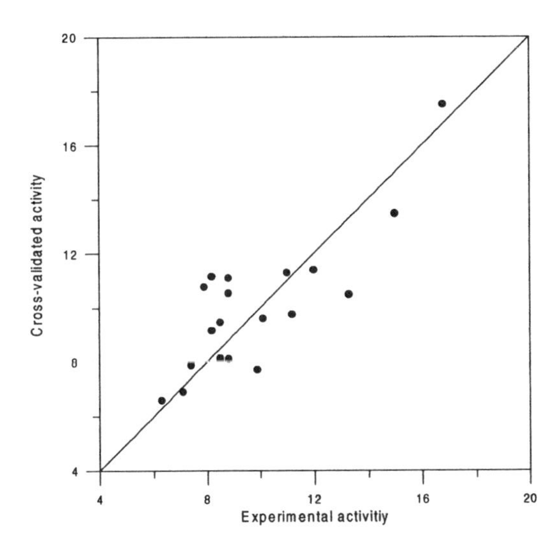

Figure 1. Experimental versus cross-validated conformational stability values for the mutated α-subunits tryptophan synthase when using the 4 PC QSAR model.

The only point close to the correct results possess q^2=0.47, a value lower than the minimal one considered to be acceptable. As a result, it has been proved that the model built relates only the QSM with the real activities, and subsequently, that quantum similarity matrices contain, in some sense, the information relevant for activity. Obviously, environment-dependent properties, such as the specific interactions with neighbours, may contribute to the stability, and a picture including only information related to the amino acid residues can become insufficient. However, the results are good enough to produce a qualitative estimation of the influence of the different mutations.

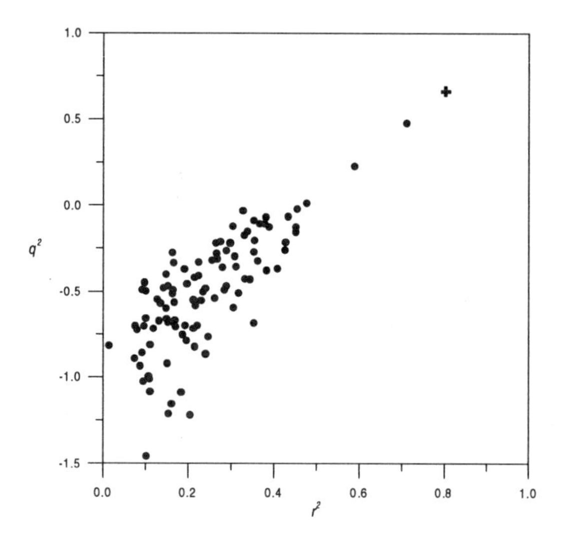

Figure 2. Randomization test for the QSAR model from SM. Optimal conditions have been maintained (4 PC), and variable selection is allowed for each randomized response. The model corresponding to real activities is marked with a cross, and those corresponding to randomized data are marked with a circle. 100 random vectors have been generated.

The same data set was analysed by other authors from other perspectives. Thus, protein thermostability was originally described with classical QSAR models within the Hansch-Fujita framework [136]. This approach assumes that biological activity in a compound is produced by a combined effect of different factors, which act independently. These factors are modeled using different physicochemical properties. Hydrophobic-like descriptors were found to be optimal for describing this kind of property [133].

Yutani et al. [133] explored the correlations between stability and hydrophobicity using two parameters: the Nozaki-Tanford index [137] and the optimal matching hydrophobicity (OMH), defined by Sweet and Eisenberg [138]. The first index quantifies the free energy of amino acid transfer from water to ethanol, and OMH is a similarity measure between the different amino acid residues obtained by constructing a correlation coefficient that compares their assigned hydrophobicities. In this case, the presence of outliers suggested that the volume of the residues and their ionization state could also play an important role in the structure-stability relationship.

On the other hand, Damborský [134] confirmed the importance of hydrophobicity to describe the stability effects of the different mutations. The author used only one descriptor, the hydrophobicity index of Rackovsky and Scheraga [139], to correlate the data. No additional descriptors improved the predictivity of the regression. This simple model achieved valuable correlation indices: $r^2 = 0.755$ and $q^2 = 0.711$.

The satisfactory results attained with QSM may indicate that steric effects, those which are mainly contained in the overlap DF integrals, can also play a role in conformational stability, as suggested by Yutani [133]. In comparison to the previous approaches, QSARs derived from SM are not so simple and accurate. However, the main interest of QSM as descriptors leans upon the uniqueness of the selection. As can be seen, even when the main factor responsible for stability is identified, different definitions of the hydrophobicity indices lead to very different results. The criterion for descriptor selection in the Hansch-Fujita analysis is one of its weakest points: there are no clear rules to a priori choose the parameters, and a stepwise selection is usually performed to select a subset of properties from huge databases. Alternatively, QS provides a simple, unique and unbiased definition of the structural descriptors, which can be used regardless of the molecular set and the type of property studied.

5.3 Inward Product and QSPR solutions under QSM formalism

Because the discrete form of the expectation values for a QOS can be written as the system of equations:

$$\mathbf{Zw} = \pi , \tag{33}$$

where $\mathbf{Z}$ is the SM, π a property vector, and $\mathbf{w}$, as commented before in Section 5.1, can be interpreted as a discrete representation of some, unknown and rather entangled operator, describing the QOS properties $\{\pi_i\}$ considered like discrete expectation values.

Usually the SM elements $\{z_{ij}\}$ are defined in such a way, see Section 5.1 as well as Definition 4 and Definition 7, that they are PD values: $\forall i,j;\ z_{ij}>0$.

The same can be said with respect the elements of the property vector: $\forall i; \pi_i>0$. This last characteristic can be seen in such a PD way, because many molecular activities come from concentration-like measures or experimental countings producing natural numbers. Even it can be hoped that, like temperature, the elements of the property vector can be scaled and shifted in such a way they become PD. Thus, if the known elements of the discrete expectation value equation belong to the appropriate VSS, nothing opposes to consider the same will also be true with the operator elements, that is: $\forall i; w_i>0$.

Accepting these preliminary considerations, the content of Sections 3.1 and 3.2 can be used, as well as the inward product definition and properties, as found in Appendix G, to set the following equalities:

$$\mathbf{Z} = \mathbf{A} * \mathbf{A}; \quad |\pi\rangle = \mathbf{p} * \mathbf{p}; \quad \mathbf{w} = \mathbf{x} * \mathbf{x}$$

where the symbol *; stands for an inward or Hadamard matrix product. This is the same as to recognise that the elements of the matrices used above are related by the equations:

$$\forall i,j;\ z_{ij}=(a_{ij})^2,\ \pi_i=(p_i)^2,\ w_i=(x_i)^2.$$

Thus, the discrete expectation value described by Equation (33) may be rewritten as:

$$(\mathbf{A} * \mathbf{A})(\mathbf{x} * \mathbf{x}) = \mathbf{p} * \mathbf{p}. \tag{34}$$

An *approximate* alternative equation seems to be precluded by the previous Equation (34), as:

$$(\mathbf{Ax}) * (\mathbf{Ax}) = \mathbf{p} * \mathbf{p},$$

which transforms into another linear system, involving the inward square root matrices of those appearing in Equation (33):

$$\mathbf{Ax} = \mathbf{p}. \tag{35}$$

The matrix **A** is a symmetric matrix, like its inward square power, the SM **Z**. Hence, there exists a set of real eigenvalues: $\{\alpha_i\}$ and orthogonal eigenvectors $\{\mathbf{u}_i\}$, which can be used to express the matrix **A** in the spectral decomposition form:

$$\mathbf{A} = \sum_i \alpha_i \mathbf{u}_i (\mathbf{u}_i)^T .$$

Molecular superposition factors can influence the SM in such a way that it becomes a nondefinite one, in general. A PD SM is obtained when, for all molecules involved, the geometries are kept oriented in the same manner. However, this is not the case when molecular superposition is active in every molecular pair entering the calculation. The result is such that some of the SM eigenvalues can be almost zero or even negative. The same happens in the attached inward square root. So, an approximate PD part of matrix **A** can be defined, employing:

$${}_a\mathbf{A} = \sum (\delta(\alpha_i > \varepsilon)) \alpha_i \mathbf{u}_i (\mathbf{u}_i)^T ,$$

where δ(*Logical expression*), is a logical Kronecker delta [102,104]. This symbol is zero if the inner logical expression is false and one in the other case; ε is a cutoff parameter.

In the same way, a residual complementary part of the above approximation can be defined as:

$${}_r\mathbf{A} = \sum (\delta(\alpha_i \le \varepsilon)) \alpha_i \mathbf{u}_i (\mathbf{u}_i)^T .$$

Both matrices add as to give the original one: $\mathbf{A} = {}_a\mathbf{A} + {}_r\mathbf{A}$. The approximate inverse of matrix ${}_a\mathbf{A}$, can be defined easily, using:

$${}_a\mathbf{A}^{\{-1\}} = \sum (\delta(\alpha_i > \varepsilon)) \alpha_i^{-1} \mathbf{u}_i (\mathbf{u}_i)^T$$

and owing to the orthonormalisation of the eigenvectors, the following relationship is obtained:

$${}_a\mathbf{A}\,{}_a\mathbf{A}^{\{-1\}} = {}_a\mathbf{A}^{\{-1\}}\,{}_a\mathbf{A} = \sum (\delta(\alpha_i > \varepsilon)) \mathbf{u}_i (\mathbf{u}_i)^T = {}_a\mathbf{P} ,$$

where the matrix ${}_a\mathbf{P}$ is the projector over the eigensubspace of the PD eigenvalues such that: $\alpha_i > \varepsilon$. The approximate inverse matrix ${}_a\mathbf{A}^{\{-1\}}$, can be used in turn to obtain an approximate solution of the linear system (35), such that:

$${}_a\mathbf{x} = {}_a\mathbf{A}^{\{-1\}} \mathbf{p} \qquad (36)$$

and in the same manner it is obtained a set of approximate inward square root properties:

$$_{\mathbf{a}}\mathbf{p}={}_{\mathbf{a}}\mathbf{A}_{\mathbf{a}}\mathbf{x}={}_{\mathbf{a}}\mathbf{P}\,\mathbf{p}, \tag{37}$$

which is clearly a projection of the inward square root of the original property vector into the chosen PD eigensubspace.

The vectors defined in Equations (36) and (37) are sufficient to produce VSS elements, associated to Equation (33) in some approximate fashion. That is:

$$_{\mathbf{a}}\mathbf{w}={}_{\mathbf{a}}\mathbf{x}*{}_{\mathbf{a}}\mathbf{x}, \tag{38}$$

and also:

$$_{\mathbf{a}}\mathbf{p}*{}_{\mathbf{a}}\mathbf{p} = |{}_{\mathbf{a}}\pi\rangle. \tag{39}$$

5.3.1 Put-one-in Procedure

To predict any unknown property for a new molecular structure, one can use a trivial procedure, which can be named: put-one-in, paraphrasing the well known statistical procedure for assessing the predictive power of a QSAR model, called leave-one-out [126]. An example of such procedure can be see in Section 5.2.1 above.

Suppose the vector: $\mathbf{z}_M$, computed calculating the QSM between the new molecule and the original molecular reference set, is known, as well as the molecular self-similarity: θ_M. An inward square root vector, $\mathbf{a}_M$ can be obtained simply as: $\mathbf{z}_M=\mathbf{a}_M*\mathbf{a}_M$; a still simple relationship provides: $\theta_M=(\alpha_M)^2$. A new augmented linear system as the one in Equation (35) can be written in two parts as follows:

$$\alpha_M x_M + (\mathbf{a}_M)^T \mathbf{x} = p_M \tag{40}$$

$$x_M \mathbf{a}_M + \mathbf{A}\mathbf{x} = \mathbf{p}. \tag{41}$$

Taking into account Equations (36) and (37), Equation (41) can be transformed into the approximate form:

$$x_M \mathbf{a}_M = \mathbf{p} - {}_a\mathbf{p},$$

so, an approximate value of x_M: ${}_a x_M$, can be obtained as:

$$ {}_a x_M = \left[(\mathbf{a}_M)^T (\mathbf{p} - {}_a\mathbf{p})\right]\left[(\mathbf{a}_M)^T \mathbf{a}_M\right]^{-1}$$

which upon substitution in (40) produces an approximate value of the square root of the unknown property:

$$ {}_a p_M = \alpha_M \, {}_a x_M + (\mathbf{a}_M)^T \, {}_a\mathbf{x},$$

and a final estimate of the original unknown property value can be obtained as:

$$ {}_a \pi_M = ({}_a p_M)^2 .$$

5.3.2 Simple numerical example as a test

A numerical example of this computational process is based on the steroid set, described in deep in the Section 6, and usually employed when new QSAR procedures are tested [140]. A *put-one-in* algorithm over the molecular set of 31 structures, with a property vector presented in *Table 9* (see below), has been analysed taking the Coulomb SM, already computed as reported in reference [61].

Put-one-in procedure has been repeated for the same steroid set, considering each time that one molecule has unknown property and the remaining 30, presently known, and taken as reference set. *Figure 3* shows the results, which present comparable features to previous QSAR models based on MQSM, as those reported in reference [61], using a unique SM. The error cut-off parameter used has been: $\varepsilon=10^{-2}$; the put-one-in correlation coefficient yielded a value of $r^2_{POI} = 0.764$.

These results and the fact that no statistical analysis is needed, point towards further development and testing of the computational structure, presented in this section. An algorithm, well adapted to the structure of QMS, as the previously described one, permits to look ahead to find out how to design the complete, self-contained theoretical framework, providing full understanding and general applicability of the connection between QSAR and QSM.

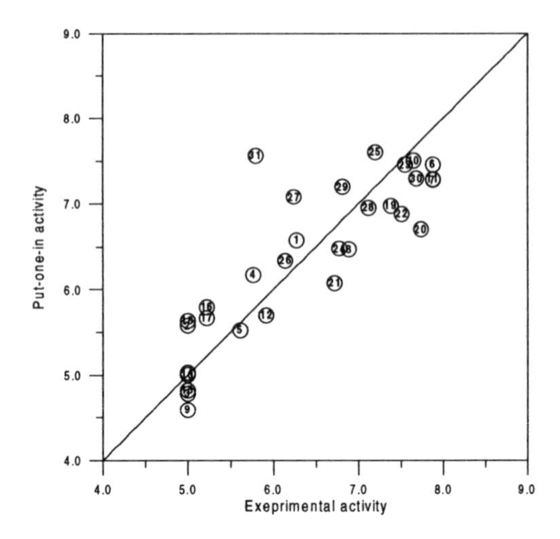

Figure 3. Put-one-in predicted versus experimental values for the 31 Cramer steroids.

5.4 Convex superposition of SM

Suppose, now, in another possible QSAR situation, that a discrete QOS is known. The associated discrete tag set: $\{\mathbf{z}_I\}$ can be supposed constructed by a convex combination of several appropriate elements belonging to the same tag set, in such a form that:

$$\mathbf{z}_I = \sum_k \alpha_k \, \mathbf{z}_I(\Omega_k) = \sum_k \alpha_k \, \mathbf{z}_I(k) \wedge K_n(\{\alpha_k\}); \tag{42}$$

as discussed in Section 5.1 above. Using Equation (27), it can be written:

$$\begin{aligned} \pi_I &\approx \mathbf{w}^T \left(\sum_k \alpha_k \, \mathbf{z}_I(k) \right) = \sum_k \alpha_k \left(\mathbf{w}^T \, \mathbf{z}_I(k) \right) \\ &= \sum_k \alpha_k \, \pi_I(k) \end{aligned} \tag{43}$$

where $\pi_I(k) \equiv \pi_I(\Omega_k)$ represents the estimated value of the considered property, using the weight operator Ω_k, obtained for the I-th QO.

5.4.1 Numerical Examples of Tuned QSAR: A toxicity study

Sometimes, the use of a single SM is not enough to ensure a good description of a molecular set. Chemical mechanisms of reaction can be too sophisticated to be represented by the information contained in only one

SM. A suitable way to extend the capacity of description of QSAR models based on QSM consists of allowing a combination of different matrices. In order to guarantee a correct behaviour of the resulting descriptors, the mixing coefficients are restricted to be convex, that is, they must fulfil the convex conditions symbolised by $K_n(\{\alpha_i\})$ and detailed in Equation (5).

Tuned QSAR, based on convex superposition of SM as commented above, has been proposed recently as a tool to obtain a well-adapted parameter matrix from several SM [41,42]. In the present example, combinations of overlap and Coulomb QSM were only permitted, but the procedure could be extended to mixtures of other types of operators. Thus, the tuned SM is expressed as:

$$\mathbf{Z}^{\text{tuned}} = \alpha\, \mathbf{Z}_{\text{OVE}} + (1-\alpha)\mathbf{Z}_{\text{COU}} \tag{44}$$

being $\alpha \in [0,1]$. Each one of these matrices combined can be interpreted as a different contribution to the molecular description. In this particular case, overlap QSM represent the steric effects and Coulomb QSM bring up information on the electrostatic interactions.

The application example of the tuned approach is focused on another QSPR problem: the establishment of quantitative relationships between molecular structure and toxicity to animal species. The major goal of this kind of analysis consists of providing simple computational models able to describe and predict the molecular toxicity produced by a set of chemicals, which could successfully substitute animal testing.

It is generally accepted that aquatic toxicity of the majority of organic chemicals is produced by a mechanism of narcosis, directly related to the accumulation of these compounds in the cell membranes. The capacity of accumulation is represented by a hydrophobicity index, such as the octanol/water partition coefficient (log P). The difference between the solubility in water and octanol, which mimics the phospholipid cell membrane [141], provides good results for these systems, as proved by different studies [142,143]. On the other hand, it has been demonstrated that for highly homogeneous series, overlap QS-SM are linearly related to log P [63]. This point constituted the basis for a previous study, where the toxicity for a set of 92 benzenes to the fish *Poecilia reticulata* was dealt with QSM, achieving valuable adjustments and predictions [69]. Together with these approaches, toxic organic compounds have been also analyzed with topological indices [144,145] and with the electron-electron repulsion energy descriptor, a type of QS-SM [68].

In particular, the molecular set used in this study is made up of 177 substituted benzenes, including anilines, phenols, halobenzenes, and nitrobenzenes. This family is divided into a training set of 135 molecules and a test set of 42 compounds, of unknown toxicity. Here, toxicity to fathead minnow (*Pimephales promelas*) was studied, which was measured by the lethal concentration (mol l^{-1}) necessary to reduce 50% the initial population of the fish, the so-called LC50 factor. In order to avoid large value differences, LC50 was scaled using the minus logarithmic form. Experimental data was taken from the work of Geiger et al. [146], reproduced in ref. [143].

The geometries of the 177 substituted benzenes were optimized at an AM1 level [147] with AMPAC 6.0 software [148], and first-order molecular DF were calculated using ASA approach. Optimal molecular alignment was found using the Algorithm 2, according to the procedure detailed in Appendix E. This leads to an exact superposition of the common benzene rings. Overlap and Coulomb QSM were transformed into Carbó indices, and the SM were transformed as explained before.

The QSAR study with a single SM yields to no significant results ($q^2 < 0.5$), hence a more sophisticated treatment is required. As stated before, the tuned QSAR models allow for a combination of matrices, which are expected to improve the molecular description. The matrix weights are interrelated by the convex constraint, and therefore only one parameter is effectively free (α coefficient in Equation (44)). Different preliminary studies determined that optimal descriptions were achieved when using 11 PC. So many parameters are required due to the smooth distribution of the variance when dealing with large data sets. In spite of this, a high number of parameters still satisfies the significance ratio inequality:

#molecules/#descriptors > 5

and therefore its use is statistically justified, obviously attending to the success in the robustness and validation tests.

To encompass different matrix distributions, the α weight was set to 0.0, 0.25, 0.50, 0.75 and 1.0 units, generating the subsequent QSAR models for each case. The results are summarised in *Figure 4*, where the evolution of the statistical parameters can be appreciated. Results derived only from either overlap or Coulomb SM are comparable, and the optimal model is obtained when both matrices are mixed with half the total weight ($\alpha = 0.5$), yielding $r^2 = 0.605$ and $q^2 = 0.521$.

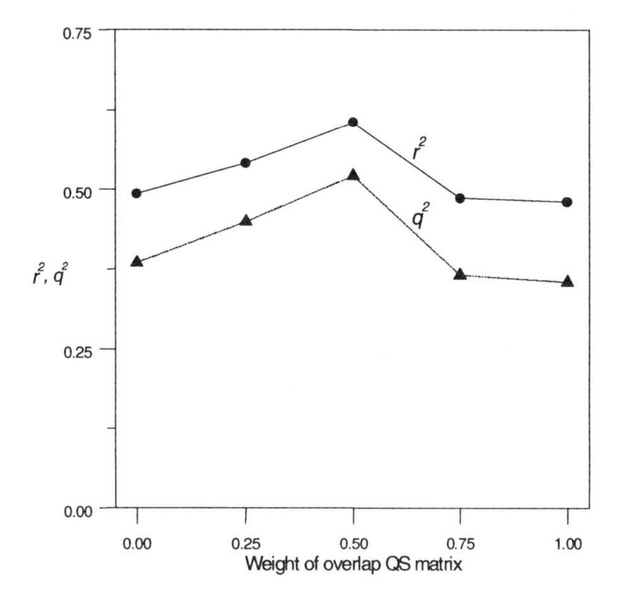

Figure 4. Multiple determination (r2) and prediction (q2) coefficients when α is changed in ¼ units. 11 PC have been used.

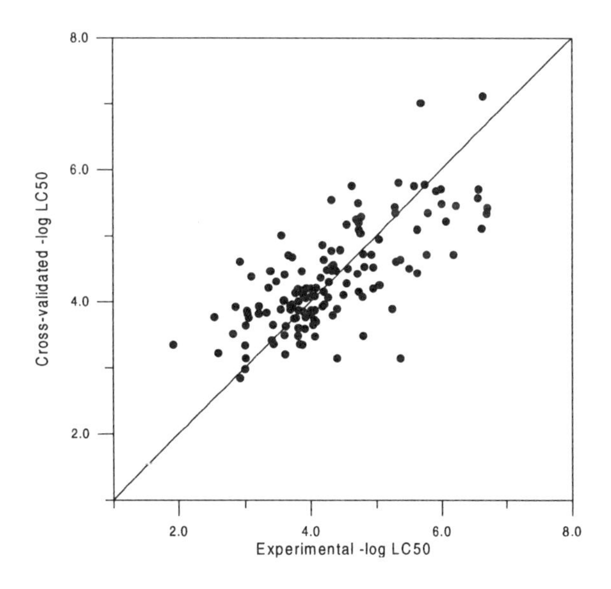

Figure 5. Experimental versus cross-validated toxicities for the 135 compounds used as a training set. Optimal tuned QSAR model has been used: 11 PC, α=0.5.

Figure 5 shows the toxicities predicted by cross-validation for the training set when the optimal model is used: $\alpha = 0.5$ and 11 PC. Even though the results are not accurate, a roughly linear trend can be observed. The presence of several outliers clearly influences in the q^2 value. This kind of analysis can provide a useful aid for a discrimination between high and low toxic agents.

These results can be compared with previous studies. The same data set was studied by Karabunarliev et al. [143] using classical descriptors, namely log P and A_{max}, the maximal acceptor superdelocalizability for the π-sites of the benzene ring. The first quantity represents the "baseline toxicity" described by hydrophobicity, and the second one reflects the likelihood nucleophilic reaction to occur prevailingly at the position of the benzene ring with maximal acceptor superdelocalizability value. The original training set was reduced to 122 chemicals, extracting those molecules poorly described by the model. The results obtained with this two-parameter regression were $r^2 = 0.831$ and $q^2 = 0.668$. Again this model is more efficient and simple than our proposal, but it has been proved that QSM, accounting for steric and electrostatic effects, can also describe properties exclusively associated to hydrophobic mechanisms with a certain accuracy. In addition, it must be emphasized again that 16 compounds were removed from the original data set, so the results were artificially overrated.

The Karabunarliev model was used to generate a set of predictions for the remaining 42 compounds, whose toxicity has not been experimentally evaluated yet. In a similar way, the optimal tuned QSAR model has been used to predict the toxic action of these pollutants, and the derived results have been compared to those reported in the aforementioned study. *Figure 6* shows a plot contrasting the values predicted by both approaches.

As it can be observed, all the predictions are roughly similar in both approaches, except for two molecules, namely benzylchloride and α,α-dichloro-*m*-xylene, which are estimated to possess very different toxicities. In both cases, tuned SM descriptors predict a high level of toxicity (6.583 and 6.096, respectively), whereas the classical model associate low toxic actions (3.771 and 4.176). Even though the experimental LC50 are still unknown, these values can be compared to the toxicity for another aquatic species, namely the guppy *Poecilia reticulata*, which have been already evaluated [149]. For this fish, the toxic action of these two compounds are among the highest ones of the family (5.510 and 6.160). Taking into account the existing correlation between the toxicities to the two species, this seems to indicate that the predictions generated by the tuned QSAR model could be more reliable than those obtained with the classical approach.

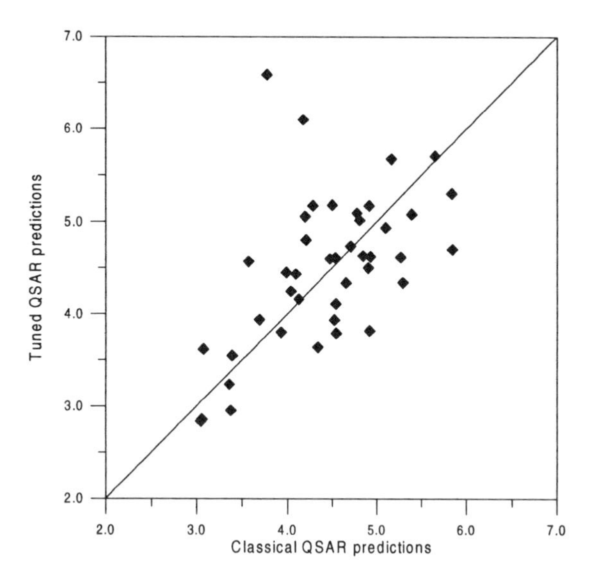

Figure 6. Tuned QSAR versus classical QSAR prediction for the 42 chemicals test set.

On the other hand, the behaviour of the remaining compounds is retrieved by the two methodologies, which are built independently and deal with different molecular aspects. This fact enhances the confidence in the qualitative general trends of toxicity for this chemical family, which might be taken into account by experimental researchers or industrial companies as a guide for future tests and product developments.

5.5 QS-SM as molecular descriptor in simple linear regression models

In the past few years, a large amount of research surveys has focused on the quantum similarity framework applied to the structure-based drug design [34-69]. This chemistry area has been rapidly extended by reason of its enormous significance in the development of new active compounds. Therefore, QSM can actively contribute in the determination of quantitative relationships between molecular structures and biological activity, as well as in the elucidation of its theoretical meaning [55] as reviewed in Section 5.1. Recently, an attempt was made to characterize simple linear QSAR models using quantum mechanical descriptors instead of hydrophobic and electronic physicochemical parameters [62-66]. This methodology is based on replacing empirical molecular descriptors used in classical QSAR analysis by appropriate QS-SM. This new QSAR approach was proven to be useful in the substitution of two empirical parameters: the octanol-water partition coefficient (log P) and Hammett σ substituent constant.

a) log P in QSAR studies can be substituted by a QSM between two molecular density functions of a given molecule immersed in two different solvents: octanol and water [62,63]. But this measure implies the computation of ab initio density functions, which in many cases is not feasible to calculate. To avoid such difficulties, a simplification can be used, where only the gas phase molecular density function is needed for the calculation. Then, the hydrophobic character of a molecule is described by an overlap-like molecular QS-SM.

b) In preceding publications [62,64], it was demonstrated that fragment QS-SM can be used to characterise molecular electronic properties. Normally, in classical QSAR studies the electronic properties provoked by substituent effects are described by the substituent constant σ coming from the Hammett equation. It has been proven that these effects also can be characterised using QS-SM of local molecular regions, corresponding to the functional group which can be identified with the molecular part which participates in a (re)active process. This methodology is founded on the basis of the recently described holographic electron density theorem [150], which exposes that any local region of a molecule contains all information about the complete molecular electron density. Consequently, the aim of the method consists of choosing an electron density fragment with non-zero volume where the predetermined molecular property would be expanded, and to compute a QS-SM for this (re)active chemical functional group. For example, a previous study consisted of characterizing substituent and solvent effects on dissociation constants of carboxylic acids by computing the fragment QS-SM of the COOH group [64].

c) On the other hand, due to the interpretation of $\langle V_{ee} \rangle$ as a QS-SM that has been given in Section 4.1.2, can also be used $\langle V_{ee} \rangle$ as a molecular descriptor. Application of this idea to predict the toxic action for a large series of organic compounds have been reported [68].

To illustrate such type of simple linear regression models, two QSAR analyses are presented: an example employing overlap-like QS-SM and the second using $\langle V_{ee} \rangle$ as molecular descriptor.

5.5.1 Predicting biological activities from QS-SM

The present application example examines the binding constants of an homologous series of n-alkyl carbamates to the cytochrome P-450 [151]. Biological activities for these compounds and molecular structures are presented in *Table 3*. The binding affinities for these molecules were found to correlate with lipophilicity [151]. Molecular geometries have been optimized using Gaussian 94 program at the HF/3-21g* level of theory. QS-SM have been computed using the promolecular ASA approach (see Appendix B for more details). A satisfactory linear regression equation that relates QS-SM and biological activity has been obtained for this molecular set:

$$\log K_S = -0.6027\theta_{AA} + 2.4287 \qquad (45)$$
$$n = 6;\ r^2 = 0.972;\ q^2 = 0.903$$

where θ_{AA} represents the standarized values of QS-SM, obtained imposing the conditions of zero mean and unit variance. Both values of scaled and shifted (θ_{AA}) and non-scaled (Z_{AA}) QS-SM are presented in *Table 3*. In addition, predicted values of binding constant K_S obtained from a leave-one-out cross-validation analysis, are listed in *Table 3*.

Table 3. Experimental and predicted activity values for the carbamate series

Molecular structures	Observed log K_s	Z_{AA}	θ_{AA}	Predicted log K_s
$CH_3(CH_2)_3OCONH_2$	3.279	372.46	-1.3522	3.212
$CH_3(CH_2)_4OCONH_2$	2.968	404.21	-0.8452	2.926
$CH_3(CH_2)_5OCONH_2$	2.699	435.96	-0.3381	2.617
$CH_3(CH_2)_6OCONH_2$	2.230	467.70	0.1690	2.347
$CH_3(CH_2)_7OCONH_2$	1.851	499.45	0.6761	2.076
$CH_3(CH_2)_9OCONH_2$	1.544	562.95	1.6903	1.168

5.5.2 Application example: $\langle V_{ee} \rangle$ as molar refractivity descriptor

This section discusses the use of Coulomb operator expectation value as a single molecular descriptor. In addition, an example of its application to QSPR is provided, and compared to previous studies where other descriptors were used.

Arising from the theoretical framework described in Section 4.1.2, the expectation value of any nondifferential operator can be considered as a particular case of QS-SM, and hence it can be used as a molecular descriptor in QSPR-QSAR studies. In particular, Coulomb operator expectation value constitutes the source of electron-electron repulsion energy ($\langle V_{ee} \rangle$). The usefulness of this value leans upon it is methodologically calculated in any common quantum chemical software package, and no specific QS software is required. $\langle V_{ee} \rangle$ may be related to two kind of effects usually employed in classical QSPR-QSAR studies: steric and electronic. Steric effects concern the influence of molecular or fragmental contributions in molecular properties, see for example [152]. Electronic effects consider the interactions between molecular orbitals and their relationship with molecular response. These electronic effects can be divided into two classes: intramolecular and intermolecular. Intramolecular ones are related to fragment contribution to the whole molecular electronic distribution, like Hammet's sigma [153]. Intermolecular descriptors monitor the interactions due to molecular orbitals from different molecules, like HOMO/LUMO energy gaps [154]. $\langle V_{ee} \rangle$ is strongly connected to molecular size, because its value depends on the number of atoms of the molecule and their kind. In addition, it can be also considered an intramolecular electronic descriptor because it describes, due to its quantum origin, the molecular electronic distribution. Thus, intramolecular orbital interactions are taken into account when $\langle V_{ee} \rangle$ is computed. A deeper discussion on these considerations will be provided after the practical example presentation. $\langle V_{ee} \rangle$, as it is based on quantum mechanical ideas, considers the molecule as a whole, and there is no need to employ fragmental or atomic contributions. When $\langle V_{ee} \rangle$ is used as a molecular descriptor, the general formulation for Equation (32) presented in Section 5.1.1 may be modified to:

$$\pi_I \cong a\langle V_{ee} \rangle + b \tag{46}$$

In this framework, $\langle V_{ee} \rangle$ has already been successfully applied in a quantitative structure-toxicity study [68], where it was able to describe the toxicological capacity of several molecular sets. The example presented to illustrate the application of $\langle V_{ee} \rangle$ in QSAR and QSPR studies consists of the description of molar refractivity (*MR*) for a large series of small organic molecules, which include alkanes, alkenes, alkines, alcohols, ethers, esters, acids, amines, ketones, aromatic compounds and polisubstituted compounds [155]. *MR* is related to the deviations of a light beam when interacting with molecular electronic structures. *MR* knowledge and prediction may be of interest in many research areas, like electronics and optical fibres.

The QSPR model using $\langle V_{ee} \rangle$ is presented in Equation (47), as well as all relevant statistical parameters. The molecular set and *MR* values were obtained from [155]. Molecular structures were considered in gas-phase and optimised at the semiempirical AM1 level using AMPAC 6.0 [148]. $\langle V_{ee} \rangle$

values were computed at the HF/3-21G* level using Gaussian 94 program [72]. In addition, *Figure 7* presents a plot showing the cross-validated *MR* versus the experimental values.

$$MR = 46.67 \times 10^{-3} \langle V_{ee} \rangle + 11.31 \tag{47}$$

$$n = 183;\ r^2 = 0.900;\ q^2 = 0.897;\ s = 2.886$$

The results obtained for this system evidence a fair correlation when $<V_{ee}>$ is used as a descriptor. The value of $r^2 = 0.900$ and the high predictive capacity observed, $q^2 = 0.897$, point out the utility of this method. The previous study [155] compared the physicochemical significance of different topological parameters. The results were considerably poorer, yielding to $r^2 = 0.824$ using the first order molecular connectivity index [156,157], $r^2 = 0.672$ using the second order molecular connectivity index [156,157] and finally $r^2 = 0.640$ using log P. As previously mentioned, $<V_{ee}>$ is related to molecular size. However, the electronic effects also play an important role in this molecular set, since many compounds possess the same empirical formula, but different connectivity, for instance heptane and 2,4-dimethylpentane. If $<V_{ee}>$ were only related to molecular size, it would not be able to differentiate between this kind of molecules and would generate an identical prediction for both. In this way, $<V_{ee}>$ contains information about the fluctuations of the electronic cloud due to structural isomers. *Figure 7* evidences these facts, and the major part of the molecules is correctly predicted within an error interval less than 10%. However, there are some molecules that are not so accurately predicted. This situation may occur because single descriptor models do not encompass the whole molecular effects responsible for a physicochemical property. In this way, $<V_{ee}>$ is presented here as a complement to be used in QSPR and QSAR analysis where steric and electronic effects are important.

To test the real predictive capacity of the model, the equations relating $<V_{ee}>$ with *MR* will be tested on a series of molecules whose property has not been included to elaborate the regression. In this case, the set has been divided into two groups: a training set consisting of 100 molecules and a test set of 83, chosen arbitrarily. The model has been recalculated for the training set and has been used to predict *MR* values of the test set. The regression obtained for the training set is presented in the following equation, including all relevant statistical parameters.

$$MR = 47.57 \cdot 10^{-3} \langle V_{ee} \rangle + 11.18$$
$$n = 100;\ r^2 = 0.905; q^2 = 0.901; s = 2.901 \tag{48}$$

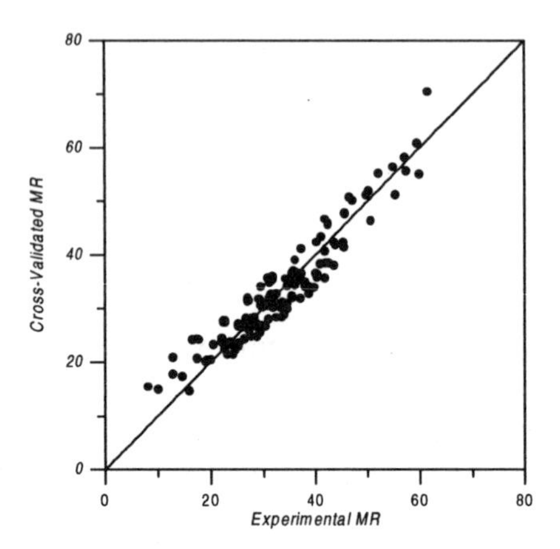

Figure 7. Cross-validated versus experimental MR plot.

The results are not substantially modified when only 100 molecules are considered and when the 83 molecules of the test set are predicted, the *MR* values obtained are fairly close to real ones. *Figure 8* presents predicted and experimental MR for both training (•) and tests sets (+).

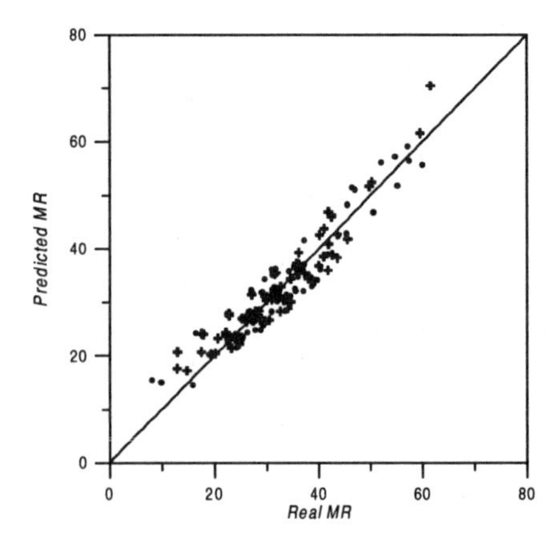

Figure 8. Experimental MR, estimations for the training set and predictions for the test set. Training set has been marked with a circle, and test set with a cross.

The correlation between real *vs.* predicted *MR* values for the test set yields to r^2=0.896. These results prove the stability of the model when, in this case, $<V_{ee}>$ is used as a molecular descriptor.

5.6 Molecular Fragment Similarity

Having demonstrated in Section 5.5 that it is possible to describe empirical properties, such as hydrophobicity and electronic effects, using appropriate QS-SM, the next step consists of developing a novel predictive QSAR approach based on molecular fragment similarity [66]. In the initial studies, and in order to validate this methodology, former knowing of the active molecular fragments responsible of biological activity was extracted from preceding classical QSAR analysis. In this way, the study was focused on searching the appropriate quantum-mechanical descriptors that can replace the empirical parameters employed to generate the classical multilinear regression equations. Here, and in order to generalise the present approach, an attempt to predict the molecular regions that produce or are responsible for the biological activity, without a priori specifications, is examined. Next, the computational scheme developed to achieve this objective, and based on two programs, is presented:

Definition of molecular fragments and related QS-SM. The present methodology can only be applied to a structurally related series of compounds. It is necessary to identify a base molecular structure in all the studied set. Then, from this common skeleton, the following molecular fragments are defined: all individual atoms, except hydrogen atoms; all bonded diads of atoms; and all consecutive triads of atoms. In addition, other fragments are localised, such as the rings that can be detected in the common structure, and the molecular fragments corresponding to the substituents which vary in each molecule. Finally, as a general descriptor of hydrophobicity, the QS-SM for the whole molecule is always considered in these studies. In the particular case that a molecule does not have a determinate molecular fragment, it is assumed that the QS-SM value for this descriptor is *zero*.

Computation of QS-SM. Both overlap-like and Coulomb-like similarity measures can be computed for all defined molecular fragments. Involved density functions in the QSM integrals are constructed using ASA approach, which is described in more rigorous detail in Appendix B. Its definition corresponds to the expression:

$$\rho_A^{ASA}(\mathbf{r}) = \sum_{a \in A} P_a \rho_a^{ASA}(\mathbf{r}) \qquad (49)$$

where the coefficient P_a represents the atom a total charge, and $\rho_a^{ASA}(\mathbf{r})$ the atomic density function. A can be understood as the whole molecule or any molecular fragment. In accordance with the semiempirical computations employed to optimise molecular geometries for the present examples, valence electron density functions are considered.

Selection of molecular descriptors. A method is needed to reduce the huge amount of data provided by all-computed fragment QS-SM. The definition of several variables provides more information about the observations, but on the other hand, a suitable procedure for summarising and analysing data are required for extracting relevant information. The QS-SM data are analyzed to reveal the fundamental factors underlying the biological activity by means of a partial-least squares (PLS) technique [122,123]. This corresponds to a variable reduction methodology which relates molecular descriptors matrix (independent variables), **X**, and the biological response variables vector, **y**, using a linear multivariate model. This technique is useful to analyze data with collinear and noisy variables.

Multilinear regression (MLR) and correlation analysis. The last step consists of generating a QSAR model. Using the selected QS-SM by the PLS technique, all possible linear relationships are constructed using a nested summation symbol (NSS), see Appendix H, algorithm. The optimal model is chosen as the linear equation which better predicts the desired biological activity.

5.6.1 QSAR of pyrimidine nucleosides as antiviral agents for human immunodeficiency virus type 1

A set of 27 pyrimidine nucleosides related to 3'-azido-3'-deoxythymidine (AZT), with a common structure shown in *Figure 9* and substituents listed in *Table 4*, was studied as antiviral agents for human immunodeficiency Virus type 1 (HIV-1) in peripheral blood mononuclear cells. The activity of these compounds, computed as median effective concentration (antiviral effect) on day 5 after infection, EC_{50} (μM) [158], was summarised in *Table 4*. The geometries of all studied compounds have been fully optimized at the semiempirical AM1 level [147] using AMPAC program [148].

The first computation step consists of defining and computing fragment overlap-like QS-SM for the training set, which is composed by the first 22 AZT derivatives listed in *Table 4*. In the AZT molecular structures, 79 descriptors have been considered: QS-SM for the whole molecule; QS-SM for all single-atom fragments different of hydrogen; bonded diads of atoms; consecutive triads of atoms belonging to the common skeleton; the two cyclic substructures (a five-member ring and a six-member ring); and the substituents X, R_5, $R_{3'}$, and $R_{5'}$ (see *Figure 9*).

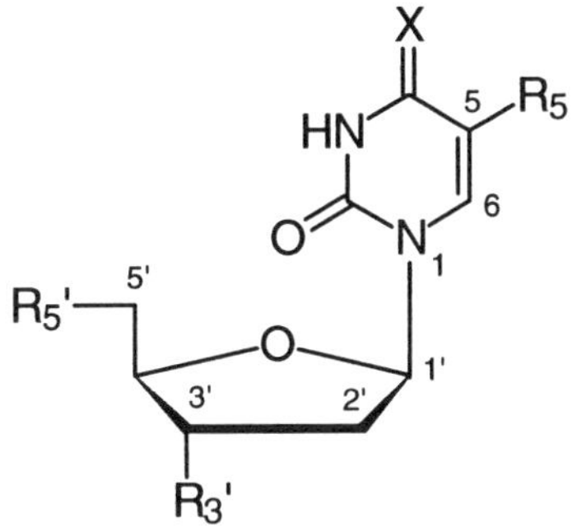

Figure 9. Common molecular structure for AZT derivatives.

Table 4. Anti-HIV-1 agents and potency in peripheral blood mononuclear cells

ID	R_5	$R_{5'}$	$R_{3'}$	X	EC_{50}	$-logEC_{50}$
1	H	OH	N_3	O	0.18-0.46[a]	0.495
2	CH_3	OH	N_3	O	0.002-0.009[a]	2.260
3	C_2H_5	OH	N_3	O	0.056-1.00[a]	0.277
4	C_3H_7	OH	N_3	O	63.0	-1.799
5	Br	OH	N_3	O	1.04	-0.017
6	I	OH	N_3	O	1.14	-0.057
7	H	OH	NH_2	O	60.0	-1.778
8	C_2H_5	OH	NH_2	O	54.9	-1.740
9	H	OH	N_3	NH	0.66-1.19[a]	0.034
10	CH_3	OH	N_3	NH	0.081-0.22[a]	0.822
11	H	OH	I	O	12.1	-1.083
12	CH_3	OH	I	O	46.3	-1.666
13	C_2H_5	OH	I	O	86.0	-1.935
14	H	OH	H	O	96.8	-1.986
15	CH_3	OH	H	O	0.17	0.770
16	C_2H_5	OH	H	O	4.90	-0.690
17	CH_3	N_3	OH	O	8.6	-0.935
18	CH_3	NH_2	OH	O	77.4	-1.889
19	C_2H_5	NH_2	OH	O	94.3	-1.975
20	H	OH	2',3'-unsaturated	O	68.3-79.9[a]	-1.870
21	CH_3	OH	2',3'-unsaturated	O	0.009-0.04[a]	1.611
22	C_2H_5	OH	2',3'-unsaturated	O	75.7	-1.879
23	CH=CHBr (E)	OH	N_3	O	>100	<-2.0
24	CH_3	OH	NH_2	O	>100	<-2.0
25	H	N_3	OH	O	>100	<-2.0
26	C_2H_5	N_3	OH	O	>100	<-2.0
27	H	NH_2	OH	O	>100	<-2.0

[a] *It has been computed the arithmetic mean*

As mentioned before, a PLS technique is used to reduce the number of available molecular descriptors and, in this way, to avoid the time-consuming combinatorial search required for all defined QS-SM. Previous to PLS transformation, each column unit vector $\{\mathbf{x}_k\}$ containing fragment QS-SM is standardized in order to obtain autoscaled variables with zero mean and unit variance. Then, it is possible to identify the fragments which are important for the modeling of response variables **y** only by seeing PLS regression coefficients, **b**. To discriminate which **X** variables are the most important in the model, it has chosen those $\{\mathbf{x}_k\}$ variables such their PLS regression coefficients b_k were larger than half the maximum value of **b** [122]. In the present example, computing a PLS analysis for 3 factors, 19 QS-SM fragments have been selected. Finally, the last step consists of the MLR computation over all possible combinations of the molecular descriptors selected in the PLS procedure using a NSS algorithm (see Appendix H), choosing at the end the correlation analysis with a bigger value of q^2. The best regression equation (4 parameters) for the training set of 22 AZT derivatives, which relates the biological activity $-\log EC_{50}$ and the selected quantum chemical descriptors:

$$
\begin{aligned}
&-\log EC_{50} = -0.6473\theta_{AA}^{C_{5'}} + 23.1445\theta_{AA}^{C_5C_6} - 23.5869\theta_{AA}^{C_5C_6N_1} \\
&-1.4764\theta_{AA}^{R_5} - 0.6831 \\
&n = 22;\; r^2 = 0.761;\; q^2 = 0.649
\end{aligned} \tag{50}
$$

Regarding Equation (50), there have been identified the molecular fragments $C_{5'}$, C_5C_6, $C_5C_6N_1$ and R_5 as the common structural patterns, responsible of biological affinity for the studied set of AZT derivatives. Regression coefficients for fragment QS-SM $\theta_{AA}^{C_5C_6}$ and $\theta_{AA}^{C_5C_6N_1}$ are very large, similar and with opposed sign. This feature indicates that it could exist as a more appropriate molecular fragment, presumably related with some association of the fragments C_5C_6 and $C_5C_6N_1$, but with a complex definition that was not considered in this study.

The present methodology serves to find common features among the studied molecules, providing a pharmacophore model. Within this procedure, it is possible to identify the common structural patterns and variations in molecular structure that explain changes in the biological affinity.

To verify the predictive power of the reported QSAR model corresponding to Equation (50), a correlation between observed and predicted values of the antiviral effect $-\log EC_{50}$ is shown in *Figure 10*. The predicted values were computed using a LOO cross-validation analysis [126], yielding a q^2 value of 0.649.

Finally, an independent validation study of QSAR model (50) was performed, consisting to predict the activity for a new subset of compounds with unknown activity. It consists of a set of 5 non-active compounds (from **23** to **27** AZT derivatives listed in *Table 4*), with EC_{50}>100 μM. In consequence, this validation study corresponds to an extrapolation. The predicted values of $-\log EC_{50}$ have been computed using Equation (50) and are listed in *Table 5*. As can be seen, except for molecule **24**, the rest of compounds have been correctly predicted as nonactive derivatives.

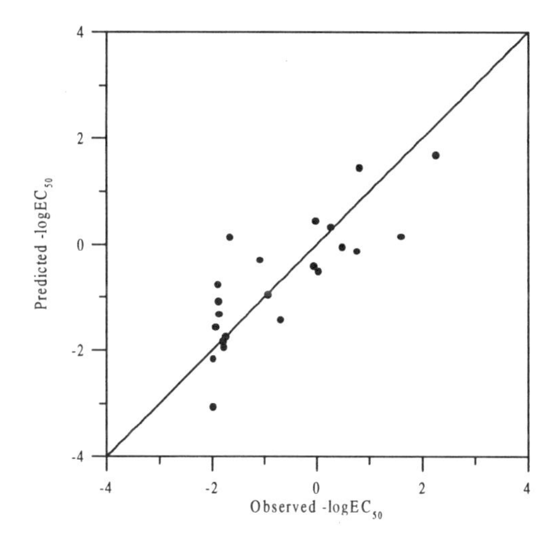

Figure 10. Observed versus LOO predicted activity for pyrimidine nucleosides derivatives

Table 5. Predicted antiviral effect for 5 nonactive compounds.

ID	Predicted $-\log EC_{50}$	Predicted EC_{50}
23	-1.925	84.14
24	-0.430	2.69
25	-2.849	706.3
26	-2.474	297.9
27	-3.023	1045

6. TOPOLOGICAL QUANTUM SIMILARITY MATRICES AND INDICES

It has been observed, when VSS were discussed, that SVSS could play an interesting role in the description of topological molecular vectors. At this moment the definition, structure, and particular QSAR use of QSM seems almost complete. However a point put into evidence at the beginning of this work has not yet been developed. The topological applications of atomic similarity matrices had not been discussed.

Everyone working in QSAR or QSPR problems will acknowledge the utility of Boolean topological matrices and the large amount of indices deduced from there by manipulating their elements [154]. However, Boolean topological matrices are such simplified molecular descriptors, that it is impossible to use them in cases where information depending on the atomic configuration in 3-dimensional space is needed. Also it is not clear how this leakage of information can affect the QSPR obtained from simplified Boolean topological matrices.

This section will try to develop the potential use of QSM ideas to obtain topological matrices where 3-dimensional information can be present.

6.1 Topological Quantum Similarity Matrices General Concept

Suppose an arbitrary molecule and known the positions: $\{\mathbf{R}_I\}$ and nature: $\{Z_I\}$ of their component atoms. This is the same as to know a simple molecular model where atomic DF (ADF): $\{\rho_I(\mathbf{r}-\mathbf{R}_I)\}$, are located at each atomic site. The quality of these ADF is not relevant for the following discussion; they can be obtained from some ASA DF set or be simply made of a single function, using squared STO or GTO, for example. A QSM matrix can be constructed from the intra-atomic relationships, in the same manner discussed for QOS.

That is, an atomic similarity matrix (ASM) is defined as:

$$\mathbf{Z}(\Omega)=\left\{z_{IJ}(\Omega)=\iint \rho_I(\mathbf{r}_1-\mathbf{R}_I)\Omega(\mathbf{r}_1;\mathbf{r}_2)\rho_J(\mathbf{r}_2-\mathbf{R}_J)d\mathbf{r}_1 d\mathbf{r}_2\right\} .$$

In this case the nature of the ADF set is arbitrary but must be homogeneous, and then the atomic similarity matrix will be symmetric. Similarly, the weight operator Ω, must be positive definite, in order to keep the whole matrix PD nature; as it is chosen in the QOS case for the same reasons. The ASM upper triangle will behave as a transformation of the topological distance vector $|\boldsymbol{D}>$, as discussed in Section 2.1.2, when the presence of SVSS is demonstrated. In the same way as with the SM attached to QOS, the ASM can be considered belonging to some matrix VSS of the adequate dimension.

The elements of the ASM can be considered entangled PD functions of the interatomic distances. As such they can represent some general procedure to generate topological representations of the attached molecules. A rich set of molecular descriptors, much more general than the simple Boolean topological matrices can be constructed in this way. Over these ASM the same manipulations, as the ones described in the Boolean case, can be performed. Thus, a large variety of new topological indices, with origin in quantum similarity grounds appear to be available.

Boolean topological matrices cannot obviously distinguish conformers or atomic configurations of the same molecule. ASM not only bear this information, but also it will provide molecular descriptors, belonging to the same molecular VSS or SVSS, for each molecular atomic configuration. Compared with Boolean topological matrices, quite complex patterns arise when ASM are considered. Therefore, a great deal of molecular information is present in the actual structures than in the simple Boolean topological matrices definition.

6.2 Topological Quantum Similarity Matrices over 1s GTO functions

As it has been stated, the traditional description of molecular structures by means of topological graphs has been widely used in QSAR-QSPR studies. Within this field, numerical correlation between quantities derived from topological graphs and physicochemical or biological properties [159], usually produces acceptable results.

It is often found, however, that the three-dimensional (3D) molecular structure can be of essential importance for describing some of the molecular properties, allowing more accurate QSAR predictive tools. Unfortunately, when topological indices (TI) derived from topological matrices (TM) are used, part of the spatial information of the molecule may be lost. Some studies focus on this problem [160,161].

The approach presented here, constitutes and extension of the classical TI definitions. One can call these indices topological quantum similarity indices (TQSI). In this sense, not only new molecular descriptors could be defined, but also a novel approach related to the QSM will be outlined. A preliminary version of this new methodology will be presented in the following sections.

The representation of a given molecular structure by a topological graph can be coded by means of the well-known attached TM. The TM elements are set to 1 if the associated atoms can be considered connected (bonded) and by 0 otherwise. The TM elements can be used afterwards in the definition of TI, which are taken as molecular descriptors, susceptible to be correlated with molecular properties.

Table 6 contains the definition of some auxiliary matrices or vectors usually employed in classical molecular topological studies.

The results obtained with the classical methodology will be compared with the present one. In the present approach, the TM elements become real numbers and for its construction it is not necessary to know whether the atoms forming a molecule are bonded or not. These matrices can be named topological quantum similarity matrices (TQSM).

Table 6. Elements related to the computation of many TI. n denotes the number of atoms in a molecule

Matrix/vector	Elements definition
T(n,n)	T_{ij}=1 if atoms i and j are bonded, or 0 otherwise
D(n,n)	D_{ij}: topological length of shortest path from atom i to atom j
V(n)	V_i: sum of entries in the i-th row or column of matrix **T**

To compute a TQSM, it is necessary to know the 3-D molecular geometry using over it a simple set of basis functions. The basis set is composed in the following discussion by means of one standard normalized 1s GTO function with exponents ζ_i and centred over every atom i, with nuclear coordinates $\mathbf{R}_i$, belonging to a molecule:

$$g(\mathbf{r}-\mathbf{R}_i,\zeta_i)=\left(\frac{2\zeta_i}{\pi}\right)^{3/4} e^{-\zeta_i|\mathbf{r}-\mathbf{R}_i|^2} \tag{51}$$

The function exponents used in this work are displayed in *Table 7.* For every atom, each exponent is defined in such a way that more of the 95% of the electronic density is contained within the spherical volume defined by the van der Waals radius. The listed values do not correspond to optimised parameters. Results concerning to this aspect will be published elsewhere.

Table 7. Atomic 1s GTO exponents used in this work

Atom	Exponent
C	0.46738
O	0.53053
F	0.54824

In the calculations presented here, all the molecular graphs are considered hydrogen suppressed. In this way, the choice which is customary in the classical approach has been followed.

Some TQSM will be defined here. In the following formulation, i and j denote atoms belonging to the same molecule. Basically, every TQSM element is computed by means of an integral which involves two normalized 1s GTO functions as defined in Equation (51) centred at atoms i and j. The most relevant TQSM elements are computed as follows:

Overlap matrix **S**: each element S_{ij} of the overlap matrix is defined as the overlap integral:

$$S_{ij} = \int g(\mathbf{r} - \mathbf{R}_i, \zeta_i)\, g(\mathbf{r} - \mathbf{R}_j, \zeta_j)\, d\mathbf{r} \tag{52}$$

Cioslowski-like matrix **C**: each element C_{ij} is defined as the square of the corresponding overlap matrix element:

$$C_{ij} = S_{ij}^2 \tag{53}$$

Coulomb electron-repulsion integral matrix **R**: Each element R_{ij} is defined as the two-electron repulsion integral:

$$R_{ij} = \int\int \left|g(\mathbf{r}_1 - \mathbf{R}_i, \zeta_i)\right|^2 \mathbf{r}_{12}^{-1} \left|g(\mathbf{r}_2 - \mathbf{R}_j, \zeta_j)\right|^2 d\mathbf{r}_1 d\mathbf{r}_2 \tag{54}$$

Gravitational integral matrix **G**: Each element is defined as the two-electron gravitational integral:

$$G_{ij} = \int\int \left|g(\mathbf{r}_1 - \mathbf{R}_i, \zeta_i)\right|^2 \mathbf{r}_{12}^{-2} \left|g(\mathbf{r}_2 - \mathbf{R}_j, \zeta_j)\right|^2 d\mathbf{r}_1 d\mathbf{r}_2 \tag{55}$$

Interatomic similarity matrix **Z**: The elements are defined as the overlap similarity between two atoms belonging to the same molecule:

$$Z_{ij} = \int \left|g(\mathbf{r} - \mathbf{R}_i, \zeta_i)\right|^2 \left|g(\mathbf{r} - \mathbf{R}_j, \zeta_j)\right|^2 d\mathbf{r} \tag{56}$$

All the TQSM are defined with null diagonal elements. This option has been considered in order to follow a similar criteria which is customary in the classical TM construction. Other choices can be adopted.

6.3 Topological Quantum Similarity Indices

Table 8 contains the recipes for calculating some relevant topological indices: Wiener (W) and Wiener Path (WPN) Numbers, Randic (χ), Schultz (MTI), Balaban (B) and Hosoya (Z) indices, Harary number (H) and the generalized connectivity indices (${}^{m}\chi_{t}$) of Kier and Hall. Most of them are described in references [58,161].

Table 8. Definition of several TI. Their mathematical structures are kept in order to build the TQSI. See text for more information.

Index	Definition
Wiener Path Number	$\mathrm{WPN} = \sum_{i}^{n}\sum_{j>i}^{n} D_{ij}$
Wiener index	W = WPN + p3, where p3 is the number of atoms separated by 3 bonds in the molecule.
Randic index	$\chi = \sum_{i}^{n}\sum_{j>i}^{n} \frac{T_{ij}}{\sqrt{V_i V_j}}$
Schultz index (Molecular topological index)	$\mathrm{MTI} = \sum_{i=1}^{n}[\mathbf{V}(\mathbf{T}+\mathbf{D})]_i$, where $[\mathbf{A}]_i$ stands for the i-th row of matrix **A**.
Harary number	$H = \sum_{i}^{n}\sum_{j>i}^{n} D_{ij}^{-2}$
Balaban index	$B = \frac{n_e}{\mu+1}\sum_{i}^{n}\sum_{j>i}^{n}\left[(D)_i (D)_j\right]^{\frac{-1}{2}}$, μ=number of cycles, n_e=number of edges, and $(D)_i$=sum of distances from vertex i.
Hosoya index	$Z = \sum_{i=0}^{n/2} p(i)$, where p(i) is the number of ways to draw i non-adjacent bonds in the molecular graph. It is also defined p(0)=1.
Generalized connectivity indices of order m and type t	${}^{m}\chi_t = \sum_{s=1}^{n_t}\prod_{i=1}^{m+1}(V_i)_s^{-1/2}$, n_t = number of connected subgraphs of type t.

As seen from the definitions in *Table 8*, usually, the TI can be defined by a sum over atomic terms which, in turn, can be obtained from the data embedded in the matrices or the vectors shown in *Table 6*. Note the special definition of the Randic index: in the numerator of the contributing terms there is present a TM element. This will be relevant when TQSI are considered.

A TQSI is constructed from the same mathematical formulae as in *Table 8*, but the matrix or vector elements are replaced by the corresponding TQSM ones. The summations run over the same integer indices (thus, the discretised molecular bond structure is also considered) but it is expected that the new indices include, in some way, information about the molecular 3-D structure.

When TQSI are considered, the Hosoya index is treated as a slightly special case. Despite the same classical contributions must be computer generated, in the present calculation, the p(i) term to be added is obtained in a special manner: it is not the number of ways to obtain i non-adjacent molecular graph edges. Instead, for every distinct form, the product of the TQSM elements attached to the selected bonds is computed. Then, the natural logarithm of this product is considered and all this term contributions are added.

Some 3-D variants of well-known topological indices can also be defined. Such indices are the 3D Wiener path number (^{3D}W), 3D Shultz index (^{3D}MTI) and the 3D Harary number (^{3D}H). Their definition is the same than the related ones appearing in *Table 8*, but the matrix of distances entering into the index computation is the real one collecting all the Cartesian distances between pairs of atoms belonging to the molecule. In addition, within the approach presented here, the Hosoya index and each of the p(i) contributions themselves are taken as molecular descriptors.

6.4 An application example: A family of steroids

To test the behaviour of the proposed methodology, it was decided to study a standard benchmark molecular set: the family of 31 steroids that bind to the corticosteroid binding globulin, which has been already studied in Section 5.3.2. Gasteiger and collaborators [162] performed an in-depth review related to all the 31 molecules in the set. All the geometries are taken as in reference [162]. Information on this and other useful items can be downloaded from the WWW site cited in [163].

During the phase of molecular indices generation, some restrictions were considered to reduce the amount of data to be analyzed:

- The generalized connectivity indices of Kier and Hall were computed only up to order 6.

- Only contributions up to order 9 are computed to obtain the Hosoya index.
- Specific QSAR studies were performed using the 5 TQSM individually. Also, an equivalent study was done using the classical TM. In this last case, the Hosoya index definition coincides with the classical one [164,165].

Table 9. Steroids CBG binding affinity: Experimental values are expressed as pKs and are taken from ref. [162]. In the last column appears the best set of cross-validated values obtained in this work. See text for more details.

Compound	ID number	Exp. Value	Cross-validated value
Aldosterone	**1**	-6.279	-6.582
Androstanediol	**2**	-5.000	-5.503
5-Androstenendiol	**3**	-5.000	-4.445
4-Androstenedione	**4**	-5.763	-6.310
Androsterone	**5**	-5.613	-5.444
Corticosterone	**6**	-7.881	-7.335
Cortisol	**7**	-7.881	-7.282
Cortisone	**8**	-6.892	-7.515
Dehydroepiandrosterone	**9**	-5.000	-4.532
11-Deoxycorticosterone	**10**	-7.653	-7.743
11-Deoxycortisol	**11**	-7.881	-8.046
Dihydrotestosterone	**12**	-5.919	-6.579
Estradiol	**13**	-5.000	-5.275
Estriol	**14**	-5.000	-4.919
Estrone	**15**	-5.000	-5.456
Ethiochonalonone	**16**	-5.225	-5.510
Pregnenolone	**17**	-5.225	-5.567
17a-Hydroxypregnenolone	**18**	-5.000	-5.448
Progesterone	**19**	-7.380	-7.08
g17a-Hydroxyprogesterone	**20**	-7.740	-6.861
Testosterone	**21**	-6.724	-6.015
Prednisolone	**22**	-7.512	-7.666
Cortisolacetat	**23**	-7.553	-7.679
4-Pregnene-3,11,20-trione	**24**	-6.779	-6.698
Epicorticosterone	**25**	-7.200	-7.519
19-Nortestosterone	**26**	-6.144	-5.690
16a,17a-Dihidroxyprogesterone	**27**	-6.247	-6.645
17a-Methylprogesterone	**28**	-7.120	-7.322
19-Norprogesterone	**29**	-6.817	-6.609
2a-Methylcortisol	**30**	-7.688	-7.707
2a-Methyl-9a-Fluorocortisol	**31**	-5.797	-5.466

Such restrictions forced to generate 57 topological indices by molecule and for each kind of TQSM. The data was originally organized in form of a (31×57) matrix. Then, some of the column descriptors were removed. This was done because some of these columns were the vector zero or were equal to another one (a special case of linear dependency) or only contained one non-zero element (this will cause problems during the process of true cross-validation). Finally, only m=36 descriptors remained attached to the TQSM and, thus, (31×36) data matrices were processed. *Table 10* collects the set of 36 indices used in this work.

Table 10. Set of topological indices used in this work.

Index Code	Symbol	Description
1	WPN	Wiener path number
2	W	Wiener index
3	^{3D}WPN	3D Wiener path number
4	χ	Randic index
5	MTI	Schultz index
6	^{3D}MTI	3D Schultz index
7	H	Harary number
8	^{3D}H	3D Harary number
9	B	Balaban index
10	^{3D}B	3D Balaban index
11	Z	Hosoya index[(a)]
12	p(1)	p(i)[(a)] is the contribution of order i to the Hosoya index
13	p(2)	
14	p(3)	
15	p(4)	
16	p(5)	
17	p(6)	
18	p(7)	
19	p(8)	
20	p(9)	
21	$^{0}\chi_{P}$	Generalized connectivity Indices
22	$^{1}\chi_{P}$	
23	$^{2}\chi_{P}$	
24	$^{3}\chi_{P}$	
25	$^{4}\chi_{C}$	
26	$^{4}\chi_{P}$	
27	$^{4}\chi_{C}$	
28	$^{4}\chi_{PC}$	
29	$^{5}\chi_{P}$	
30	$^{5}\chi^{C}$	
31	$^{5}\chi_{CH}$	
32	$^{5}\chi_{PC}$	
33	$^{6}\chi_{P}$	
34	$^{6}\chi_{C}$	
35	$^{6}\chi_{CH}$	
36	$^{6}\chi_{PC}$	

[a] *As it is explained in the text, the definition of this index differs when TQSM or classical TM are being considered.*

To perform a comparative analysis, the number of descriptors for each kind of topological matrix had to be the same. This allowed some TQSI to enter into the set of classical ones. Such indices are the 3-D mentioned above: those that have the codes 3, 6, 8, and 10 in *Table 10*. Despite of this extension, the number of indices attached to the TM was only m=35. The difference of one unit with respect to the number of TQSI is found because, in the classical approach, the Randic index and the $^1\chi_P$ connectivity one are equivalents. This coincidence is not found when TQSM are used due to the definitions shown in *Table 8*.

In the search for the linear models no pre-treatment related to the PCA, PLS or other linear or non linear techniques was performed. For each descriptor matrix, all possible combinations of indices entering in a given linear correlation equation are generated. From a chosen set of m=35 or m=36 indices, this procedure performs "m over k" calculations, where k is the number of active variables present in the regression model. In the results presented here, the index k varies from 1 up to 9. The whole procedure is performed applying a nested summation symbol (NSS) algorithm codification (see Appendix H). To speed up the calculations, for every descriptor set combination it is calculated the attached value of the q^2 coefficient [166]. Once the optimal combination is extracted, a true c-v computation is made and the related value of r^2 is reported.

Tables 11 to *16* contain the relevant results obtained when all the indicated combinations of indices were performed. In each table, the first column shows the number of descriptors, k, entering into the linear model. The second one presents the value of r^2 obtained in the c-v procedure. Also, the statistical significance arising from the Snedecor F-test for the linear regression attached to the c-v is presented. Finally, the linear model equation is reported. In each equation, the symbol for each descriptor index must be identified in *Table 10* and computed using the corresponding topological matrix. In the linear models, the dependent variable **y** stands for the adjusted value of the activity, which is expressed in the same logarithmic scale as it appears in *Table 9*. The comparison of *Table 11* against *Tables 12* to *16* reveals that the results arising from the classical indices are poorer than those obtained from the TQSM. Some aspects corroborate this conclusion: first, the values of r^2 for the c-v data are almost always better in the case of TQSI. Secondly, the corresponding statistical significances are slightly better than those arising from the data of the classical topological matrix. Finally, in *Table 11* the maximum value for r^2 is obtained when a set of 6 descriptors is optimised (r^2=0.617), whereas for any kind of TQSM a higher value is obtained for the same number of descriptors. In addition, when TQSI are being considered, the maximum value of r^2 is reached when using 8 or 9 selected descriptors. The enhanced values obtained using TQSM cannot be associated to the fact that the set of TQSI gives one more column descriptor than the classical one.

Table 11. Data coming from the linear models obtained using the classical TM.

K	r^2 in c-v	Significance p	Linear model equation
1	0.3485	0.00047	$y = -0.631710\,\chi + 0.584650$
2	0.5539	<0.00001	$y = -6.80513\,\chi + 0.109315\,p(2) + 36.3320$
3	0.5711	<0.00001	$y = -0.00299609\,MTI + 0.408869\,H + 4.80190 \cdot 10^{-5}\,p(9) - 11.9241$
4	0.5860	<0.00001	$y = -5.80747\,\chi + 0.954499\,H + 2.08753 \cdot 10^{-5}\,p(9) - 6.18881\,{}^{4}\chi_{C} + 12.9966$
5	0.5941	<0.00001	$y = -0.00115122\,p(4) + 7.19691 \cdot 10^{-5}\,p(9) - 8.99523\,{}^{4}\chi_{C} + 1.29308\,{}^{6}\chi_{P} + 0.415011\,{}^{6}\chi_{PC} - 12.2709$
6	0.6174	<0.00001	$y = -0.259245\,WPN + 0.125887\,{}^{3D}WPN + 0.0567972\,MTI - 0.0296704\,{}^{3D}MTI + 1.36417 \cdot 10^{-5}\,Z + 2.21348\,{}^{0}\chi_{P} - 24.4812$
7	0.6048	<0.00001	$y = -0.243320\,W + 0.133003\,{}^{3D}WPN + 0.0546371\,MTI - 0.0313233\,{}^{3D}MTI + 0.590032\,H + 5.22155 \cdot 10^{-5}\,p(9) + 0.285499\,{}^{6}\chi_{PC} - 15.0907$
8	0.5923	<0.00001	$y = -0.205512\,W + 0.107321\,{}^{3D}WPN - 3.71141\,\chi + 0.0504903\,MTI - 0.0253692\,{}^{3D}MTI + 1.07973\,{}^{2}\chi_{P} - 2.29896\,{}^{5}\chi_{PC} + 1.67127\,{}^{6}\chi_{CH} + 17.9713$
9	0.5966	<0.00001	$y = -0.212278\,W + 0.133312\,{}^{3D}WPN + 0.0511613\,MTI - 0.0313169\,{}^{3D}MTI - 3.77090\,{}^{3}\chi_{P} - 1.29217\,{}^{4}\chi_{C} - 4.25158\,{}^{4}\chi_{P} + 5.42344\,{}^{6}\chi_{P} + 1.23000\,{}^{6}\chi_{PC} + 24.7631$

Table 12. Data coming from the linear models obtained using the TQSM S.

K	r^2 in c-v	Significance p	Linear model equation
1	0.3959	0.00015	$y = -0.985932\,\chi + 1.45934$
2	0.5145	0.00001	$y = -3.92190\,\chi + 0.0516229\,p(3) + 18.8701$
3	0.5489	<0.00001	$y = -0.159189\,WPN + 0.150734\,W + 0.0397568\,p(8) - 5.21990$
4	0.5880	<0.00001	$y = -0.0270046\,WPN + 1.84832\,p(1) + 0.275361\,p(9) + 1.96336\,{}^{0}\chi_{P} - 39.5464$
5	0.6959	<0.00001	$y = -0.0272842\,{}^{3D}WPN - 64.2096\,{}^{3D}B + 0.0590476\,p(7) + 6.59182\,{}^{0}\chi_{P} - 0.524068\,{}^{5}\chi_{P} - 18.9161$
6	0.7432	<0.00001	$y = 0.0760581\,{}^{3D}WPN - 18.8259\,\chi - 0.0371972\,{}^{3D}MTI + 1.94512\,{}^{3D}H - 21.8325\,B + 2.02084\,p(2) + 102.182$
7	0.7773	<0.00001	$y = 0.0704124\,{}^{3D}WPN - 23.6715\,\chi - 0.0353800\,{}^{3D}MTI - 20.5103\,B + 2.37113\,p(2) + 1.66830\,{}^{1}\chi_{P} - 1.54703\,{}^{6}\chi_{CH} + 111.567$
8	0.8087	<0.00001	$y = 0.115799\,W - 0.0725991\,MTI + 2.98281\,H - 42.2718\,B + 0.0507752\,p(5) - 1.75628\,{}^{3}\chi_{P}$

K	r^2 in c-v	Significance p	Linear model equation
			$- 0.626142\ {}^4\chi_P + 0.616826\ {}^5\chi_C + 50.1490$
9	0.8295	<0.00001	$y = -0.0104394\ {}^{3D}MTI + 6.04928\ H - 19.2501\ {}^{3D}H + 0.0384960\ p(5) + 4.84038\ {}^2\chi_P - 1.81503\ {}^3\chi_P - 5.37187\ {}^4\chi_c - 2.11893\ {}^4\chi_P + 0.334864\ {}^5\chi_{PC} - 57.0143$

Table 13. Data coming from the linear models obtained using the TQSM C.

K	r^2 in c-v	Significance p	Linear model equation
1	0.3657	0.00031	$y = -0.684306\ \chi + 0.709345$
2	0.5395	<0.00001	$y = -3.69140\ \chi + 8.92520{\cdot}10^{-4}\ p(3) + 22.6735$
3	0.6446	<0.00001	$y = -7.05924\ \chi + 0.0316608\ p(2) - 0.00211777\ {}^6\chi_P + 38.5669$
4	0.6742	<0.00001	$y = -0.0316447\ MTI + 3.74207{\cdot}10^{-5}\ Z - 1.25643{\cdot}10^{-4}\ p(8) - 0.0166959\ {}^4\chi_P + 18.0008$
5	0.7590	<0.00001	$y = -0.00855952\ {}^{3D}WPN + 0.00117431\ p(3) + 0.310672\ {}^2\chi_P - 0.0407307\ {}^4\chi_P - 0.183158\ {}^4\chi_C - 12.2813$
6	0.7825	<0.00001	$y = -0.0226653\ W + 0.00173207\ p(3) + 0.319495\ {}^2\chi_P - 0.0413886\ {}^4\chi_P - 0.185162\ {}^4\chi_C - 0.0278816\ {}^6\chi_{CH} - 9.28960$
7	0.8375	<0.00001	$y = -0.0218409\ WPN + 0.00127794\ p(3) + 0.675492\ {}^2\chi_P - 0.0807284\ {}^3\chi_P - 0.266165\ {}^4\chi_c - 0.0596890\ {}^4\chi_P + 0.00122358\ {}^6\chi_{PC} - 15.2253$
8	0.8464	<0.00001	$y = -0.0240972\ WPN + 0.00128227\ p(3) + 0.926355\ {}^2\chi_P - 0.127675\ {}^3\chi_P - 0.467621\ {}^4\chi_c - 0.0720036\ {}^4\chi_P + 0.173544\ {}^4\chi_C + 0.00188091\ {}^6\chi_{PC} - 18.1988$
9	0.8431	<0.00001	$y = -0.0260143\ WPN - 1.79510\ p(1) + 0.0520044\ p(2) + 1.05743\ {}^2\chi_P - 0.126139\ {}^3\chi_P - 0.533776\ {}^4\chi_c - 0.0704954\ {}^4\chi_P + 0.241000\ {}^4\chi_C + 0.00186549\ {}^6\chi_{PC} + 9.93050$

Table 14. Data from coming from the linear models obtained using the TQSM R.

K	r^2 in c-v	Significance p	Linear model equation
1	0.4175	0.00009	$y = 20.6953\ {}^6\chi_{CH} - 9.97379$
2	0.4990	0.00001	$y = -13.0677\ \chi + 0.225179E\text{-}05\ p(9) + 27.1318$
3	0.5626	<0.00001	$y = -0.164318\ WPN + 0.156238\ W + 4.28854{\cdot}10^{-6}\ p(9) - 5.72758$
4	0.5963	<0.00001	$y = -0.00839460\ MTI + 0.896334\ H + 2.09245{\cdot}10^{-4}\ p(5) - 327.450\ {}^5\chi_{CH} + 9.71717$
5	0.6408	<0.00001	$y = -0.00909252\ MTI + 0.857664\ H + 2.27305{\cdot}10^{-4}\ p(5) - 4.60778\ {}^5\chi_p - 303.646\ {}^5\chi_{CH} + 26.2894$
6	0.6893	<0.00001	$y = -253.264\ \chi - 89.4934\ p(1) + 5.14361\ p(2) - 0.313297\ p(3) + 0.0130563\ p(4) + 82.0409\ {}^1\chi_P + 766.623$
7	0.7024	<0.00001	$y = -231.600\ \chi - 5.81039\ {}^{3D}H - 92.6729\ p(1) + 5.42491\ p(2) - 0.327338\ p(3) + 0.0135724\ p(4) + 67.8383\ {}^1\chi_P + 844.941$

K	r^2 in c-v	Significance p	Linear model equation
8	0.7062	<0.00001	$y = 0.232832\ \mathrm{WPN} - 184.876\ \chi - 0.0266580\ \mathrm{MTI} + 8.94562\ \mathrm{H} - 202.978\ {}^{3D}\mathrm{B} - 14.9381\ p(1) + 1.22936\cdot10^{-5}\ p(9) - 30.3715\ {}^{4}\chi_C - 609.066$
9	0.7271	<0.00001	$y = -0.00339147\ {}^{3D}\mathrm{WPN} - 275.264\ \chi - 97.2917\ p(1) + 5.14996\ p(2) - 0.233272\ p(3) + 6.95570\cdot10^{-4}\ p(7) - 6.60697\cdot10^{-4}\ p(8) + 2.97656\cdot10^{-4}\ p(9) + 91.1601\ {}^{1}\chi_P + 835.572$

Table 15. Data coming from the linear models obtained using the TQSM G.

K	r^2 in c-v	Significance p	Linear model equation
1	0.4141	0.00009	$y = -2.06963\ \chi + 4.26957$
2	0.5523	<0.00001	$y = -4.47291\ \chi + 0.304791\mathrm{E}{-05}\ p(6) + 15.6531$
3	0.5629	<0.00001	$y = -0.164231\ \mathrm{WPN} + 0.156152\ \mathrm{W} + 2.57014\cdot10^{-6}\ p(9) - 5.72507$
4	0.6142	<0.00001	$y = -0.0224323\ \mathrm{W} + 0.944439\mathrm{E}{-06}\ \mathrm{Z} + 0.0199121\ p(2) - 3.32269\ {}^{5}\chi_{CH} + 1.50648$
5	0.6546	<0.00001	$y = -0.0248421\ \mathrm{MTI} + 0.00343997\ p(3) + 2.66000\cdot10^{-6}\ p(9) - 0.106670\ {}^{4}\chi_P - 4.14119\ {}^{5}\chi_{CH} + 21.0160$
6	0.7397	<0.00001	$y = -0.0357416\ \mathrm{MTI} + 0.00483513\ p(3) + 1.41100\ {}^{2}\chi_P - 0.388262\ {}^{4}\chi_P - 1.80206\ {}^{4}\chi_C - 0.102283\ {}^{6}\chi_P + 7.86770$
7	0.7896	<0.00001	$y = -0.0710493\ \mathrm{WPN} + 0.820787\ \mathrm{H} + 3.76747\cdot10^{-4}\ p(4) + 4.49753\ {}^{2}\chi_P - 2.40651\ {}^{4}\chi_c - 0.381931\ {}^{4}\chi_P - 0.228846\ {}^{6}\chi_P - 79.9399$
8	0.8098	<0.00001	$y = -0.0618643\ \mathrm{MTI} + 11.9419\ {}^{3D}\mathrm{H} + 1.07113\cdot10^{-4}\ p(5) + 4.44355\ {}^{2}\chi_P - 2.75441\ {}^{4}\chi_c - 0.209319\ {}^{4}\chi_P - 0.270982\ {}^{6}\chi_P + 0.228220\ {}^{6}\chi_C - 109.385$
9	0.8273	<0.00001	$y = -0.0723889\ \mathrm{MTI} + 0.00484849\ {}^{3D}\mathrm{MTI} + 12.3278\ {}^{3D}\mathrm{H} + 1.08335\cdot10^{-4}\ p(5) + 4.42308\ {}^{2}\chi_P - 2.72847\ {}^{4}\chi_C - 0.236230\ {}^{4}\chi_P - 0.265736\ {}^{6}\chi_P + 0.206697\ {}^{6}\chi_C - 108.875$

Table 16. Data coming from the linear models obtained using the TQSM Z.

K	r^2 in c-v	Significance p	Linear model equation
1	0.3636	0.00033	$y = -0.681200\ \chi + 0.682621$
2	0.5538	<0.00001	$y = -5.31344\ \chi + 0.00682478\ p(2) + 25.5152$
3	0.6166	<0.00001	$y = -7.13080\ \chi + 0.00950417\ p(2) - 0.325926\mathrm{E}{-06}\ {}^{6}\chi_{CH} + 39.3384$
4	0.6907	<0.00001	$y = -1.54425\ \mathrm{MTI} + 1.19686\cdot10^{-5}\ \mathrm{Z} - 4.01527\cdot10^{-5}\ p(8) - 0.908229\mathrm{E}{-06}\ {}^{4}\chi_P + 17.0733$
5	0.7749	<0.00001	$y = -0.0189161\ \mathrm{WPN} + 4.16776\cdot10^{-4}\ p(3) + 9.66286\cdot10^{-4}\ {}^{2}\chi_P - 0.246881\mathrm{E}{-05}\ {}^{4}\chi_P - 1.14067\cdot10^{-5}\ {}^{4}\chi_C - 14.4497$
6	0.7698	<0.00001	$y = -0.00999634\ \mathrm{WPN} - 0.00410123\ {}^{3D}\mathrm{WPN} + 3.90562\cdot10^{-4}\ p(3) + 9.86920\cdot10^{-4}\ {}^{2}\chi_P - 2.50266\cdot10^{-6}\ {}^{4}\chi_P - 1.17757\cdot10^{-5}\ {}^{4}\chi_C - 14.6329$
7	0.8296	<0.00001	$y = -0.0215443\ \mathrm{W} + 3.99694\cdot10^{-4}\ p(3) + 0.00205049\ {}^{2}\chi_P - 3.16723\cdot10^{-5}\ {}^{3}\chi_P$

K	r^2 in c-v	Significance p	Linear model equation
			$- 1.15062\cdot10^{-4}\ {}^4\chi_c - 3.62817\cdot10^{-6}\ {}^4\chi_P$ $+ 1.45325\cdot10^{-9}\ {}^6\chi_{PC} - 17.6581$
8	0.8447	<0.00001	$y = -0.0242555\ W + 4.03051\cdot10^{-4}\ p(3) + 0.00297847\ {}^2\chi_P$ $- 5.68511\cdot10^{-5}\ {}^3\chi_P - 2.22784\cdot10^{-4}\ {}^4\chi_c$ $- 4.54913\cdot10^{-6}\ {}^4\chi_P + 1.33432\cdot10^{-5}\ {}^4\chi_C$ $+ 2.48843\cdot10^{-9}\ {}^6\chi_{PC} - 21.2394$
9	0.8445	<0.00001	$y = -0.0219862\ WPN + 3.45924\cdot10^{-4}\ p(3)$ $+ 0.00282118\ {}^2\chi_P - 5.79523\cdot10^{-5}\ {}^3\chi_P$ $- 2.06971\cdot10^{-4}\ {}^4\chi_c - 4.64288\cdot10^{-6}\ {}^4\chi_P$ $+ 1.18230\cdot10^{-5}\ {}^4\chi_C + 2.52533\cdot10^{-8}\ {}^5\chi_P$ $+ 2.43975\cdot10^{-9}\ {}^6\chi_{PC} - 19.5797$

This is so because the pair of descriptors χ (Randic) and ${}^1\chi_P$ (a connectivity index) appear only simultaneously in a few linear models and the best ones (those from *Table 13*) do not contain this pair acting in a synergic manner nor any of the specified indices enter in the best models (k=5 to 9). *Figure 11* shows the evolution of the value of r^2 with respect to the number of descriptors entering into the linear model. In the figure, it has been represented the numerical values coming from *Table 11* (the classical topological approach), *Table 13* (the best results obtained with TQSI) and *Table 14* (the worst results obtained using TQSI). Some of the aspects discussed in the previous paragraph are clearly visualised in the figure. The labels identify the original TQSM used to compute the indices.

Another characteristic to be mentioned: In the linear models related to a specific matrix, most of the selected indices appear many times. This can be understood in the sense that, when using this kind of descriptors, there is a set of them well conditioned and oriented with respect to the studied molecular property or activity.

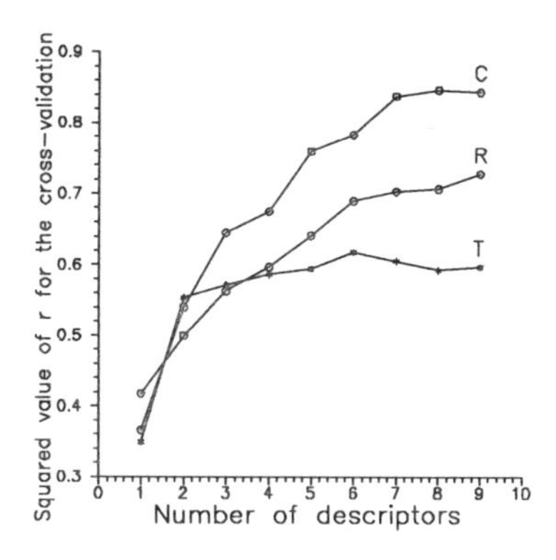

Figure 5. Graphical representation of some of the results contained in tables 6-11. The labels identify the original TQSM used to compute the indices

Also, a statistical test has been performed in order to check the degree of molecular information contained into the indices. The idea was to perform 100 additional systematic and equivalent studies but this time using 31×36 random matrices, acting as TQSM, with uniformly distributed value elements. In this numerical experiment, up to k=6 descriptors entered into the linear models. *Table 17* contains the most relevant data coming from this Monte Carlo simulation. The first column indicates the number of random descriptors entering into the linear model. The second column lists the maximum value of r^2 for the cross-validation obtained for every group of descriptors. The third column shows the minimal values found for the statistical significance. In general, the values of the second and third columns do not arise from the same particular computation. In the fourth column it can be found the mean values of r^2 obtained for every group of descriptors. Finally, the last column indicates the related values of s'=s/10, being s the standard deviation of the 100 values of r^2. Thus, s' is the dispersion parameter used in the t-Student statistics, in order to estimate the interval of definition for the mean value of r^2.

Table 17. Statistical data related to the t-Student parameter in the study of a Monte Carlo test. Information in each row come from 100 data measures. See text for more information.

K	Maximal value of r^2	Minimal value of p	Mean value of r^2	S'=s/10
1	0.2115	0.00924	0.0862	0.0050
2	0.3611	0.00035	0.1753	0.0079
3	0.4447	0.00004	0.2424	0.0089
4	0.5955	<0.00001	0.3070	0.0099
5	0.6558	<0.00001	0.3628	0.0100
6	0.7229	<0.00001	0.4141	0.0103

In all the cases, the Student's t-statistics reveals that the means appearing in *Table 17* are essentially distinct from the values of r^2 shown in *Tables 11* to *16*. The statistical confidence level is always greater than 99.999%. This result demonstrates that the data in *Tables 11* to *16* are statistically significant, predictive and contain substantial molecular activity information.

Also, because the number of random tests is considerable, for every value of k the distribution of the obtained values for r^2 must approach a gaussian distribution. *Table 18* shows the results obtained when a gaussian fitting is performed over every group of 100 data. The first column indicates the number of random descriptors entering into the linear model. Then, the adjusted mean and standard deviations are written. Finally, the values of r^2 for the linear fittings in a probability plot are reported.

In all the cases, as it can be seen from the inspection of *Tables 11* to *16*, the obtained values of r^2 for the cross-validation using TI or TQSI are found on the right of the fitted gaussian distributions. The cumulated probability lying on the left of the normal curve is always greater than 99.5% and, in many cases, including the most relevant ones, such a probability is superior to 99.99%.

Table 18. Statistical data related to gaussian fittings for the values of r2 for cross-validation obtained in the Monte Carlo test. See text for more information.

K	Mean value of r^2	Adjusted Value of σ	Value of r^2 in the fitting
1	0.0862	0.0512	0.9619
2	0.1753	0.0800	0.9709
3	0.2424	0.0903	0.9823
4	0.3070	0.1001	0.9854
5	0.3628	0.1012	0.9838
6	0.4141	0.1044	0.9843

Eventhough all the topologically-related prediction studies are statistically significant, the worst case is found precisely when TI are used. The predictions arising from the classical method give the smallest values of r^2. Then, such values are always closer to the predefined intervals of statistical confidence.

The authors consider that the Monte Carlo test presented here, together with the attached statistical study, is in this case even stronger than the well-known randomization test procedure. In a standard randomisation test, the vector elements containing the molecular property values are randomly swapped. Then, a c-v calculation is done and the related resulting parameters are reported. A word of caution must be said with respect to this kind of procedures: usually it is used a predefined linear model which was, in some sense, optimized to give good results when the correct property vector is worked out. Within this approach, it is probable that the reordering of the molecular property values will drive to the prediction of corrupted activity values in the process of c-v. On the contrary, the Monte Carlo test reported in this work demands not only a stronger and stable linear model, but also a better set of molecular descriptors. This is so because this kind of simulation test is a full randomization procedure. It must be noted that in this statistical experiment not only some parameters are changed, but also new linear models are searched trying to fit the molecular data with corrupted starting descriptor values. Thus, entire freedom was given to the system in order to find the best result in every test. This idea was considered in the randomisation test presented here. 100 randomly sorted property vectors have been generated and the system tried to correlate each corrupted data vector using the TQSM **C**. The procedure was unable to find a significant value of r^2 for the c-v. The largest value found was r^2=0.5490 when using k=6 descriptors. The results are displayed in *Figure 12*.

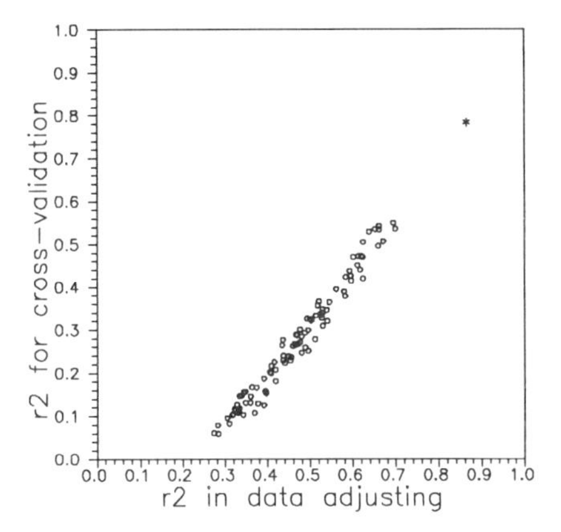

Figure 12. Randomization test results for the model of 6 descriptors of Table 13. 100 randomized points were calculated as it is explained in the text. The asterisk signals the correct model result.

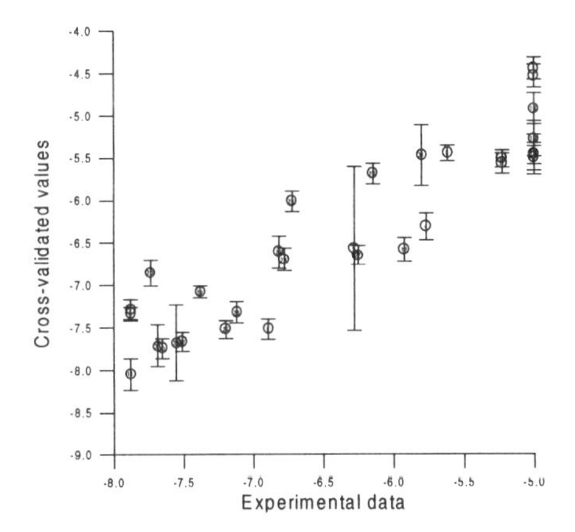

Figure 13. Results for a leave-two-out procedure. See text for more information

Figure 13 shows the results obtained in a leave-two-out procedure when the best model of 8 descriptors in *Table 13* is employed. Circles signal the mean c-v value property in the 30 tests performed over each molecule. The bars mark the error estimation. It was set equal to 3s, being s the estimation of the standard deviation for each group of 30 values when they are adjusted to a gaussian distribution. Note the special behaviour of molecule number 1.

It bears the maximal error, that is, it is attached to the most sensible result to be correlated.

Figure 14 shows the c-v value properties obtained employing the model of 8 descriptors coming from the TQSM C. The related numerical values can be found in *Table 9* and *Table 13*.

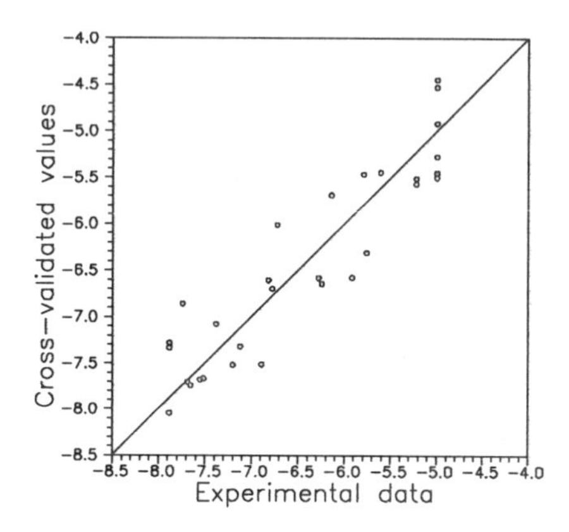

Figure 14. Cross-validation results using the linear model of 8 descriptors coming from the TQSM C. The attached value of r^2 is 0.8464

In general, the results presented in *Tables 11* to *16* of this work are as accurate, or better in some cases, than the previous ones published by other authors. For example, M. Wagener et al. in ref. [162], using a combination of Kohonen and feedforward neural networks, obtained a value of q^2=0.63, when the complete set of 31 steroids as in the present work is studied. R.D. Cramer et al. in ref. [140], have studied a reduced set of 21 molecular structures and provide a result of q^2=0.662 using CoMFA methodology working with two optimal PLS components. On the other hand, A. C. Good *et al.* in ref. [5] employed the same set of 21 steroids. These authors present a set of results computed using various methodologies, the best q^2 value was 0.761. B. D. Silverman and D.E. Platt [167], for the same reduced 21 steroid family, employ CoMMA method. The obtained q^2 value using two PLS parameters was 0.674. The method presented here is able to give a value of r^2 for the c-v equal or superior to 0.9 when the set of the first 21 molecules is considered.

7. CONCLUSION

A general discussion on the formal structure of QSM and the surrounding problems has been carried out. Several definitions design in an accurate manner the notion of QO and, in this way, the tagged set structure of QOS naturally appears. The attached formalism becomes a rich source of new concepts, such as the KE DF and other DF forms, which are based on extended Hilbert spaces. This formalism leads towards a matrix formulation of energy expectation values. The new density functions forms can be described and visualised too, producing useful suggestive pictures of the molecular structure environment.

Discretization of QO descriptions, via general definitions of QSM, also conduces to the definition of SM and discrete QOS. The connection of all the involved tag sets to some VSS, allows the analysis of several aspects of QSM as well as of the QSPR. In this way, it can be said that QSM n-dimensional vectors, associated to a given molecular structure, can be considered, from a quantum mechanical optic, as an unbiased way to represent microscopic systems in a discrete manner. In this sense, SM and convex combinations of SM have lead to satisfactory descriptions of molecular properties in different environments. In this QSM context, the user has no way to choose arbitrarily the parameters and, doing so, possibly interact with the resulting model. Empirical QSAR parameters, whatever their origin, shall be considered as examples of more or less successful attempts to simulate such QSM vectorial description.

New mathematical devices like the fragment QSM or the TSM leading to the definition of TQSI can be used as a possible source of QO descriptors. Therefore, these new descriptors can be viewed as another of the consequences of the present framework. They can be employed as a new source of quantum similarity indices computation, which are easily defined and possessing an interesting potential behaviour. In addition, it has been proved that they possess a high descriptive power that make them susceptible to be used in a QSAR-QSPR environment.

The same can be said in reference to Coulomb energy expectation values, $<V_{ee}>$, which being QS-SM can be used as such to obtain Hansch-type relationships. In the present work, an application example is shown. leading to valuable results. The results indicate that $<V_{ee}>$, a method based on electron densities instead of the conventional atomic-fragment contribution approaches, may constitute in some cases a suitable way of complementing or quantifying QSPR-QSAR in general.

Topological matrices, based on interatomic QSM, appear to be a noticeable source of molecular parameters too, as well as an interesting generalisation of the classical molecular topology by means of the introduction of quantum information to the corresponding topological indices.

An extended inward (hyper-)matrix product with precisely defined properties has been introduced. Extended inward matrix products can be also related to classical matrix products, via trivial relationships involving nested summation symbols. This operation can be used to describe a new way to obtain QSAR from QSM.

ACKNOWLEDGMENTS

Thanks are due to *Prof. Dr.* K. Sen, because some parts of this paper have been developed through daily interaction with him by of one of us (R.C.-D.) at Hyderabad University (India).

Lively discussions on the many facets of the problems contained in this work were also carried out with *Profs.* P.G. Mezey (University of Saskatchewan, Saskatoon) and R. Ponec (Institute of Chemical Process Fundamentals, Czech Academy of Sciences), during their visits to the IQC in Girona in 1998-1999. Their remarks and comments helped improve several aspects of the manuscript. Discussions about molecular toxicity and supply of experimental data are also acknowledged to Dr. E. Benfenati (Istituto Mario Negri, Milan).

One of us (X.G.) wants to acknowledge the University of Girona a predoctoral fellowship.

This work has been partially financed by the CICYT Research Project #SAF 96-0158 and sponsored by the European Commission contract #ENV4-CT97-0508, as well as the Foundation María Francisca de Roviralta.

APPENDIX A. TAGGED SETS EXAMPLES

A1. Boolean Tagged Sets

Tagged sets may be considered as sets such that their elements and any kind of coherent information to describe them are *simultaneously* taken into account. The simplest of tagged sets can be defined whenever the tag set part elements can be transformed into or expressed by n-dimensional Boolean or bit strings. Every n-bit string could be easily associated with any of the 2^n vertices of a n-dimensional unit length hypercube, H_n. Thus, any set of objects, possessing some kind of information attached to them, can be structured as a tagged set, using the appropriate n-dimensional cube vertices as the tag set elements [41,81].

Moreover, there appears to be a characteristic feature, which will reappear throughout this chapter, associated to the definition of any hypercube vertex tag set. Boolean tags are formed by unit length n-dimensional cube vertices, which because the bit-like nature of their components can be considered, directed and included into a positive definite hyper-quadrant section belonging to some n-dimensional space. It is also obvious that other tagged sets can be transformed into a Boolean form: Consider a tag set made by n-tuples of rational numbers as a quite common and general example. The nature of the molecular information precludes the possibility of easily transform chemical tagged sets into Boolean molecular tagged sets and inversely. In general, the effortless transformation of a tagged set into a Boolean one is propitiated by two circumstances: a) the natural intrinsic *Positive Definite* (PD) character of the experimental or theoretical information gathered into the tag set:, and b) the peculiar structure of modern electronic computational tools, whose information is processed in terms of Boolean strings and translated into integer or rational numbers.

From the above point of view, Boolean tagged sets can be considered as a sort of canonical form, which can describe, in some ultimate way, any kind of discrete rational information orderly attached to a chosen object set.

A2. Functional Tagged Sets

Until now, tagged sets in this Appendix have been supposed implicitly constructed employing n-dimensional vector-like tag sets. However, there is no need to circumscribe tag set parts to finite-dimensional spaces.

Another crucial point has to be considered before going ahead in the description of tagged sets. It is related to Boolean tagged sets, and appears when a ∞-dimensional hypercube vertex subset is taken as the tag set part. Then, a parallelism between the ∞-dimensional vertices and the elements of the [0,1] segment acting as tags naturally appears. Finally, it must be considered the possible use of Boolean matrices of arbitrary dimension ($m \times n$), or still more generally: Boolean hyper matrices, as sound candidates to belong to the tag set part. All these multiple possibilities, associated to Boolean tagged sets, open the way to consider the possible definition of still more general tagged sets.

As already explained, tag sets can also be made of elements coming from ∞-dimensional spaces. Any function space can be considered belonging to the ∞-dimensional class of spaces. Moreover, among all the possible function families, possessing homogeneous properties, the most appealing candidate, from the molecular point of view, corresponds to a subset of some probability density functional space.

Two reasons point towards this kind of choice. First, probability DF (PDF) are normalizable. They are PD functions too, yielding values within the (0,1) open segment, and they, thus, may behave as some isomorphic infinite-dimensional limit form of a Boolean tag set. Second, according to the interpretation given by von Neumann [85] or Bohm [86], PDF formed by the squared module of state wavefunctions, constitute mathematical elements attached to the descriptive behaviour of quantum systems. Recent [87], and not so recent [83], discussions signal towards this descriptive role of the quantum DF too. It seems that, in this case from a quantum mechanical perspective, a PDF must be necessarily used, if one is willing to take into account the whole information attached to a given molecular structure.

From the preceding ideas, PDF tagged sets appears to be the natural ∞-dimensional extension connected to the discrete n-dimensional Boolean tagged sets to be used in quantum chemical applications. Actually, there is no need to search for any new mathematical structure: Definition 1, as it is, still holds, even when any ∞-dimensional space subset is employed as the tag set part.

A3. Boolean Tagged Ensembles

From all things said in the previous sections, a summary can be structured in terms of a new general definition of Boolean tagged sets. The next definition will deal with them.

Definition 9: Boolean Tagged Ensembles

Suppose known a collection, $C = \{T_I\}$, of tagged sets bearing the same cardinality. That is: $\forall T_I \in C \rightarrow \# (T_I) = n$.

Suppose known a unit n-dimensional hypercube, $\mathrm{H}_n = \{\mathbf{v}_p\}$.

A Boolean tagged ensemble, E, is a tagged set whose object set is C, and whose tag set is H_n :

$$\mathrm{E} = C \times H_n = \{\forall \mathrm{Z}_{I;p} \in \mathrm{E} \mid \exists T_I \in C \wedge \exists \mathbf{v}_p \in \mathrm{H} \Rightarrow Z_{I;p} = (T_I; \mathbf{v}_p)\}.$$

The definition of Boolean tagged ensembles can be viewed as some kind of recursive property attached to the tagged set definition. The notion of Boolean tagged ensemble can be easily generalised to various kinds of tagged ensembles:

Definition 10: Tagged Ensembles

A tagged ensemble is a tagged set whose object set elements are tagged tets. □

APPENDIX B. ASA FITTING

Although a recent paper [97] gives the complete details of ASA fitting, algorithm developments and new atomic fitting tables will be of help to the readers interested into applications of the ASA DF.

Essentially, the ASA fitting algorithm can be divided into three well defined parts: a) generation of ASA exponents using *even-tempered* geometric sequences [98]; b) optimization of coefficients using an elementary Jacobi rotation (EJR) technique [101]; c) exponent optimization refinement using a Newton method [168].

Optimal sets of ASA coefficients and exponents are obtained by minimizing the function:

$$\varepsilon^{(2)} = \int \left|\rho(\mathbf{r}) - \rho^{\mathrm{ASA}}(\mathbf{r})\right|^2 \mathrm{d}\mathbf{r}, \tag{57}$$

which corresponds to the common definition of the quadratic error integral function between ab initio, $\rho(\mathbf{r})$, and ASA, $\rho^{\mathrm{ASA}}(\mathbf{r})$, electronic density functions, subject to the convex conditions described in Equations (4) and (5). Substituting the ASA density function defined in Equation (1) and using a matrix notation, the function $\varepsilon^{(2)}$ can be written now as:

$$\varepsilon^{(2)} = \mathrm{Z} + \mathbf{w}^{\mathrm{T}}\mathbf{Z}\mathbf{w} - 2\,\mathbf{b}^{\mathrm{T}}\mathbf{w}, \tag{58}$$

where: $\mathrm{Z}=\int|\rho(\mathbf{r})|^2 \mathrm{d}\mathbf{r}$, can be interpreted as an ab initio quantum self-similarity measure, see Section 4.1 for more details, and the elements of the matrix $\mathbf{Z}=\{\mathrm{Z}_{ij}\}$ as well as these of vector $\mathbf{b} = \{\mathrm{b}_i\}$ are given respectively, by the integrals:

$$\mathrm{Z}_{ij} = \int \left|\varphi_i(\mathbf{r})\right|^2 \left|\varphi_j(\mathbf{r})\right|^2 \mathrm{d}\mathbf{r} \tag{59}$$

$$\mathrm{b}_i = \int \left|\varphi_i(\mathbf{r})\right|^2 \rho(\mathbf{r})\,\mathrm{d}\mathbf{r}. \tag{60}$$

As has been explained in Section 3.2, the set of PD real coefficients $\{w_i\}$ can be substituted employing a complex coefficient set, $\{x_i\}$, using a discrete generating rule as in Equation (3). Thus, considering only the generating rule with real coefficients the function $\varepsilon^{(2)}$ transforms into the expression:

$$\varepsilon^{(2)} = Z + \sum_{i,j\in a} x_i^2 x_j^2 Z_{ij} - 2\sum_{i\in a} x_i^2 b_i \, . \tag{61}$$

Variation of quadratic error integral function employing EJR has been slightly modified with respect to the methodology described in a previous paper [96]. When a EJR is applied over a vector [169], an orthogonal transformation is performed, which can be identified as $\mathbf{J}_{pq}(\alpha)$ and described by the equations:

$$\begin{aligned} \dot{x}_p &\leftarrow c\, x_p - s\, x_q \\ \dot{x}_q &\leftarrow s\, x_p + c\, x_q \; , \end{aligned} \tag{62}$$

where only the elements p and q of the vector **x** are modified. The symbols c and s, appearing in Equation (62) determine the cosine and sine of the EJR angle α. If an EJR $\mathbf{J}_{pq}(\alpha)$ is applied over the appropriate elements of the quadratic error equation, the variation of $\varepsilon^{(2)}$ respect the active pair of elements {p,q} is easily evaluated [96], giving as a result a quartic equation with respect to s and linear in c:

$$\delta\varepsilon^{(2)} = E_{04}s^4 + E_{13}cs^3 + E_{02}s^2 + E_{11}cs \; , \tag{63}$$

where the sine and cosine polynomial coefficients are defined as:

$$\begin{aligned} E_{04} &= \theta_{pq}\left[\left(x_p^2 - x_q^2\right)^2 - 4x_p^2 x_q^2\right] \\ E_{13} &= 4\theta_{pq}\left(x_p^2 - x_q^2\right)x_p x_q \\ E_{02} &= 4\theta_{pq} x_p^2 x_q^2 - 2\left(x_p^2 - x_q^2\right)G \\ E_{11} &= -4x_p x_q G \\ &\text{and} \\ G &= \sum_{i\neq p,q} x_i^2\left(Z_{pi} - Z_{qi}\right) + x_p^2 Z_{pp} - x_q^2 Z_{qq} - \left(x_p^2 - x_q^2\right)Z_{pq} - b_p + b_q \\ \theta_{pq} &= Z_{pp} + Z_{qq} - 2Z_{pq} \end{aligned} \tag{64}$$

The procedure to obtain the optimal rotation angle can be greatly improved using straightforward Taylor expansions [170] in order to replace the s and c expressions, present in Equation (62). For small values of the EJR angle α, up to third order, there can be written:

$$\begin{aligned} c &= \cos(\alpha) \approx 1 - \frac{\alpha^2}{2} + \theta(\alpha^3) \\ s &= \sin(\alpha) \approx \alpha\left(1 - \frac{\alpha^2}{3!}\right) + \theta(\alpha^3) \end{aligned} \tag{65}$$

In consequence, it is possible to obtain a new $\delta\varepsilon^{(2)}$ expression, where the former c and s values can be substituted by α polynomials. Then, Equation (63) may be rewritten as:

$$\delta\varepsilon^{(2)} = \alpha^3 a + \alpha^2 b + \alpha\, c\,, \tag{66}$$

where: $a = E_{13} - 2/3E_{11}$, $b = E_{02}$ and $c = E_{11}$. If the stationary point null gradient condition is taken into account on Equation (66), a second order equation in α is obtained, and the Hessian provides the minimum condition. The optimal angle α_+ coincides with the root associated with the positive sign of the discriminant, and finally the optimal cosine and sine are given by:

$$c^* \approx 1 - \frac{\alpha_+^2}{2} \;\wedge\; s^* \approx \alpha_+\left(1 - \frac{\alpha_+^2}{6}\right) \tag{67}$$

Using this optimal solution into the EJR transformation, as defined in Equation (62), a simple expression is obtained for the coefficient set variation:

$$\begin{aligned} \dot{x}_p &\leftarrow \left(1 - \frac{\alpha_+^2}{2}\right) x_p - \alpha_+\left(1 - \frac{\alpha_+^2}{6}\right) x_q \\ \dot{x}_q &\leftarrow \alpha_+\left(1 - \frac{\alpha_+^2}{6}\right) x_p + \left(1 - \frac{\alpha_+^2}{2}\right) x_q \end{aligned} \tag{68}$$

To illustrate the applicability of the described algorithm, an application example is presented for an atomic basis set. This study is an extension of previous works [96,97] where two fitted basis sets are examined in great detail: a 3-21G basis set for atoms H to Kr [96] and a Huzinaga basis set for atoms H to Rn [97]. Here the fitting algorithm is applied to a 6-21G basis set [171,172] for atoms H to Ar. As in the previous works, the first step consisted of computing RHF energies and density functions using ATOMIC program [173]. *Table 19* lists computed RHF energies and normalized QS-SM for atoms H to Ar. From this ab initio density functions it has been computed the ASA density functions for different number of fitted atomic functions. The main results are presented in *Table 20*: fitting quadratic error ($\varepsilon^{(2)}$), relative error in the computation of Z_{AA} and relative error in the computation of one-electron potential energy, V(**r**).

Table 19. RHF energy and normalised QS-SM values for 6-21G basis set

Atomic Symbol	Electronic State	HF	Z_{AA}
H	^2S	−0.496979	0.03939
He	^1S	−2.835680	0.18801
Li	^2S	−7.430628	0.34934
Be	^1S	−14.569304	0.52605
B	^2P	−24.516722	0.70022
C	^3P	−37.658318	0.88175
N	^4S	−54.340752	1.07053
O	^3P	−74.699104	1.26727
F	^2P	−99.229572	1.47337
Ne	^1S	−128.275846	1.69055
Na	^2S	−161.84135	1.93227
Mg	^1S	−199.59492	2.17429
Al	^2P	−241.85261	2.42037
Si	^3P	−288.82535	2.67091
P	^4S	−340.68404	2.92643
S	^3P	−397.46296	3.18636
Cl	^2P	−459.43273	3.45095
Ar	^1S	−526.76040	3.72017

Table 20. Fitting results for 6-21G basis set for atoms H to Ar

		No. Fitted Atomic Functions				
		2	3	4	5	6
H	$\varepsilon^{(2)}$	1.46E-04	7.24E-07	2.94E-07	1.46E-07	6.08E-08
	% Z_{AA}	-2.572	-0.077	0.009	0.039	0.013
	%V(**r**)	-0.950	-0.056	-0.001	0.031	0.010
He	$\varepsilon^{(2)}$	1.09E-03	4.75E-06	5.20E-07	7.48E-07	3.99E-08
	% Z_{AA}	-3.113	-0.073	-0.002	0.038	0.010
	%V(**r**)	-1.284	-0.040	-0.003	0.045	0.012
Li	$\varepsilon^{(2)}$	1.10E-02	1.18E-03	1.27E-04	1.80E-05	5.61E-06
	% Z_{AA}	-3.734	0.156	-0.010	-0.012	-0.002
	%V(**r**)	-6.609	0.005	-0.039	-0.071	-0.128
Be	$\varepsilon^{(2)}$	2.06E-02	1.51E-03	1.47E-04	2.16E-05	5.55E-06
	% Z_{AA}	2.196	0.172	-0.010	-0.021	-0.026
	%V(**r**)	0.279	0.348	-0.059	-0.111	-0.150
B	$\varepsilon^{(2)}$	2.67E-02	1.90E-03	2.04E-04	4.68E-05	1.88E-05
	% Z_{AA}	2.701	0.077	-0.069	-0.061	-0.004
	%V(**r**)	1.896	0.145	-0.237	-0.307	0.020
C	$\varepsilon^{(2)}$	3.26E-02	2.38E-03	3.08E-04	1.00E-04	2.30E-05
	% Z_{AA}	2.574	0.011	-0.123	-0.118	0.001
	%V(**r**)	2.042	-0.044	-0.389	-0.444	0.070
N	$\varepsilon^{(2)}$	3.85E-02	2.97E-03	4.75E-04	1.91E-04	3.11E-05
	% Z_{AA}	2.356	-0.125	-0.218	-0.191	0.001

		No. Fitted Atomic Functions				
		2	3	4	5	6
	%V(**r**)	1.901	-0.224	-0.535	-0.562	0.054
O	$\varepsilon^{(2)}$	4.49E-02	3.78E-03	7.69E-04	2.61E-04	3.48E-05
	% Z_{AA}	2.059	-0.270	-0.321	0.011	0.002
	%V(**r**)	1.643	-0.435	-0.704	0.068	0.056
F	$\varepsilon^{(2)}$	5.16E-02	4.78E-03	1.17E-03	3.11E-04	4.37E-05
	% Z_{AA}	1.754	-0.408	-0.441	-0.003	0.007
	%V(**r**)	1.387	-0.613	-0.831	-0.006	0.056
Ne	$\varepsilon^{(2)}$	5.85E-02	5.98E-03	1.68E-03	3.68E-04	5.30E-05
	% Z_{AA}	1.452	-0.600	-0.556	-0.001	0.009
	%V(**r**)	1.148	-0.770	-0.952	0.003	0.049
Na	$\varepsilon^{(2)}$	6.79E-02	8.63E-03	3.66E-03	5.19E-04	1.88E-04
	% Z_{AA}	0.601	-1.265	-1.132	-0.119	-0.093
	%V(**r**)	-0.956	-2.784	-2.857	-0.880	-0.761
Mg	$\varepsilon^{(2)}$	7.95E-02	1.33E-02	4.38E-03	6.77E-04	2.76E-04
	% Z_{AA}	-0.206	-1.937	0.016	-0.040	-0.035
	%V(**r**)	-2.582	-4.306	-0.748	-0.522	-0.434
Al	$\varepsilon^{(2)}$	9.46E-02	2.06E-02	4.77E-03	7.21E-04	2.80E-04
	% Z_{AA}	-1.030	-2.587	0.116	-0.008	-0.018
	%V(**r**)	-4.136	-5.753	-0.229	-0.273	-0.245
Si	$\varepsilon^{(2)}$	1.13E-01	3.09E-02	5.00E-03	7.07E-04	2.57E-04
	% Z_{AA}	-1.795	-3.168	0.144	-0.001	-0.012
	%V(**r**)	-5.420	-6.907	0.023	-0.144	-0.162
P	$\varepsilon^{(2)}$	1.36E-01	4.43E-02	5.18E-03	6.85E-04	2.23E-04
	% Z_{AA}	-2.468	-3.636	0.133	-0.005	-0.012
	%V(**r**)	-6.462	-7.795	0.128	-0.090	-0.122
S	$\varepsilon^{(2)}$	1.64E-01	6.09E-02	5.35E-03	6.64E-04	2.14E-04
	% Z_{AA}	-3.023	-3.949	0.124	-0.006	-0.016
	%V(**r**)	-7.319	-8.460	0.147	-0.087	-0.114
Cl	$\varepsilon^{(2)}$	1.96E-01	7.89E-02	5.50E-03	6.42E-04	1.75E-04
	% Z_{AA}	-3.448	2.054	0.117	-0.005	-0.015
	%V(**r**)	-7.951	1.118	0.150	-0.084	-0.125
Ar	$\varepsilon^{(2)}$	2.32E-01	8.25E-02	5.65E-03	6.21E-04	1.65E-04
	% Z_{AA}	-3.726	2.029	0.113	-0.001	-0.019
	%V(**r**)	-8.371	1.320	0.147	-0.085	-0.121

Coefficients and exponents for this basis set of ASA functions can be downloaded from [174].

APPENDIX C. EXTENDED DENSITY FUNCTIONS

This appendix gives more details on other possible forms of extended density functions. Electrostatic potentials are presented, and several extended DF follow: quadrupole, angular momentum, and mass variation with velocity.

C1. Electrostatic Potentials

A typical example of the scheme discussed above may be constituted by the electronic part of Electrostatic Molecular Potentials (eEMP), first described and used by Bonaccorsi, Scrocco and Tomasi [175]. eEMP evaluated at the position **R** in 3-dimensional space, V(**R**), computed over first order DF, $\rho(\mathbf{r})$, is defined using Equation (8) as:

$$\Omega(\mathbf{r}) = |\mathbf{r}-\mathbf{R}|^{-1} \wedge V(\mathbf{R}) = \int |\mathbf{r}-\mathbf{R}|^{-1} \rho(\mathbf{r}) d\mathbf{r} . \quad (69)$$

Not taking into account the electron charge sign, eEMP acts as a PD distribution, with maxima located at molecular nuclei. A similar form of the eEMP when compared with the DF must be expected. This particular aspect will be later discussed, and several examples given.

C2. Quadrupole DF

These previous findings allow to think of the possible use of another set of elements in extended Hilbert spaces (EHS), see Section 3.4, made with the extended wavefunction part multiplied by the position vector, as: $|\chi\rangle = |\Psi; \mathbf{r}\Psi\rangle$. The new extended functions are the quantum mechanical position companions of the former momentum ones: $|\Phi\rangle = |\Psi; \nabla\Psi\rangle$. As for the extended functions $|\Phi\rangle$, a density and a projector can be also described for $|\chi\rangle$. A quadrupole DF (QDF) can be thus defined as

$$q(\mathbf{r}) = |\mathbf{r}|^2 \rho(\mathbf{r}),$$

forming part of the DF attached to this new breed of extended functions:

$$|\chi|^2 = \rho + |\mathbf{r}|^2 \rho = (1+|\mathbf{r}|2)\rho = \rho(\mathbf{r}) + q(\mathbf{r})$$

Also, this new extended DF can be obtained as the trace of the projector:

$$|\chi\rangle\langle\chi| = \begin{pmatrix} 1 & |\mathbf{r}\rangle \\ \langle\mathbf{r}| & |\mathbf{r}\rangle\langle\mathbf{r}| \end{pmatrix} \rho = \begin{pmatrix} \rho & |\mu\rangle \\ \langle\mu| & \mathbf{Q} \end{pmatrix}.$$

which corresponds to a matrix with dipole moment distributions in the off-diagonal elements and, moreover:

$$\mathrm{Tr}|\mathbf{Q}| = q(\mathbf{r}).$$

C3. Angular Momentum DF

In the same context it may be interesting to study how angular momentum could be introduced in extended wavefunction schemes. Defining the antisymmetric matrix:

$$\mathbf{e}(\mathbf{r}) = \begin{pmatrix} 0 & -z & y \\ z & 0 & -x \\ -y & x & 0 \end{pmatrix}$$

then it is easy to see that quantum mechanical angular momentum in matrix form can be obtained as:

$$\mathbf{r} \times \nabla \Psi \equiv \mathbf{e}(\mathbf{r}) \nabla \Psi = \mathbf{L}\Psi ,$$

where $\mathbf{L}$ is the angular momentum operator. Thus, the matrix Λ=Diag(1,$\mathbf{e}(\mathbf{r})$) transforms the extended wavefunctions into angular momentum ones:

$$\Lambda |\Phi\rangle = \Lambda |\Psi, \nabla\Psi\rangle \equiv |\Psi, \mathbf{r} \times \nabla\Psi\rangle .$$

At the same time, one can define angular momentum DF (AMDF) using the same arguments as before, taking the extended part of the resultant extended wavefunction:

$$\lambda(\mathbf{r}) = |\mathbf{r} \times \nabla\Psi|^2 = \mathrm{Tr}(\mathbf{r} \otimes \mathbf{r}) \mathrm{Tr}((\nabla\Psi) \otimes (\nabla\Psi)) - |\mathrm{Tr}(\mathbf{r} \otimes (\nabla\Psi))|^2 .$$

If Ψ is taking the form of an 1s GTO function, preparing in this way the ASA structure of such extended DF, the corresponding AM DF can be written as:

$$\lambda(\mathbf{r}) = (4\alpha^2)\{\mathrm{Tr}(\mathbf{r} \otimes \mathbf{r}) \mathrm{Tr}((\mathbf{r} - \mathbf{R}) \otimes (\mathbf{r} - \mathbf{R})) - |\mathrm{Tr}(\mathbf{r} \otimes (\mathbf{r} - \mathbf{R}))|^2\} |\Psi|^2 .$$

C4. Mass Variation with Velocity DF

When dealing with relativistic Hamiltonians, a new operator appears because of several approximate procedures allowing the simplification of Dirac's equation. See, for example, the treatise of Bethe and Salpeter [176], as well as the Moss discussion [177], the McWeeny's book [178], or the work of Almlöf and Gropen [179].

The so-called mass-velocity term, is mainly related to the fourth power of momentum, and thus with the operator ∇^4. A diagonal operator gives the adequate operator structure:

$$\upsilon = \mathrm{Diag}\left(1;-\nabla^2\right) = \mathrm{Diag}(1;\Delta)$$

acting on the extended wavefunctions, in this way, using Green's first theorem:

$$\begin{aligned}\left\langle\nabla^4\right\rangle &= \langle\Phi|\upsilon|\Phi\rangle = 1 - \int(\nabla\Psi)^*\left(\nabla^2\right)(\nabla\Psi)\,\mathrm{d}V \\ &= 1 + \int\left(\nabla^2\Psi\right)^*\left(\nabla^2\Psi\right)\mathrm{d}V = 1 + \int(\Delta\Psi)^*(\Delta\Psi)\,\mathrm{d}V \\ &= 1 + \int|\Delta\Psi|^2\,\mathrm{d}V.\end{aligned}$$

The three integrals appearing in the last sequence terms are the ones, which according to Bethe and Salpeter [176] must be used whenever mass-velocity integral terms have to be evaluated. So, the corresponding mass-velocity (MV) DF part may be written as: $\nu(\mathbf{r})=|\Delta\Psi|^2$.

The corresponding ASA MV DF form could be written knowing the ASA gradient. The laplacian of a 1s GTO function located at the coordinates $\mathbf{R}_\mathrm{I}$, it is easily found to be:

$$\Delta\varphi_i(\mathbf{r}) = 2\alpha_i\left(2\alpha_i|\mathbf{r}-\mathbf{R}_\mathrm{I}|^2 - 3\right)\varphi_i(\mathbf{r}).$$

Then the ASA MV DF could be expressed as the sum:

$$\upsilon(\mathbf{r}) = \sum_i \omega_i|\Delta\varphi_i(\mathbf{r})|^2 = 4\sum_i \gamma_i\left(2\alpha_i|\mathbf{r}-\mathbf{R}_\mathrm{I}|^2 - 3\right)^2|\varphi_i(\mathbf{r})|^2,$$

where, as before in KE DF: $\gamma_i=\omega_i\alpha_i^2$. It can be easily seen that MV DF appears, according to the well-known properties of 1s GTO, as the square of a d_{x2}-type function.

APPENDIX D. VISUALISATION EXAMPLES OF DF

The new set of varied DF found in the previous section can be studied in the same way as the well-known eDF maps are represented in the plots used in quantum chemistry [24]. Three-dimensional maps of isodensity surfaces can be generated with available computational techniques [213,214]. This corresponds to follow several steps, some of them so trivial that appear to be irrelevant in a study as the present one. The representation process starts with the evaluation of DF grids, enveloping the molecular coordinates, which can origin wire frame structures related with the isodensity values. After that, they can be rendered and rotated in space as virtual objects, until some adequate point of view is found. Finally, the chosen object snapshot can be manipulated, represented on a screen and, if necessary, printed into a paper surface. The processing detail, the computational techniques and the required programs and data will be briefly explained. All the necessary items are available to the interested reader and permit to generate surfaces of his own [213-218]. A comparison of extended DF for some assorted molecular structures are given to provide a limited sample of the available possibilities [217,218], in order to visually show the many facets of molecular surfaces.

D1. Visualisation examples of several molecular extended DF

Here, several examples of molecular structures and DF, as discussed in the previous sections, are given to illustrate the possibilities of such DF generalisation based on the structure and opportunities given by the EH spaces.

Although the isodensity surfaces can be generated for any available DF, in all the examples given in the following sections, the plotted functions are generated under the ASA framework, using the function structures previously described. The conditions discussed in Section 2.3 hold for all the displayed DF in the next sections. That is, whichever DF, obtained by a convex superposition, must have the associated basis set made by normalised DF. This kind of fundamental property has been followed in all the cases shown below. Such examples contain a sample of the possible extended molecular DF, as described in the preceding sections, that is: ASA examples of eDF, KE, Q, AM, MV and eEMP are provided. The molecular structures chosen are: estradiol, dioxine, AZT, and phenylalanine. The corresponding molecular structures, used in the following pictures, have been constructed and optimised in gas-phase using AMPAC 6.0 [148] under the AM1 Hamiltonian. The density functions have been constructed according to Promolecular ASA approach:

$$\rho_A^{promolec}(\mathbf{r}) = Z_A^{-1} \sum_{a \in A} Z_a \rho_a(\mathbf{r} - \mathbf{R}_a)$$

In that way, each hydrogen has been fitted to 1 GTO, each carbon, nitrogen and oxygen to 3 GTOs and chlorine to 4 GTOs. For each molecule, every isodensity surface is plotted at four isodensity levels.

The procedure used to construct a sufficient DF total surface representation involves the following steps:

Algorithm 3: Construction of the Molecular Surfaces

- Translate the molecular origin onto the molecular centre of charges.
- Define a grid, large enough to envelop the molecule. Use a grid spacing, sufficiently dense as to obtain a smooth representation.
- Compute the value of the involved DF at each point of the grid.
- The resulting set of grid coordinates plus function values, is used to construct and render the total DF surface.
- Plot the surface. □

The surface construction step has been performed here using the marching cubes algorithm (MCA) [213-217]. The MCA can construct triangles from a 3-D cloud of points, defining a wireframe surface, which can be later filled up or, as in the following examples, rendered. The MCA Fortran 90 source code can be obtained from our web site [217]. The resulting wireframe models, calculated in this way and presented here, have been rendered and plotted afterwards by means of the GiD program [218]. A comparison between ASA and ab initio eDF surfaces has already been made [219] and will not be repeated here.

D2. ASA eDF visualisation

The first example presented, along *Figures 15* to *18* consists into the representation of the ASA eDF for the four molecular structures chosen.

As observed from the figures, all systems are accurately described when low and high isodensity levels are considered, reaching a quality quite close to the obtained using ab initio procedures (see [24]). However, due to the nature of the promolecular approach used, the description of the intermediate levels, which coincide with bond formation steps, are not so accurately described. Although this inconvenience is omitted when working in QSM, the main user of ASA DF, because the density is greatly concentrated around atoms, and the contribution of bond density in similarity measures is negligible.

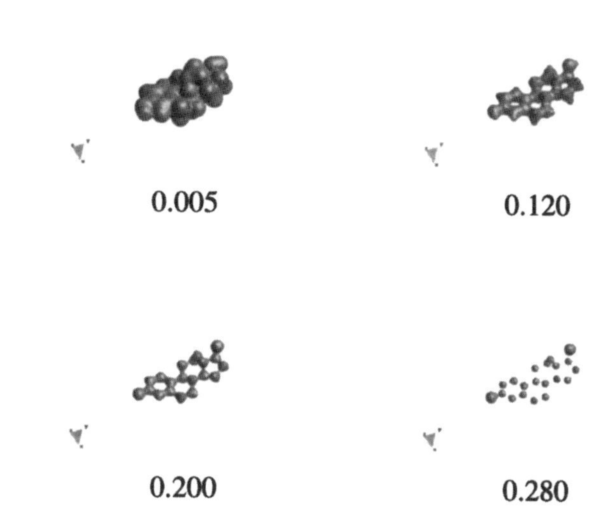

Figure 15. Graphical representation of ASA eDF calculated at four isodensity levels for the estradiol molecule.

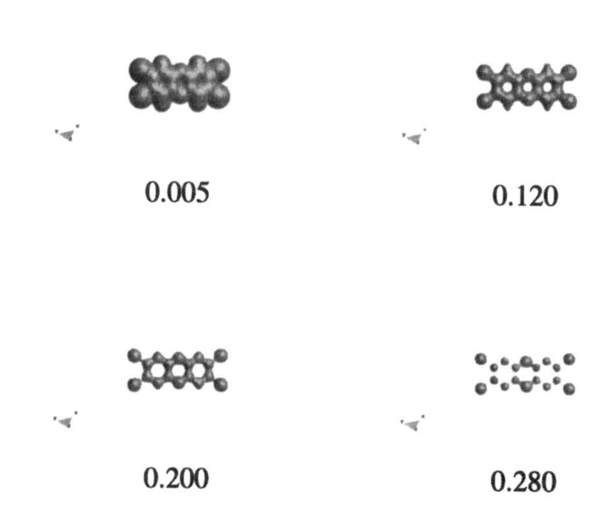

Figure 16. Graphical representation of ASA eDF calculated at four isodensity levels for the dioxine molecule.

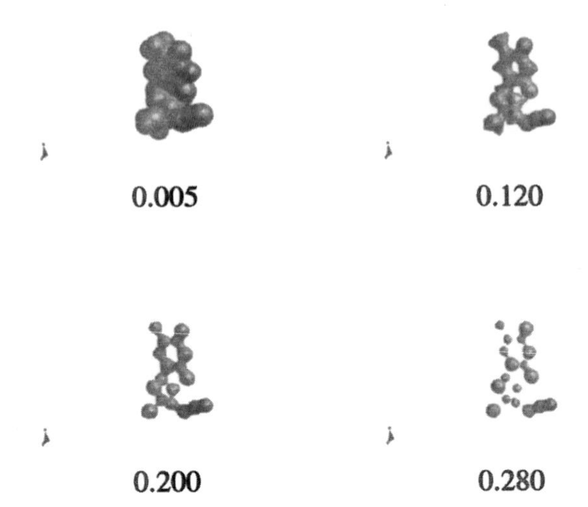

Figure 17. Graphical representation of ASA eDF calculated at four isodensity levels for the AZT molecule.

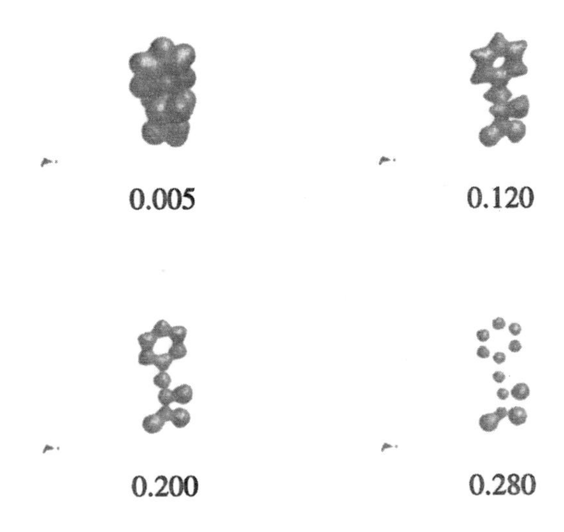

Figure 18. Graphical representation of ASA eDF calculated at four isodensity levels for the phenylalanine molecule.

D3. ASA KE DF visualisation

The next presented example, corresponding to *Figures 19* to *22*, involves the visualisation of KE DF. Similarly to the previous four ASA eDF surfaces, the representation is made over the same molecular set, and in the same manner, the corresponding isodensity surfaces are plotted at four levels.

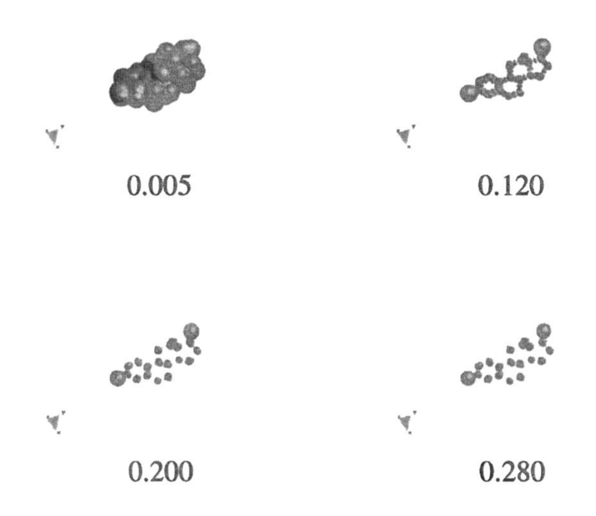

Figure 19. Graphical representation of ASA KE DF calculated at four isodensity levels for the estradiol molecule.

Taking the eDF as a comparison model, KE DF presents similar forms at low isodensity values, enveloping the molecular shape, and the same can be said at high function values, whereas only heavy atoms are represented. However, when the intermediate values are considered, an interesting difference arises, due to the decrease of eDF along the interatomic distance, KE DF increases creating an extreme between bonded atoms where the electron momenta reach lower values. This situation is easy to understand, because the nearer any electron is to an atom, the faster it will move because of the increment of nuclear attraction.

Another remarkable difference, which is difficult to notice from the present figures, is that, and according to ASA KE DF equation, the value of the function in the atomic coordinates is zero, just where eDF reaches a maximal value.

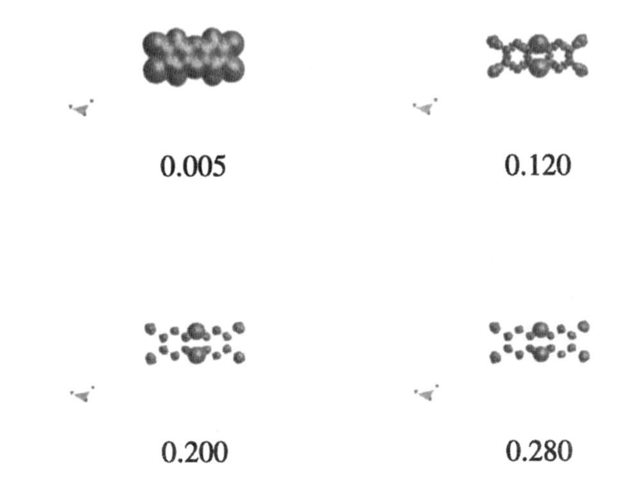

Figure 20. Graphical representation of ASA KE DF calculated at four isodensity levels for the dioxine molecule.

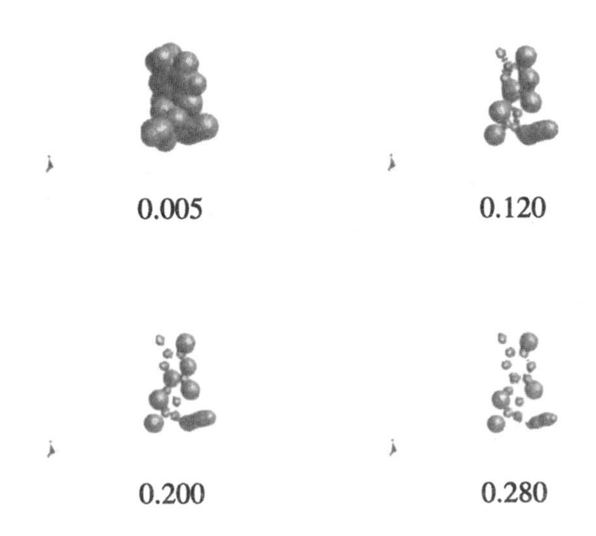

Figure 21. Graphical representation of ASA KE DF calculated at four isodensity levels for the AZT molecule.

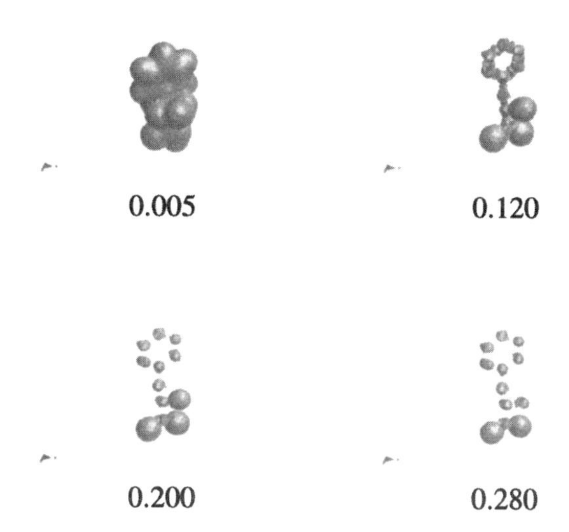

Figure 22. Graphical representation of ASA KE DF calculated at four isodensity levels for the phenylalanine molecule.

D4. ASA QDF visualisation

The next EH DF representation example presented, in *Figures 23* to *26*, involves the visualisation of QDF. Surface representation, similar to the previous figures, is based the same molecular set as before, plotted at four isodensity levels. Quadrupole moments provide a second-order approximation to the total electron distribution, and indeed, to the molecular shape, as they are the second order derivatives of the energy with respect to an applied electric field. Quadrupole-like DF arises from this idea and provides a new vision of the molecular shape from the point of view of a pre-established centre. In the presented examples, for each structure, all molecular coordinates have been translated at their centre of charges, in order to provide a sound reference for reproducibility purposes, as QDF is not invariant upon translation.

The following figures show how at high values of QDF the regular molecular shape is split from the chosen molecular centre of charges. The function begins to have a structured form around the heavy atoms, present surrounding the centre of charges. At low values of QDF, the shape becomes as regular as the eDF forms.

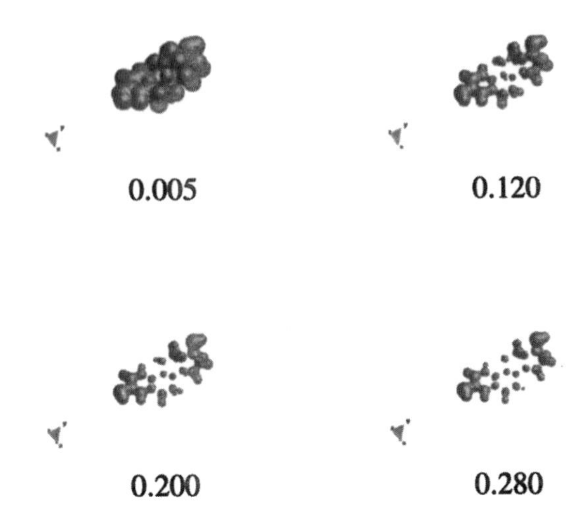

Figure 23. Graphical representation of ASA QDF calculated at four isodensity levels for the estradiol molecule.

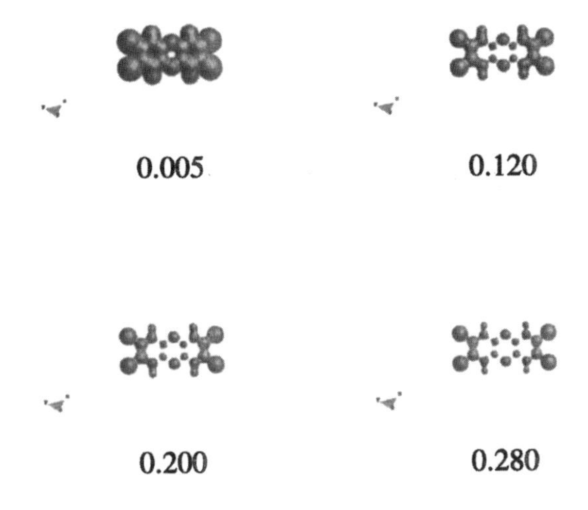

Figure 24. Graphical representation of ASA QDF calculated at four isodensity levels for the dioxine molecule.

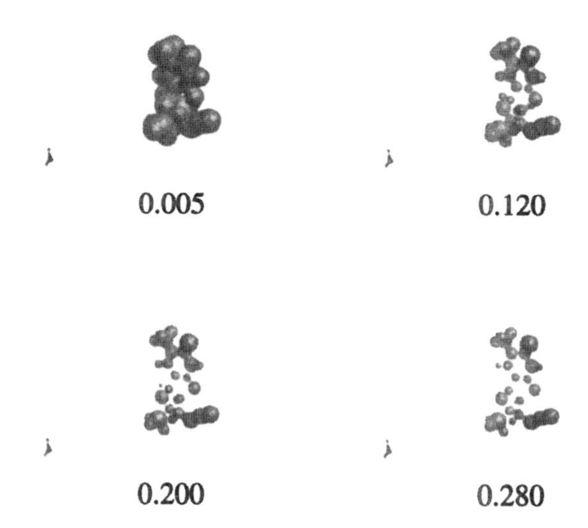

Figure 25. Graphical representation of ASA QDF calculated at four isodensity levels for the AZT molecule.

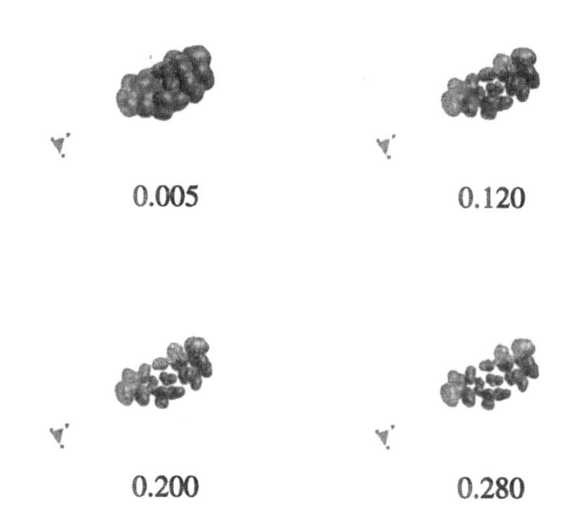

Figure 26. Graphical representation of ASA QDF calculated at four isodensity levels for the phenylalanine molecule.

D5. ASA AM DF visualisation

Figures 27 to *30* correspond to the visualisation of AM DF. Similar to those examples, the representation consists of the visualisation of the same set of molecules as before, plotted also at four isodensity levels.

To understand the physical interpretation of this new kind of DF, it is important to consider what one can interpret with molecular AM. AM corresponds to the momentum, associated with the rotational motion of a molecular object. The greater the AM, the stronger the braking force needed to bring it to rest. Combining this concept with the existence of molecular inertial axis, it can be seen how the major part of AM DF is gathered around the edges of the axis involving less mass translations. This point of view appears more obvious when high values of AM DF surfaces are observed. When low AM DF values are analysed, it is easy to grasp that the shape of the AM DF becomes uniformly distributed, as the contributions from the rest of the axis are added, adopting finally an eDF-like shape, as in all other extended DF studied so far.

As in the QDF cases, the AM DF is not invariant upon translations, and the present images have been obtained upon translation of every molecular atomic coordinate set into the centre of charges.

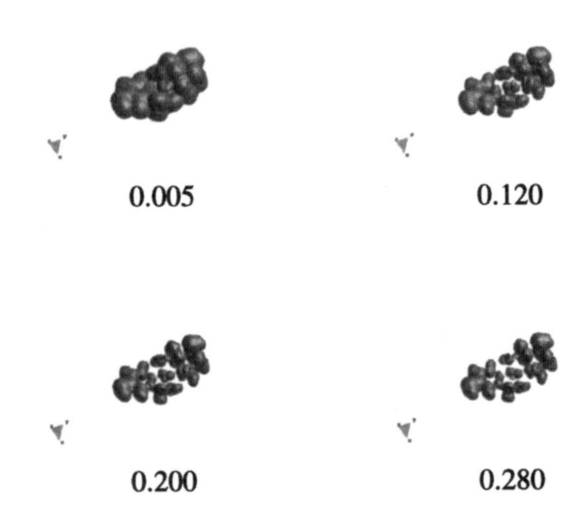

Figure 27. Graphical representation of ASA AM DF calculated at four isodensity levels for the estradiol molecule.

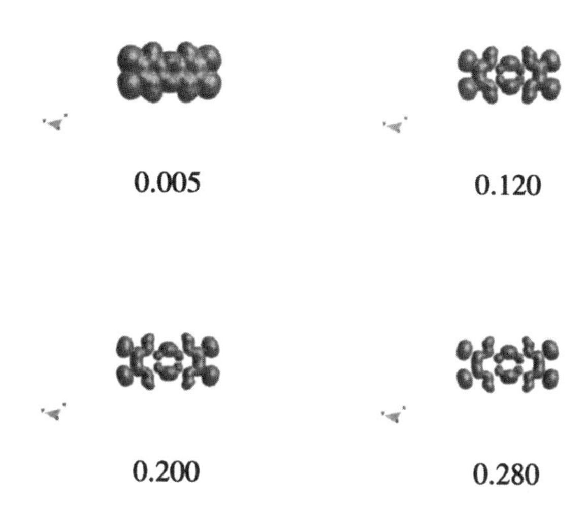

Figure 28. Graphical representation of ASA AM DF calculated at four isodensity levels for the dioxine molecule.

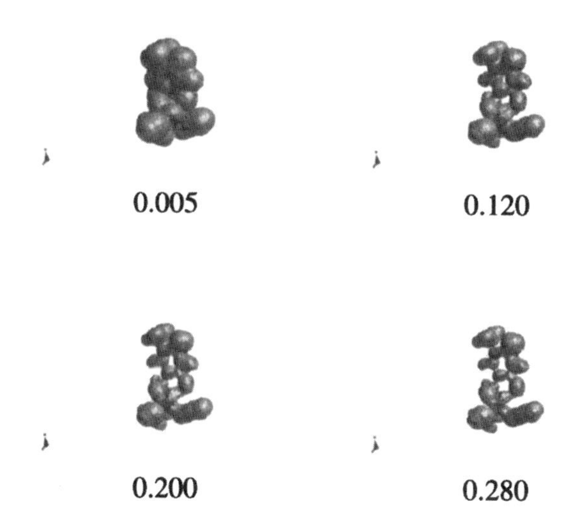

Figure 29. Graphical representation of ASA AM DF calculated at four isodensity levels for the AZT molecule.

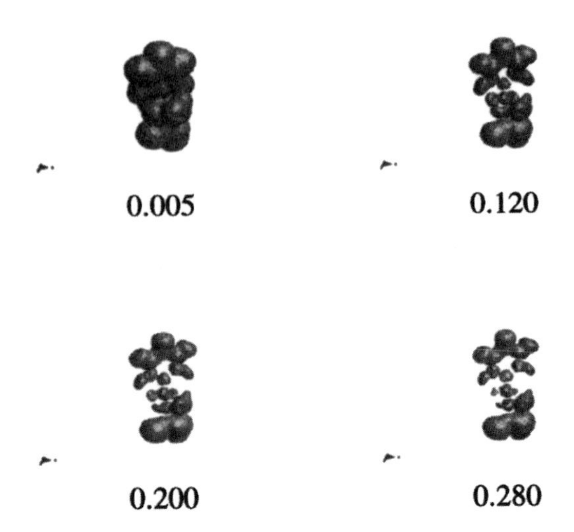

Figure 30. Graphical representation of ASA AM DF calculated at four isodensity levels for the AZT molecule.

D6. ASA VMV DF visualisation

The next example presented, from *Figures 31* to *34*, consists in the representation of MV DF. The figures involve the same set of molecules and molecular surfaces have been also plotted at four levels.

This new kind of DF arises, using the relativistic idea of the effect of MV DF. Once the definition of the KE DF is understood, it is assumed that the zones on the molecule where the electrons achieve more KE, or more momentum, are the ones located around the atoms. The next ones are the bond regions, with a minimum located at the middle of the bond. Finally, the zones with less KE are the most external ones.

According to the relativistic Breit Hamiltonian, the EH MV DF values are more accused as the momentum of the involved object, in this case an electron, increases. In this way, the zones where KE increases will present a high value of MV variation, which will be mostly localised surrounding the atoms of the molecule. Away from atomic centres, the value of MV DF decreases, and it turns most similar to eDF due to the KE in these points is low and more homogeneous in space.

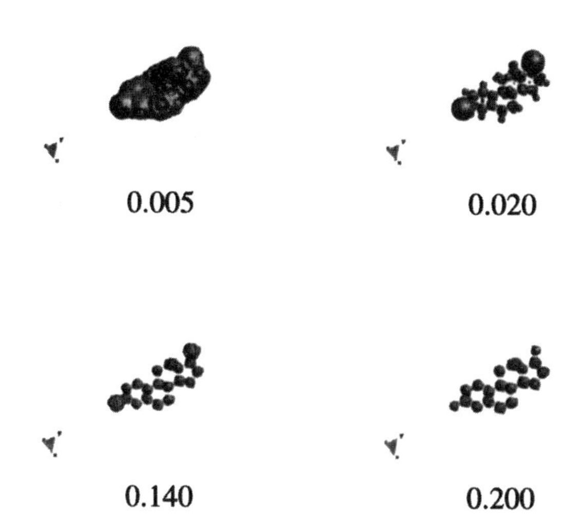

Figure 31. Graphical representation of ASA VMV DF calculated at four isodensity levels for the estradiol molecule.

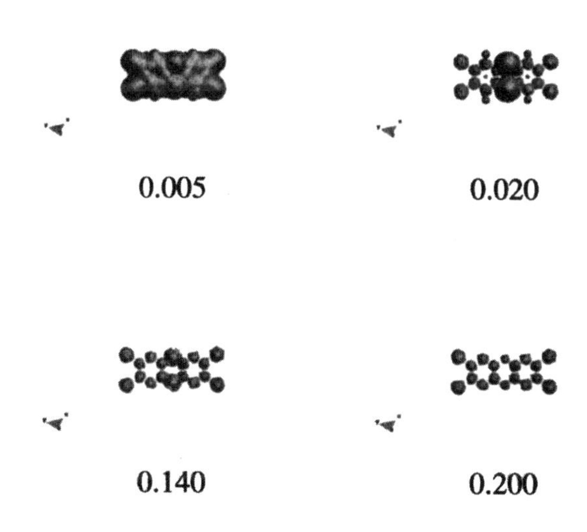

Figure 32. Graphical representation of ASA VMV DF calculated at four isodensity levels for the dioxine molecule.

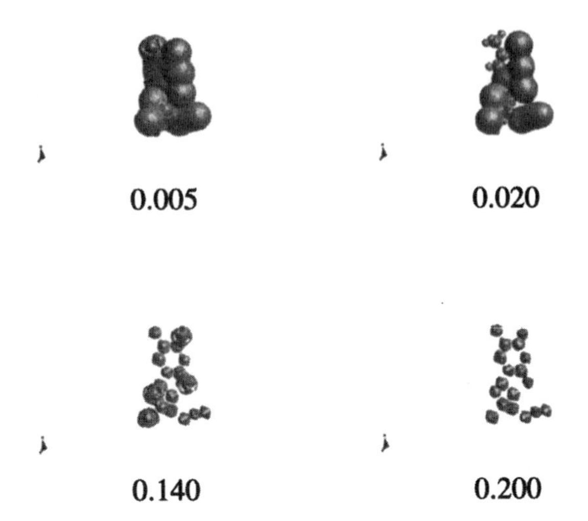

Figure 33. Graphical representation of ASA VMV DF calculated at four isodensity levels for the AZT molecule.

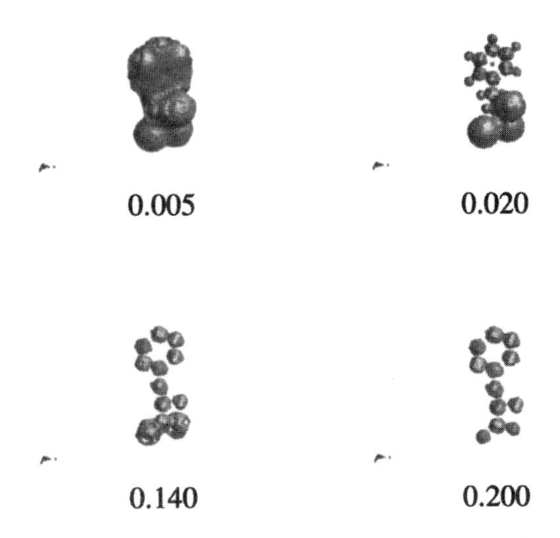

Figure 34. Graphical representation of ASA VMV DF calculated at four isodensity levels for the phenylalanine molecule.

D7. ASA eEMP visualisation

The last example presented, in *Figures 35* to *38*, consists of the representation of the eEMP for the same previous molecules at four isopotential levels. From the graphical examples of the eEMP, it can be seen that at low eEMP values all representations adopt a spherical shape, whereas at high values only heavy atoms are plotted, as occurs in eDF and KE DF representations. At intermediate values, the plots are more uniform than the respective representation of previous examples. However, these eEMP plots show the fuzzyness of the molecular shape when low values are considered.

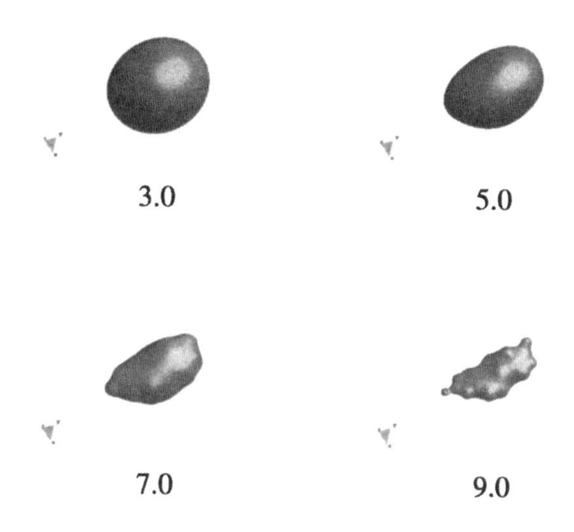

Figure 35. Graphical representation of ASA eEMP calculated at four isodensity levels for the estradiol molecule.

APPENDIX E. MOLECULAR SUPERPOSITION

In this appendix, a new method for molecular alignment based on topological and geometrical features is presented. First, it is explored a discussion about some methodologies including QSM where superposition plays an important role. A brief discussion on the algorithm used so far to find the optimal superposition within a QSM framework is given. Next, a new approach is described and detailed. Finally, an assorted set of examples is presented to illustrate the usefulness and the performances of the new algorithm.

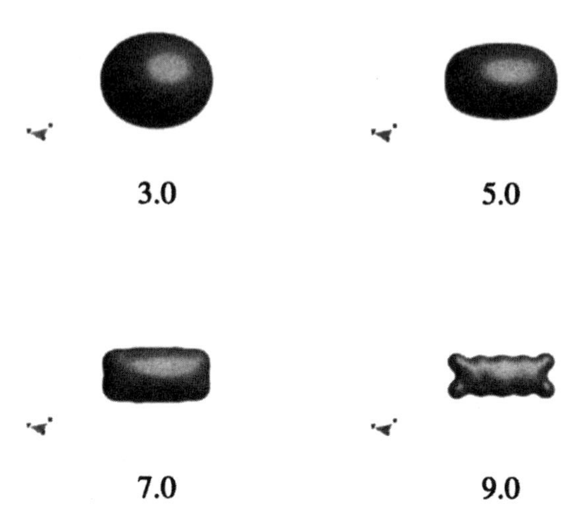

Figure 36. Graphical representation of ASA eEMP calculated at four isodensity levels for the dioxine molecule.

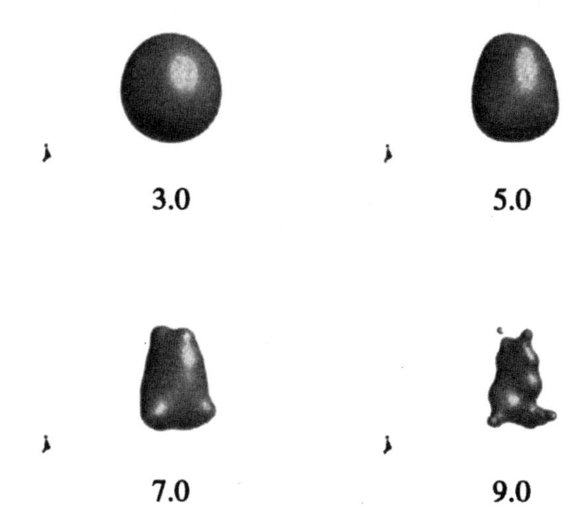

Figure 37. Graphical representation of ASA eEMP calculated at four isodensity levels for the AZT molecule.

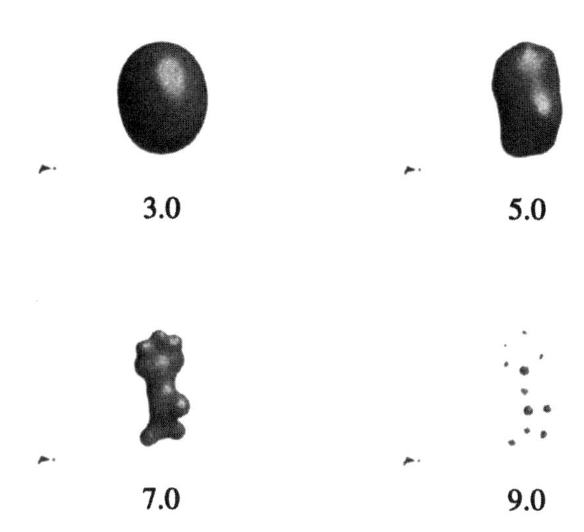

Figure 38. Graphical representation of ASA eEMP calculated at four isodensity levels for the phenylalanine molecule.

E1. The Molecular Alignment Problem

Optimal molecular alignment constitutes a major problem in several theoretical and computational disciplines and it is a wide studied field [180-191]. For example, any technique based on obtaining information from direct comparison of two or more 3-D molecular structures arrives at the conclusion that the results highly depend on the way the studied objects are located in space. Some 3-D-QSAR models, see for example [140,192,193], are based on the calculation of force fields at each point of a grid surrounding molecules. Little variations on spatial location of molecules may modify the values of the evaluated function at these grid points. Another problem where alignment constitutes the most important step is structural recognition in 3-D databases, see for example [194-206], which in some cases, identification can be even restricted to molecular fragments. Many studies on pattern recognition have been published [207-210], which conclude that one of the most important factors to be considered is the speed of the superposition method.

QSM also needs a solution for such a problem. As it has been commented in Section 4.2.2, the QSM values depend on the relative position in space of two arbitrary studied molecules. So far, this problem can be solved considering that the optimal alignment involved the maximisation of the associated QSM of both molecules [77] using the PD of the measures. Even if this solution usually superimposes both structures according to their structural features, the method is extremely sensitive to the presence of heavy atoms and tends to superimpose this kind of atoms regardless of backbone atoms. This fact can be explained by the existence of higher density peaks located at heavy atoms. *Figure 39* illustrates this behaviour.

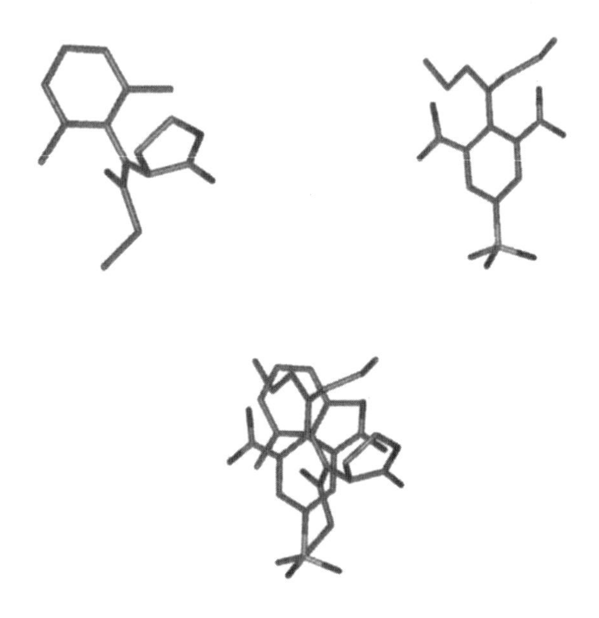

Figure 39. Alignment of ofurace and orylazin according to the maximal similarity rule.

Figure 39 evidences that superposing heavy atoms do not need to agree with the chemical intuition.. The heaviest atom in the first molecule ofurace (CAS number: 058810-48-3) is a chlorine ($Z_{Cl\text{-}Cl} \approx 984$ au), whereas in the second orylazin (CAS number: 019044-88-3) it is a sulphur ($Z_{S\text{-}S} \approx 805$ au). The maximal similarity algorithm superposes these two atoms ($Z_{Cl\text{-}S} \approx 889$ au), and aligns the rest of the molecule trying to approach the pairs of atoms which will increase the value of QSM as much as possible. If the rest of atoms were considered, no other possibilities may arise due to the small contributions to overall similarity that light atoms can achieve, for example: $Z_{C\text{-}C} \approx 31$ au, $Z_{O\text{-}O} \approx 81$ au or $Z_{C6H6\text{-}C6H6} \approx 190$ au.

E2. Molecular alignment restricted to common skeleton recognition

To avoid the loss of chemical information produced by heavy atom dependence alignment, an alternate superposition approach based simply on topological and geometrical features is presented. In this case, there is no need to employ QSM during the process and the sensitivity to heavy atoms is removed. In addition, the use of DF is avoided during the alignment process, thus saving computational time.

The method is based on comparison of interatomic distances and atomic numbers, providing a way to align molecular pairs with low computational cost and ensuring the automatic recognition of common skeletons according to chemical intuition. This procedure is slightly connected to the maximum isomorphic subgraph problem studied in pattern recognition and statistics [117-119].

E3. Algorithm details

In this section, a detailed description of the topo-geometrical superposition algorithm (TGSA), as described in Algorithm 2, is provided, and some aspects are graphically plotted to provide a better explanation. *Figure 40* shows a pair of molecules selected to illustrate the algorithm: glutamine and histidine, two essential amino acids, were chosen for this example. Such molecules present an obvious common skeleton (-$CHNH_2COOH$), which has to be superimposed.

The needed information consists only of atomic numbers and molecular coordinates. The first step defines pairs of atoms, called atomic diads. Diads are defined only if the involved pair of atoms are bonded. This criterion has been adopted using a common accepted definition for common skeleton, consisting of coincident sets of atoms in a group of molecules. Bonded atoms can be defined in several ways. If a molecular modeling software package is used, the program usually supplies this information. In the present case, two atoms are considered to be bonded if interatomic distance is below 1.1 times the sum of the van der Waals radii of the involved atoms.

The total number of bonds establishes the number of atomic diads that need to be taken into account in the alignment process. Hydrogen atoms were not included throughout the calculations to save computational time.

When all diads have been defined for both molecules, every diad of one molecule is compared to all diads of the other using the similarity of their interatomic distances within an arbitrary pre-established threshold. It has been fixed to a 5% in bond length. The threshold is included to take into account the fluctuations in interatomic distances due to the presence of different substitutions within a molecule. This procedure allows us to discard a considerable amount of bonds which do not belong to the common skeleton. *Figure 41* presents an example of one of the selected diads, corresponding to a pair of carbons.

Figure 40. Molecular structures of glutamine and histidine.

Figure 41. Graphical example of a pair of similar diads. The light colored bonds are located between the carboxylic carbon and the chiral carbon.

Once the diad choice is achieved, the next step is to add a new atom to a diad, in order to construct atomic triads. Triads are generated by adding a bonded atom to at least one atom of a chosen diad, obtaining, in geometrical terms, a triangle. The vertices of this triangle correspond to atoms, and two or three sides correspond to chemical bonds. The reason for creating triads is such that when an object is to be placed at a given position in space, three points of reference are needed, three atoms in this case. *Figure 42* shows one of the constructed triads over the previous molecules. The third atom corresponds to an oxygen, which is bonded to one of the carbons of the previously selected diad. In this case, the triad is formed by two effective bonds. However nothing opposes the participation of three bonds, as would happen if epoxy groups were selected.

All triangles obtained for one molecule are compared to all triangles obtained for the second one. The distances between the atoms of the diad to the new atom are compared within the threshold tolerance. If the three distances of both compared triangles are similar, both triads are considered similar, too, and are subsequently tested for superposition. The rest of triads are automatically discarded. The case presented in *Figure 42* satisfies the test of distances, and both molecules would be superposed.

However, two opposite situations can occur: no pair of triads or several pairs are selected, situations corresponding to heterogeneous and homogeneous sets of molecules, respectively. The first case can be easily solved: The preceding steps are repeated as many times as needed, increasing the value of the threshold each time until some triad is chosen.

Figure 42. Graphical example of a pair of selected triads. In this case, the third selected atom is the carboxylic oxygen.

The second situation requires the definition of a procedure to discard triads. Here, the adopted criterion consists of choosing the pair of triads that superposes the major number of atoms with equal atomic charge. Atoms are considered to be superposed if they are as close as a threshold distance, chosen to be 0.30 au. In addition, if more than one pair of triads superpose the same number of atoms, the selected one will be the triad that minimises the sum of distances between the superposed atoms.

At the final step, both molecules are aligned by means of the selected triads of their respective molecular structures. No information about their respective density functions has been required so far. At this point MQSM can be computed. *Figure 45* presents the final molecular alignment. Superposing the pair of carbons and the oxygen atom of both systems leads also to superpose three other atoms: the acidic oxygen, the nitrogen, and another carbon of the side chain.

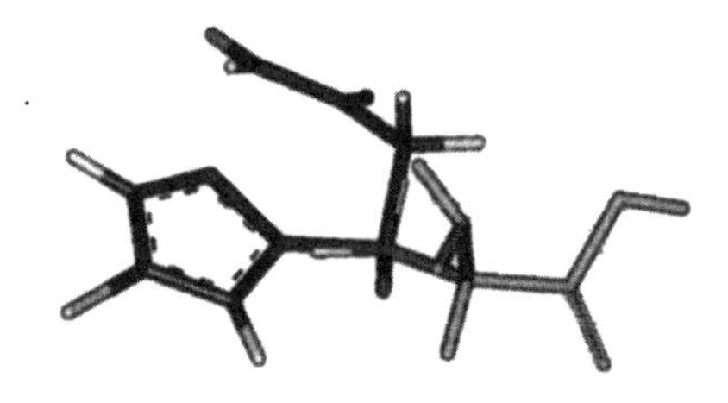

Figure 45. Final molecular alignment (superposed atoms marked for clarity).

E4. Molecular Alignment Examples

Figures 46 to *48* present three examples of molecular alignment using TGSA. The selected cases constitute sufficiently complex situations as they involve steroids, anilides and inhibitors of benzamidine. Hydrogen atoms have been suppressed from figures for clarity. In each figure, a discussion about the solution provided by the method is given.

Figure 46 shows the TGSA molecular alignment of two steroids: aldosterone and androstanediol [140]. These structures have a large common skeleton, made up of three six-member rings and a five-member one. Thus, there exists a great amount of comparable diads and triads located at different molecular sites in both molecules. For example, an external and a central six-membered ring have a similar triad which is discarded. This clearly demonstrates the usefulness of the triad selection procedure. The application of the TGSA to these molecules, within the atom superposition threshold of 0.30 au, considered 11 atoms to be superposed, located in the central part of the molecules.

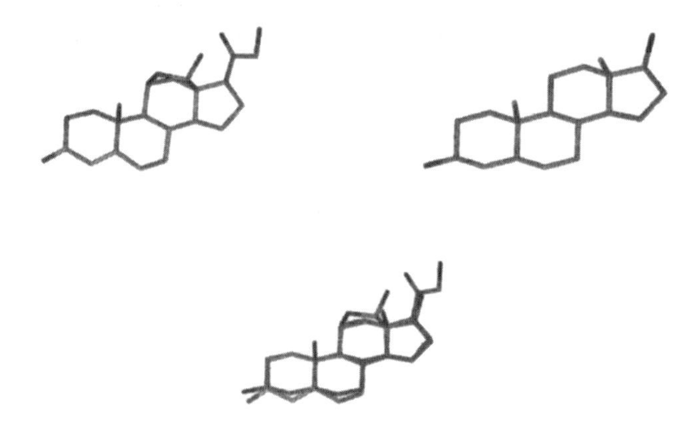

Figure 46. Alignment of two steroids: aldosterone and androstanediol.

Figure 47 presents the alignment of the pair of molecular structures analysed before [211] using the maximal similarity algorithm. In this case, when using TGSA, the alignment is modified by the recognition of the common benzene ring instead of superposing chlorine and sulphur atoms, which would be obtained using QSM criteria. In this example, seven atoms have been found to be superposed, the six carbons contained in the benzene ring and a nitrogen bonded to it.

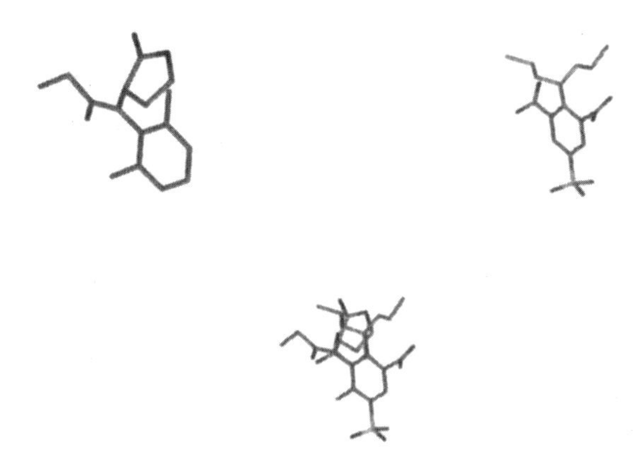

Figure 47. Alignment of ofurace and orylazin (corrected example).

Figure 48 presents the alignment of two inhibitors of benzadimine [212]. These molecules share a large common skeleton formed by two benzene rings linked by a four-member chain and a diaziridine ring, linked by a carboxyl group to the chain. The relevance of this example leans upon the computation time performance of TGSA, as will be later discussed. In this case, 15 atoms have been superimposed from alignment of the diaziridine ring, located at the upper right side substructure. Again, the efficiency of the triad selection method is demonstrated, due to existence of alternative alignments based on benzene rings or linked chains superpositions. However, they have not been chosen by TGSA because they would have matched a minor number of atoms.

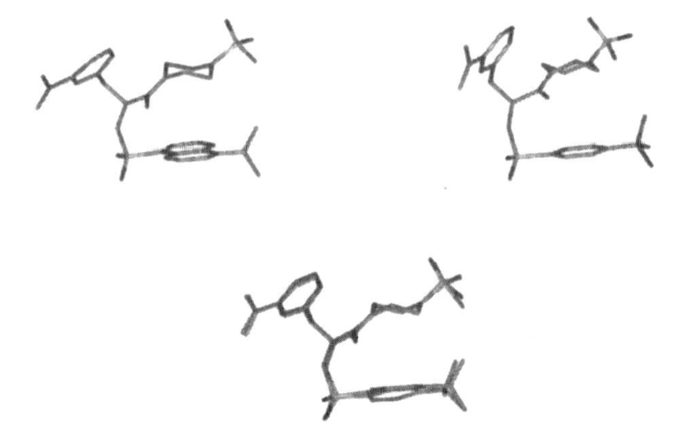

Figure 48. Alignment of two inhibitors of the benzamidine.

E5. TGSA Performance

This section is intended to provide a qualitative guide on the real implementation of the proposed procedure. In order to provide information about the performance of TGSA, 4 series of compounds were chosen:

- The set of the 20 essential amino acids.
- The set of 31 steroids studied by Cramer [140].
- A set of 17 anilides, extracted from [211], currently under study for correlation search between MQSM and aquatic toxicity.
- A set of 88 inhibitors of benzadimine studied by Böhm et al. [212].

Table 21 presents the relevant data for the efficiency assessment of the algorithm: number of molecules (N), mean number of atoms per molecule (A_M), and time, evaluated in seconds, needed to perform all alignments (T). The total number of alignments to be carried out is $N(N-1)/2$.

Table 21. Data for molecular alignment of several systems using TGSA. Computed with an Intel Pentium running at 200 MHz with 64 Mb of RAM.

System	N	A_M	T (sec)
Aminoacids	20	10	12.58
Anilides	17	21	59.97
Steroids	31	24	565.51
Inhibitors of benzadimine	88	36	9941.41

These timings point out the computational efficiency of TGSA. Even if times presented in *Table 21* are not directly comparable because of the differences in molecular systems, the TGSA efficiency can be clearly appreciated. Pairwise alignment of the amino acid set is completed in less than 13 seconds, achieving a superposition based on the ($-CHNH_2COOH$) group in all cases. The steroid set is processed in about 9½ minutes and in this case optimal alignments are not based on a single substructure as the amino acids. This is because different bond types are present, and even they share a common four-ring structure; the alignment is based in the most similar ring in both structures. The anilide group is the most heterogeneous set tested in this study: However, satisfactory overlays are also obtained. All structures share a benzene ring, which is the basis for their alignment. In this case, computational processing concluded in less than a minute, owing to their heterogeneity led to few diad and triad comparisons. The last test, the inhibitors of benzadimine, constitutes the largest studied set. Full pairwise alignment, 3828 superpositions, needed about 3.75 hours to be completed in a Pentium 200MHz PC machine. Considering the size and number of molecules involved, as well as the processor, it can be accepted as a sufficiently good result.

APPENDIX F. GENERALISED SCALAR PRODUCTS IN N-DIMENSIONAL VSS

Other interesting questions appear when thinking on the peculiar structure of QOS and their discrete descriptor spaces forming the tag set elements. When manipulation of SM elements is studied, and similarity indices defined, the problem of generalisation and meaning is raised. Similarity indices extension, taking into account the PD nature of tag sets VSS, will be discussed here.

F1. Permanent of a matrix

If the structure of SM and of discrete QOS is observed, as well as the presence of VSS, containing the discrete representation of DF, then there appears the possibility to obtain single parameters summarising, somehow, the SM in part or as a whole into a unique scalar. In order to accomplish such task, a generalisation of the scalar product may be proposed in *n*-dimensional VSS. This needs a definition:

Definition 11: Permanent of a (n×n) Matrix

Known a (n×n) matrix: $\mathbf{Z}=\{\mathbf{z}_I\}=\{z_{JI}\}\wedge\{\mathbf{z}_I\in C_n(\mathbf{R}^+)\}$, the Permanent of $\mathbf{Z}$: $Per|\mathbf{Z}|$, can be computed as:

$$Per|\mathbf{Z}|=\sum{}_n(\mathbf{i})\,P(\mathbf{i})\vartheta(\mathbf{i}),$$

where, $\mathbf{i}=\{i_p\}$, is the index vector of the nested summation symbol: $\Sigma_n(\mathbf{i})$, [103]; P $(\mathbf{i})=0$ if $\exists i_p=i_q$ in the index vector, and is 1 otherwise; finally:

$$\vartheta(\mathbf{i})=\prod_{k=1}^{n} z_{k\,i_k}\ .\ \square$$

Thus, the permanent is like a determinant but without considering index permutation parity. For more details see [220].

F2. Generalised Scalar Product of n Vectors

Defined as above, $Per|\mathbf{Z}|$ permits to be considered a scalar product involving *n* vectors, the columns of $\mathbf{Z}$, in *n*-dimensional VSS. The operation is distributive with respect to the sum of a vector on any column of the matrix, and multiplying a column by a scalar, the permanent is scaled by the same amount. The column order is immaterial. The same properties apply to the matrix rows.

F3. Generalised Norm of a Vector

A norm, associated to the kind of generalised scalar product, in *n*-dimensional VSS can be easily defined as the permanent of a matrix with the same column repeated *n* times:

$$NPer(\mathbf{z}) = Per|(\mathbf{z}, \mathbf{z}, \ldots \mathbf{z})| = (n!)\left(\prod_{k=1}^{n} z_k\right).$$

This makes clear that only in VSS, where vector elements are completely PD, this kind of norm can be defined as a PD computational device.

F4. Generalised Carbó Similarity Index

The construction of generalised scalar products and norms, involving several discrete DF representations, may be used to define the associated similarity indices, in the way, so many times discussed in the literature, a manipulation, which started when the initial definition of QSM [1] was given. See references [50,52,75] for a general review on the subject.

The so-called Carbó index involving two QO, defined by means of a cosine-like formulation:

$$r_{AB} = z_{AB}(z_{AA} z_{BB})^{-\frac{1}{2}} \quad (70)$$

where the SM elements of dimension (2×2) are used. The elements of this matrix can be also ordered as:

$$\mathbf{Z}_2 = \begin{pmatrix} z_{AA} & z_{AB} \\ z_{AB} & z_{BB} \end{pmatrix} = (\mathbf{z}_A \quad \mathbf{z}_B). \quad (71)$$

Then, using the generalised scalar product and norm, defined in Sections 4.4.2 and 4.4.3, a Carbó index may be defined. It can be based in this two QO matrix representations, according to Equation (70) and taking into account the origin of the SM elements, as:

$$
\begin{aligned}
R_{AB} &= Per|\mathbf{Z}_2|\left(NPer|\mathbf{z}_A|NPer|\mathbf{z}_B|\right)^{-\frac{1}{2}} \\
&= Per|(\mathbf{z}_A \quad \mathbf{z}_B)|\left(Per|(\mathbf{z}_A \quad \mathbf{z}_A)|Per|(\mathbf{z}_B \quad \mathbf{z}_B)|\right)^{-\frac{1}{2}} \\
&= \frac{1}{2}\left(z_{AA}z_{BB} + z_{AB}^2\right)\left[\left(z_{AA}z_{AB}\right)\left(z_{AB}z_{BB}\right)\right]^{-\frac{1}{2}} \\
&= \frac{1}{2}\left(r_{AB}^{-1} + r_{AB}\right)
\end{aligned}
\tag{72}
$$

The interesting result consists in that for two QO this generalised Carbó index, R_{AB}, becomes an arithmetic average of the classical Carbó index and its inverse. As inverse cosine-like similarity indices behave as distance indices, one can conclude that generalised two QO similarity Carbó indices behave as averaged cosine-like and distance-like indices. Three or more QO similarity indices computed in the same way, may behave in a more complex fashion. It is evident that they constitute a new bred of molecular descriptors to be taken into account.

APPENDIX G: INWARD OR HADAMARD MATRIX PRODUCTS AND GENERALISED DF

G1. Introduction

This Appendix will present a generalisation of a simple matrix operation, already defined and discussed in a previous paper [221], the *inward or Hadamard* matrix product (IMP). IMP leads to a new kind of matrix relationships, which reveals useful in quantum chemical applications and particularly in DF construction. The basic IMP related to vector spaces has been known for a long time [222,223] and even it is implemented as an intrinsic function in Fortran 95 language [224].

Initially, an IMP has been defined as an internal composition operation involving the elements of a matrix (m×n)-dimensional space, yielding another matrix of the same dimensions. This previous structure can be connected with isomorphic diagonal matrix representations, which can have some interest in quantum chemistry [225]. Moreover, IMP definition has been shown to allow writing the usual Rayleigh-Schrödinger perturbation theory (R-S PT) [226,227] with a new compact formalism. Such kind of matrix product can be easily introduced as a general matrix operation in high level languages like Fortran, as it has been earlier analysed [221].

The adjective *inward* has been used here to distinguish it from the usual matrix product. This denomination has been chosen because of the easy form it takes, when compared with the classical matrix product. Albeit simply defined, the IMP can be associated to most usual product properties, as can be found in the following sections.

G2. Inward matrix Products

In this section a detailed structure of IMP will be presented as a mean to prepare the application tools. All the arguments can apply to both matrix and hypermatrix spaces. In order to stress this characteristic property, along the exposition the prefix (hyper-) will be placed in front of the word matrix in parentheses.

G2.1. Definitions and Properties

Definition 12: Inward Matrix Product

Consider any arbitrary (hyper-)matrix space over a field: $\boldsymbol{M}_{(\times\mathbf{n})}$. Let **A**, **B** $\in \boldsymbol{M}_{(\times\mathbf{n})}$. An inward product of the (hyper-)matrix pair is a closed operation, resulting in a new (hyper-)matrix $\mathbf{P} \in \boldsymbol{M}_{(\times\mathbf{n})}$, and symbolised by: $\mathbf{P} = \mathbf{A} * \mathbf{B}$, whose elements are defined by the algorithm:

$$\forall(\mathbf{i}): \mathrm{p}(\mathbf{i}) = \mathrm{a}(\mathbf{i})\mathrm{b}(\mathbf{i}). \square$$

Above, the elements of the (hyper-)matrices are identified by means of an index vector $(\mathbf{i}) \equiv (i_1; i_2; \ldots i_p)$. Also, the (hyper-)matrix space dimension is given by the symbol: $(\times\mathbf{n}) \equiv (n_1 \times n_2 \times \ldots n_p)$. The notation follows the earlier one employed when dealing with nested sum structures [102-105].

Properties of IMP

The interest in defining such a IMP stems from the possibility to attach to it the most usual features of a multiplication composition rule. The following properties can be attached to the inward product:

Let be: **A, B, C**, ... $\in \boldsymbol{M}_{(\times\mathbf{n})}$. Inward products defined over them are:

a) Distributive with respect matrix sum:

$$\mathbf{A} * (\mathbf{B} + \mathbf{C}) = \mathbf{A} * \mathbf{B} + \mathbf{A} * \mathbf{C}.$$

b) Associative:

$$\mathbf{A}*\mathbf{B}*\mathbf{C}=\mathbf{A}*(\mathbf{B}*\mathbf{C})=(\mathbf{A}*\mathbf{B})*\mathbf{C}.$$

c) Commutative:

$$\mathbf{A}*\mathbf{B}=\mathbf{B}*\mathbf{A}.$$

d) An Inward Unit Element exists, the *unity* matrix, **1**, such that:

$$\mathbf{1}*\mathbf{A}=\mathbf{A}*\mathbf{1}=\mathbf{A}.$$

e) Using the real multiplication unit it can be defined as:

$$1=\{\ 1(i)=1\ ;\ \forall(i)\}.$$

f) Existence of an inward inverse: Iff $\mathbf{A}=\{a(\mathbf{i})\} \wedge \forall(\mathbf{i})$: $a(\mathbf{i}) \neq 0\}$; then **A** can be called inwardly inversible or regular. A new matrix defines the inward inverse of a matrix **A**: $\mathbf{A}^{[-1]}=\{a^{[-1]}(\mathbf{i})\}$ with elements, which are computed as follows: $\forall(\mathbf{i}): a^{[-1]}(\mathbf{i})=(a(\mathbf{i}))^{-1}$. This definition produces the sequence of equalities: $\mathbf{A}*\mathbf{A}^{[-1]}=\mathbf{A}^{[-1]}*\mathbf{A}=\mathbf{1}$. Note that inwardly inversible matrices are these associated to SVSS spaces, whose elements are constructed as (hyper-)matrices.

G3. Applications of Inward Products

Having described the main features of the (hyper-)matrix inward product, several application examples will be given.

G3.1. Scalar product of two (hyper-)matrices

The IMP of two (hyper-)matrices can be trivially related to the scalar product of two (hyper-)matrices. If such a scalar product is defined, using a nested summation symbol [102-105], as:

$$\langle\mathbf{A}|\mathbf{B}\rangle=\sum(\mathbf{i})a(\mathbf{i})b(\mathbf{i}).$$

Then, using the auxiliary generalisation of Definition 3: $\langle\mathbf{A}\rangle=\Sigma(\mathbf{i})a(\mathbf{i})$, [102-105] to symbolise the sum of all elements of a given (hyper-)matrix, the following equality can be easily written:

$$\langle \mathbf{A}|\mathbf{B}\rangle = \langle \mathbf{A} * \mathbf{B}\rangle .$$

G3.2. Trace of the product of two matrices

As it is usually defined, the trace of a square (n×n)-dimensional matrix corresponds to the algorithm: $\mathrm{Tr}(\mathbf{A}) = \sum_i a_{ii}$. Then, using the auxiliary definition of the transpose of a given matrix as: $\mathbf{B}^T = \{b_{ij}^{(T)} = b_{ji}\}$, the following equality is obtained:

$$\mathrm{Tr}(\mathbf{AB}) = \langle \mathbf{A} * \mathbf{B}^T \rangle .$$

G3.3. Generalised Density Functions

IMP features permits to express DF in a compact and extremely general form. Previous work [44], as the arguments already sketched in Section 3.1, may be invoked to define a matrix **W** as a convex coefficient (hyper-)matrix of arbitrary dimensions, a characteristic which can be symbolised, as before, by the symbol: (×n). Besides $\langle \mathbf{W} \rangle = 1$, all elements of **W**are chosen real and positive. Such a (hyper-)matrix can be formed by constructing a complex matrix **X**, with the same dimension as **W**, such that: $\mathbf{W} = \mathbf{X}^* * \mathbf{X}$. In this way it can be assured that:

$$\mathbf{W} = \{\forall \omega \in \mathbf{W} \rightarrow \exists \chi \in \mathbf{X}:\ \omega = \chi^* \chi = |\chi|^2 \in \mathbf{R}^+\}.$$

This special choice can be also described by the convex conditions, see Section 3.2, on **W** [44]:

$$K_{\{\times n\}}(\mathbf{W}) = \{\forall \omega \in \mathbf{W} \rightarrow \omega \in \mathbf{R}^+ \wedge \langle \mathbf{W} \rangle = 1\}.$$

In all the previous definitions and in what follows, (hyper-)matrix elements are written without subindices, this is so because within IMP definitions, the involved (hyper-)matrix elements have to bear the same subindex set. Define now a (hyper-)matrix **P** with a dimension equal to the one of **W**, and containing as elements, normalised PD multivariate functions of a position variable vector **Z** of arbitrary dimension. It can be formally written: $\int \mathbf{P}(\mathbf{Z})d\mathbf{Z}=\mathbf{1}$, where the symbol **1** is used as a unity (hyper-)matrix of the appropriate dimension too. More specifically, one can compactly build up the convenient structure of the (hyper-)matrix function **P** as:

$$\mathbf{P} = \left\{\forall \mathbf{p}(\mathbf{Z}) \in \mathbf{P} \rightarrow \forall \mathbf{Z} : \mathbf{p}(\mathbf{Z}) \in \mathbf{R}^{+} \wedge \int \mathbf{p}(\mathbf{Z}) d\mathbf{Z} = 1\right\}.$$

This can be accomplished by constructing an IMP: $\mathbf{P} = \Psi^{*} * \Psi$, then:

$$\mathbf{P} = \left\{\forall \mathbf{p} \in \mathbf{P} \rightarrow \exists \varphi \in \Psi : \mathbf{p} = |\varphi|^{2}\right\}.$$

A normalised density function of the position variables vector **Z** can be expressed as an IMP:

$$\rho(\mathbf{Z}) = \langle \mathbf{W} * \mathbf{P} \rangle \Rightarrow \int \rho(\mathbf{Z}) d\mathbf{Z} = 1.$$

G3.4. Diagonal Form of the IMP

Another property of IMP can be now given. The definition of IMP given above, becomes also understandable from the point of view of the matrix product of two diagonal matrices. If the matrices involved in the IMP are previously transformed into an isomorphic diagonal form, for example by using the column reordering rule:

$$\forall \mathbf{A} \in M_{(m\times n)} \Rightarrow \mathrm{Diag}(\mathbf{A}) = \mathrm{Diag}(a_{11}, a_{21}, \ldots, a_{m1}, a_{12}, \ldots a_{mn}) \in D_{(mn)}$$

Then, IMP are transformed into diagonal matrix products in the isomorphic diagonal matrix space $D_{(mn)}$:

$$\mathbf{A},\mathbf{B} \in M_{(m\times n)} \rightarrow \mathbf{A} * \mathbf{B} \in M_{(m\times n)} \Leftrightarrow$$
$$\mathrm{Diag}(\mathbf{A}), \mathrm{Diag}(\mathbf{B}) \in D_{(mn)} \rightarrow \mathrm{Diag}(\mathbf{A})\mathrm{Diag}(\mathbf{B}) \in D_{(mn)}$$

An alternative property will hold when a row reordering rule is used instead of the previously defined column one. This will be the same as to apply the column reordering on the transposes that is:

$$\forall \mathbf{A}^{T} \in \boldsymbol{M}_{(n\times m)} \Rightarrow \mathrm{Diag}(\mathbf{A}^{T}) = \mathrm{Diag}(a_{11}, a_{12}, \ldots, a_{1n}, a_{21}, \ldots a_{mn}) \in \boldsymbol{D}^{*}_{(nm)},$$

then the diagonal vector space:$\boldsymbol{D}^{*}_{(nm)}$, under some circumstances can act as a dual of : $\boldsymbol{D}_{(nm)}$.

However, this kind of property of diagonal matrix spaces is a consequence of natural way matrices may be associated to diagonal forms, by means of IMP.

In the next section, where the IMP is generalised, a natural way to proceed to form (hyper-)diagonal matrices will be developed.

G4. Extended Inward Matrix Product

A trivial extension of IMP can be envisaged when considering the possibility to involve (hyper-)matrices, belonging to different matrix spaces. Here, any (hyper-)matrix **A** can be described by an element set written symbolically by: $\{a(\mathbf{p})\}$, where the index vector **p**, represents all the needed set of indexes to appropriately describe any of the (hyper-)matrix elements. Any (hyper-)matrix element, can be also expressed with a partition in the index vector as, for example: $A=\{a(\mathbf{p})\}\equiv\{a(\mathbf{i},\mathbf{j})\}$, provided that: $\mathbf{p}=\mathbf{i} \oplus \mathbf{j}$. This kind of formalism can be easily connected to tensor analysis, see for example [100], and thus could be seen as the appropriate tool to establish a close relationship between matrix and tensor algebras.

a) Then, an IMP of a (hyper-)column vector by a many element (hyper)-matrix , can be understood as a composition rule defined by means of the symbolic expression: $\mathbf{v}*\mathbf{A}=\{v(\mathbf{i})(\mathbf{i},\mathbf{j})\}$. The resultant dimension of this extended IMP coincides with the one attached to the (hyper-)matrix **A.** Defined in this fashion, *extended* IMP (EIMP) are to be considered non-commutative external composition operations involving at least two (hyper-) matrix spaces.

In this case, a simple application can be described. When both **v** and **A** are associated to the trivial dimensions (nx1) and (nxn) respectively, then their Hadamard product can be defined as follows: $\mathbf{Z} = \mathbf{v}*\mathbf{A} \rightarrow Z_{ij} = v_i A_{ij}$. Alternatively, if **A** is partitioned by means of the constituent column vectors as in the expression: $\mathbf{A} = (\mathbf{a}_1, \mathbf{a}_2, \ldots \mathbf{a}_n)$, then also one can write the following inward product matrix: $\mathbf{Z} = (\mathbf{v}*\mathbf{a}_1, \mathbf{v}*\mathbf{a}_2, \ldots \mathbf{v}*\mathbf{a}_n)$. This property has far reaching consequences, as choosing the matrix $\mathbf{A}=\mathbf{I}_n$, then one has the description of the isomorphism between column vectors and diagonal matrices, but constructed in a natural way. Indeed, according to the previous definitions and discussion, it can be written:

$$\mathbf{v}*\mathbf{I}=(\mathbf{v}*\mathbf{e}_1,\mathbf{v}*\mathbf{e}_2,\ldots,\mathbf{v}*\mathbf{e}_n)$$
$$=\mathrm{Diag}(v_1,v_2,\ldots,v_n)=(\mathbf{e}_1*\mathbf{v},\mathbf{e}_2*\mathbf{v},\ldots,\mathbf{e}_n*\mathbf{v})=\mathbf{I}*\mathbf{v},$$

which is not a property of the unit matrix, but a general property of inward products. This result is valid for row vectors too, the difference will be on the partition of the matrix **A**, but the result will appear to be the same.

b) Take two (hyper-)matrices belonging to different dimension matrix spaces: $\mathbf{A}=\{a\{\mathbf{i},\mathbf{p}\}\}\wedge\mathbf{B}=\{b\{\mathbf{j},\mathbf{p}\}\}$, where an index vector partition has been previously made, such as to have the index vector **p** with a common dimension in both arrays. The non-commutative EIMP of the (hyper-) matrix pair can be defined as: $\mathbf{C}=\mathbf{A}*\mathbf{B}=\{c(\mathbf{i},\mathbf{p},\mathbf{j})=a(\mathbf{i},\mathbf{p})b(\mathbf{p},\mathbf{j})\}$. However, in this case the dimension of the resultant extended inward product (hyper-) matrix will be a composite product of the respective involved index vector dimensions. As a consequence one can formally write the EIMP dimension as: $\mathrm{Dim}(\mathbf{A}*\mathbf{B})=\mathrm{Dim}(\mathbf{i})\times\mathrm{Dim}(\mathbf{p})\times\mathrm{Dim}(\mathbf{j})$.

c) Generalised conventional (hyper-)matrix products can be generated from EIMP, just using a nested summation symbol (NSS), $\Sigma(\mathbf{p})$, [102-105], affecting the index vector involved into the EIMP definition. See Appendix H, for more details on NSS. These generalised products can be easily associated to contracted tensor products too. This means that it can be written:

$$\mathbf{D}=\mathbf{A}:\mathbf{B}=\{d(\mathbf{i},\mathbf{j})=\sum(\mathbf{p})c(\mathbf{i},\mathbf{p},\mathbf{j})=\sum(\mathbf{p})a(\mathbf{i},\mathbf{p})b(\mathbf{p},\mathbf{j})\}.$$

However, this disposition can be changed at will, because index partition can be produced, among other choices, in the following way: $\mathbf{A}=\{a\{\mathbf{i},\mathbf{p}\}\}\mathrm{U}\ \mathbf{A}=\{a\{\mathbf{i},\mathbf{p}\}\}\wedge\mathbf{B}=\{b\{\mathbf{j},\mathbf{p}\}\}$, for example. Then, the associated inward product will change the structure of the (hyper-)matrix final form accordingly, as well as the contraction result.

APPENDIX H. THE NESTED SUMMATION SYMBOL (NSS)

A NSS is a linear operator attached to an arbitrary number of nested sums, that is, an NSS represents a set of summation symbols where the number of them can be variable [102-105]. A simple NSS operator notation is:

$$\sum_n(\mathbf{j}=\mathbf{i},\mathbf{f}), \tag{73}$$

where the implied n-dimensional vectors are:

$$\boldsymbol{j} = (j_1, j_2, \dots, j_n) \;, \boldsymbol{i} = (i_1, i_2, \dots, i_n) \text{ and } \boldsymbol{f} = (f_1, f_2, \dots, f_n).$$

The index n is called the *dimension of the NSS*.

The operator (73) stands for the following set of n summation symbols:

$$\sum_{j_1=i_1}^{f_1} \sum_{j_2=i_2}^{f_2} \dots \sum_{j_n=i_n}^{f_n} \; .$$

In this context, the operator performs all the sums involved in the generation of all the distinct possible combinations of the vector $\boldsymbol{j}$ indices. In this formulation, the k-th index of the vector $\boldsymbol{j}$ runs over the values from i_k to f_k, incrementing its value one unit by every step. This last condition can be relaxed allowing every index to vary with a particular step length. In this case, a step vector $\boldsymbol{s}=(s_1, s_2, \dots, s_n)$ can enter as a NSS argument:

$$\sum_n (\mathbf{j} = \mathbf{i}, \mathbf{f}, \mathbf{s}) \tag{74}$$

It is straightforward to relate the NSS formalism with some combinatorial objects. For example, if it is defined the unit vector as the n-dimensional vector $\mathbf{1}$=(1,1,...,1), a NSS as

$$\sum_n (\mathbf{j} = \mathbf{1}, m\mathbf{1}, \mathbf{1})$$

will generate all the m^n variations with repetition.

Also, it is possible to control if some particular forms of the $\boldsymbol{j}$ vector are accepted or not. This can be specified by means of a *logical function* [102-105], $L(\boldsymbol{j})$, which, usually, depends on the $\boldsymbol{j}$ vector. For every form of the vector $\boldsymbol{j}$, the logical function returns a value of true (1) or false (0). Only in the first case, the actual form of the $\boldsymbol{j}$ vector must be considered. The proposed notations in order to explicit the use of the logical function are:

$$\sum_n (\mathbf{j} = \mathbf{i}, \mathbf{f}, \mathbf{s}) L(\mathbf{j}) \equiv \sum_n (\mathbf{j} = \mathbf{i}, \mathbf{f}, \mathbf{s}; L(\mathbf{j})).$$

Using this general notation, and depending on the particular definition of the logical function, more combinatorial objects can be constructed with a NSS, as it will be seen below.

Besides of its algebraic applications [102-105], a NSS has a computational implementation called a generalized nested do loop (GNDL) [102-105]. The GNDL algorithm constitutes the link between the mathematical notation of the NSS and the computer codification of this operator. As a short example, the FORTRAN 95 codification of the operator outlined in Equation (74) corresponds to the listing given in the Module 1 below. There, the operator is codified as a routine which possesses, as arguments, the integer values defining the vectors $\boldsymbol{i}, \boldsymbol{f}$ and $\boldsymbol{s}$. Also, the NSS dimension is wanted. In the algorithm kernel, a call to a user-supplied function, called JOB, is performed. This routine depends on the concrete application of the algorithm and will eventually use the actual from of vector $\boldsymbol{j}$.

Particular iterative codes for the generation of combinations and permutations have been published elsewhere [102-105]. Here, the corresponding recursive implementations are supplied in Module 2 and Module 3 listings. Module 2 constitutes the codification of a combinations without repetition generator. There, every time the user-supplied routine JOB is called, the vector c(1:n) contains a new combination to be processed. The combinations are generated in lexicographical order. In the code, the logical function $L(j)$ is codified in an embedded fashion: at every recursive call, there is a DO-END DO sentence (see RCOMBI subroutine) which runs over a specific range of integer values, which, in turn, depend of the previous c(:) vector indices. Something similar can be said respect to the Module 3. This module generates all the permutations without repetition of a set of n objects. The codification of the logical function is achieved using a logical vector, L(:), which codifies the storage of the objects already entering in a particular permutation. In this way, the object repetitions are avoided.

Due to the particular object oriented features of the new FORTRAN 90/95 standards, the GNDL can be implemented in the algorithms in a transparent way with respect to the programmer. Module 4 shows an example of it. There, a module is defined which implements a NSS capable of simulate the codification of an arbitrary number of FORTRAN DO instructions. The programmer, only needs to define the NSS in the main program and a simple cyclic sentence is used to call the JOB routine while the GNDL generates successive active forms of the $\boldsymbol{j}$ vector. The module contains the NSS internal related parameters and drives the $\boldsymbol{j}$ vector forms creation.

From the definition of the NSS operator and the codification of the GNDL, it is obvious to see that this kind of structures can be easily programmed in a parallel manner. Of course, if a module similar to the one appearing in Program 1 is implemented in some high level programming language, many benefits could be transferred to the new generations of programmers and codes.

Several applications of NSS have been explored [102-105]. In general, substantial benefits were obtained in the fields of computational chemistry, perturbation theory, numerical calculus, programming techniques, and so on. Here it has been used in several places, mainly in Sections 5.6 and 6.4.

Modules and Programs as Examples of NSS

Module 1. A basic NSS codification of operator (74).

```
!-----------------------------------------------
module GNDL
  integer counter
contains
   subroutine basic_NSS(n,i,f,s)
   integer j(n),i(n),f(n),s(n)
   j(:)=i(:) ! First form
   k=n
   do while (k>0)
      if ((j(k)-f(k))*s(k)>0) then
         j(k)=i(k); k=k-1
      else
         call JOB(n,j) ! Application call
         k=n
      end if
      if (k>0) j(k)=j(k)+s(k) ! Step increment
   end do
   end subroutine basic_NSS
end module GNDL
!-----------------------------------------------
```

Module 2. A recursive codification of a combinations generator. This program is called by means of the sentence call recursive_combinations(m,n) which will start the generation of the ${}^{m}C_{n}$ combinations.

```
!-------------------------------------------------------------------
module combinations
 integer, allocatable :: c(:),f(:)
 integer counter
CONTAINS
subroutine recursive_combinations(m,n)
 allocate(c(0:n),f(n))
 c(0)=0 ! Sentinel
 do i=1,n; f(i)=m-n+i; end do ! Initialization
 counter=0
 call rcombi(n,1) ! First call. Computation launch
 END subroutine recursive_combinations
recursive subroutine rcombi(n,L) ! Recursive generation of level L
  do i=c(L-1)+1,f(L)
     c(L)=i
     if (L<n) then; call rcombi(n,L+1) ! Generates next level
              else; call JOB(n,C(1))   ! Calls application
      end if
   end do
   END subroutine rcombi
end module combinations
!-------------------------------------------------------------------
```

Module 3. A recursive codification of a permutations generator. This program is called by the sentence call recursive_permutations(n) which will start the generation of the $n!$ permutations.

```
!----------------------------------------------------------------------
module permutations
   integer, allocatable :: P(:)
   integer counter
contains
   subroutine recursive_permutations(n)
   logical*1 L(n) ! Logical vector of incompatibilities
   allocate(p(n))
   do i=1,n; p(i)=i; end do ! Initializes
   counter=0
   L(:)=.true.
   call rper(n,1,L) ! Launch the computation
   END subroutine recursive_permutations
recursive subroutine rper(n,Level,L)
 logical*1 L(n),Lcopy(n)
   do i=1,n
      if (L(i)) then ! Index value available
         P(Level)=i
         if (Level<n) then
            Lcopy(:)=L(:); Lcopy(i)=.false. ! Unavailable value
            call rper(n,Level+1,Lcopy)       ! Generates next level
         else
            call JOB(n,p) ! Call application
         end if
      end if
   end do
END subroutine rper
end module permutations
!----------------------------------------------------------------------
```

Program 1. Implementation of a built-in NSS capability available to the programmer.

```
!-------------------------------------------------------------------
MODULE NSS_implementation
integer, allocatable :: j(:),i(:),f(:),p(:) ! NSS related arguments
integer :: n,k                              ! and parameters
logical :: flag
CONTAINS
subroutine initialize_NSS(pro_i,pro_f,s)
 integer, dimension(:), intent (IN) :: pro_i,pro_f
 integer, dimension(:), intent (IN), OPTIONAL :: s ! Step vector
 n=SIZE(pro_i) ! NSS dimension
 allocate(j(n),i(n),f(n),p(n))
 if (n/=SIZE(pro_f)) stop "NSS dimension problems: check f vector."
  i(:)=pro_i(:)
  j(:)=i(:)
  f(:)=pro_f(:)
  if (PRESENT(s)) then  ! Step vector
   if (n/=SIZE(s)) stop "NSS dimension problems: check s vector.."
    p(:)=s(:)
  else
    p(:)=1 ! Default option: unit step.
  end if
  do k=1,n
   if ((j(k)-f(k))*p(k)>0) then !Index out of range
     write(*,'(" Error: index number",i3," is out of ramge.")') k
     stop
  end if
  end do
  k=n  ! k signals the actual active summation symbol
  flag=.true.
end subroutine initialize_NSS
function SSA() result (RUN) ! NSS control procedure
 logical RUN ! NSS is yet active?
 RUN=.false.
2 if (flag) then
  if (k>0) then
     if ((j(k)-f(k))*p(k)>0) then ! Index out of range:
        j(k)=i(k)  ! Resets this level
        k=k-1      ! If possible, previous level will
     goto 1        ! be actived.
     end if
```

```
    end if
    RUN=.true.    !
    flag=.false. ! A new active form of j vector was generated
    k=n          ! The JOB will be called
    RETURN       !
  else
    flag=.true.
  end if
  1 if (k>0) then
      j(k)=j(k)+p(k) ! Step increment
      if (.not.RUN) goto 2
        flag=.true.
   else
      RUN=.false.           ! NSS terminates,
      deallocate (j,i,f,p)  ! Deallocates data:
      RETURN                ! makes them available for further use.
   end if
  end function SSA
  END MODULE NSS_implementation
  !---------------------------------------------------------------
  PROGRAM test_NSS_implementation
     USE NSS_implementation
     call initialize_NSS((/0,0,0/),(/1,1,1/)) ! NSS declaration
     do while (SSA())        !
        print '(3i3)', j     ! Implemented NSS usage
     end do                  !
     print *
     call initialize_NSS((/0,0,0,0/),(/2,2,2,2/),(/2,2,2,2/))
     do while (SSA())        !
        print '(4i3)', j     ! Implemented NSS usage
     end do                  !
     print *
     ! A bad definition example: innermost loop never ends
     call initialize_NSS((/0,0,0,0/),(/2,2,2,2/),(/2,2,2,-2/))
     do while (SSA())        !
        print '(4i3)', j     ! This will not be called
     end do                  !
  END
  !---------------------------------------------------------------
```

REFERENCES

1. R. Carbó, M. Arnau, and L. Leyda, *Int. J. Quantum Chem.* **17**, 1185 (1980).
2. E. E. Hodgkin and W. G. Richards, *Int. J. Quantum Chem.* **14**, 105 (1987).
3. C. Burt, W. G. Richards, and P. Huxley, *J. Comput. Chem.* **11**, 1139 (1990).
4. A. C. Good, E. E. Hodgkin, and W. G. Richards, *J. Chem. Inf. Comput. Sci.* **32**, 188 (1992).
5. A. C. Good, S.-S. So, and W. G. Richards, *J. Med. Chem.* **36**, 433 (1993).
6. C. Good and W. G. Richards, *J. Chem. Inf. Comput. Sci.* **33**, 112 (1993).
7. D. L. Cooper and N. L. Allan, *J. Comput.-Aided Mol. Des.* **3**, 253 (1989).
8. D. L. Cooper and N. L. Allan, *J. Am. Chem. Soc.* **114,** 4773 (1992).
9. D. L. Cooper, K. A. Mort, N. L. Allan, D. Kinchington, and C. McGuidan, *J. Am. Chem. Soc.* **115**, 12615 (1993).
10. N. L. Allan and D. L. Cooper, *Top. Curr. Chem.* **173**, 85 (1995).
11. J. Cioslowski and E. D. Fleishmann, *J. Am. Chem. Soc.* **113**, 64 (1991).
12. J. B. Ortiz and J. Cioslowski, *Chem. Phys. Lett.* **185**, 270 (1991).
13. J. Cioslowski and M. Challacombe, *Int. J. Quantum Chem.* **S25**, 81 (1991).
14. J. Cioslowski and S. T. Mixon, *Can. J. Chem.* **70**, 443 (1992).
15. J. Cioslowski and A. Nanayakkara, *J. Am. Chem. Soc.* **115**, 1213 (1993).
16. J. Cioslowski, B. B. Stefanov, and P. Constans, *J. Comput. Chem.* **17**, 1352 (1996).
17. R. Ponec and M. Strnad, *Collect. Czech. Chem. Commun.* **55**, 896 (1990).
18. R. Ponec and M. Strnad, *Collect. Czech. Chem. Commun.* **55**, 2583 (1990).
19. R. Ponec and M. Strnad, *J. Phys. Org. Chem.* **4**, 701 (1991).
20. R. Ponec and M. Strnad, *J. Math. Chem.* **8**, 103 (1991).
21. R. Ponec and M. Strnad, *Int. J. Quantum Chem.* **42**, 501 (1992).
22. R. Ponec, *J. Chem. Inf. Comput. Sci.* **33**, 805 (1993).
23. R. Ponec, *Overlap Determinant Method in the Theory of Pericyclic Reactions*, Lecture notes in Chemistry 65, Springer-Verlag, Berlin (1995).
24. P. G. Mezey, *Shape in Chemistry: An Introduction to Molecular Shape and Topology*, VCH, New York (1993).
25. P. G. Mezey, *Top. Curr. Chem.* **173**, 63 (1995).
26. C. Lee and S. Smithline, *J. Phys. Chem.* **98**, 1135 (1994).
27. C. Amovilli and R. McWeeny, *J. Mol. Struct. (Theochem)* **227**, 1 (1991).
28. A. C. Good, *J. Mol. Graph.* **10**, 144 (1992).
29. R. Benigni, M. Cotta-Ramusino, F. Giorgi, and G. Gallo, *J. Med. Chem.* **38**, 629 (1995).
30. J. D. Petke, *J. Comput. Chem.* **14**, 928 (1993).
31. A. Riera, *J. Mol. Struct. (Theochem)* **259**, 83 (1992).
32. F. Fratev, O. E. Polansky, A. Mehlhorn, and V. Monev, *J. Mol. Struct.* **56**, 245 (1979).
33. F. Fratev, V. Monev, A. Mehlhorn, and O. E. Polansky, *J. Mol. Struct. (Theochem)* **56**, 255 (1979).
34. M. A. Johnson and G. Maggiora (eds.), *Concepts and Applications of Molecular Similarity*, John Wiley, New York (1990).
35. R. Carbó (ed.) *Molecular Similarity and Reactivity: From Quantum Chemical to Phenomenological Approaches*, Kluwer, Dordrecht (1995).
36. K. Sen (ed.), *Molecular Similarity. Top. Curr. Chem.* Vols. 173 and 174, Springer-Verlag, Berlin (1995).
37. P. M. Dean (ed.), *Molecular Similarity in Drug Design*, Blackie Academic & Professional, London (1995).

38. R. Carbó-Dorca and P. G. Mezey (eds.), *Advances in Molecular Similarity*, Vol. 1, JAI Press, Greenwich (1996).
39. R. Carbó-Dorca and P. G. Mezey (eds.), *Advances in Molecular Similarity*, Vol. 2, JAI Press, Greenwich (1998).
40. R. Carbó and L. Domingo, *Int. J. Quantum Chem.* **23**, 517 (1987).
41. R. Carbó-Dorca, *J. Math. Chem.* **23**, 353 (1998).
42. R. Carbó-Dorca, *J. Math. Chem.* **23**, 365 (1998).
43. R. Carbó-Dorca, in *Advances in Molecular Similarity*, Vol. 2, R. Carbó-Dorca and P. G. Mezey (eds.), JAI Press, Greenwich (1998).
44. R. Carbó-Dorca and E. Besalú, *J. Mol. Struct. (Theochem)* **451**, 11 (1998).
45. R. Carbó and B. Calabuig, *Comput. Phys. Commun.* **55**, 117 (1989).
46. R. Carbó and B. Calabuig, *J. Mol. Struct. (Theochem)* **254**, 517 (1992).
47. R. Carbó and B. Calabuig, in *Computational Chemistry: Structure, Interactions and Reactivity*, Vol. A., S. Fraga (ed.) Elsevier, Amsterdam (1992).
48. R. Carbó, B. Calabuig, L. Vera, and E. Besalú, *Adv. Quantum Chem.* **25**, 253 (1994).
49. R. Carbó and B. Calabuig, *Int. J. Quantum Chem.* **42**, 1681 (1992).
50. R. Carbó and E. Besalú, in *Molecular Similarity and Reactivity: From Quantum Chemistry to Phenomenological Approaches*, R. Carbó (ed.) Kluwer, Dordrecht (1995).
51. R. Carbó and B. Calabuig, *J. Chem. Inf. Comput. Sci.* **32**, 600 (1992).
52. R. Carbó-Dorca, E. Besalú, L. Amat, and X. Fradera, in *Advances in Molecular Similarity*, Vol. 1, R. Carbó-Dorca and P. G. Mezey (eds.) JAI Press, Greenwich (1996).
53. D. Robert and R. Carbó-Dorca, *J. Math. Chem.* **23**, 327 (1998).
54. D. Robert and R. Carbó-Dorca, *Nuovo Cimento* **A 111**, 1311 (1998).
55. R. Carbó, E. Besalú, L. Amat, and X. Fradera, *J. Math. Chem.* **18**, 237 (1995).
56. X. Fradera, L. Amat, E. Besalú, and R. Carbó-Dorca, *Quant. Struct.-Act. Relat.* **16**, 25 (1997).
57. M. Lobato, L. Amat, E. Besalú, and R. Carbó-Dorca, *Scientia Gerundensis* **23**, 17 (1998).
58. M. Lobato, L. Amat, E. Besalú, and R. Carbó-Dorca, *Quant. Struct.-Act. Relat.* **16**, 465 (1997).
59. D. Robert and R. Carbó-Dorca, *J. Chem. Inf. Comput. Sci.* **38**, 620 (1998).
60. L. Amat, D. Robert, E. Besalú, and R. Carbó-Dorca, *J. Chem. Inf. Comput. Sci.* **38**, 624 (1998).
61. D. Robert, L. Amat, and R. Carbó-Dorca, *J. Chem. Inf. Comput. Sci.* **39**, 333 (1999).
62. R. Ponec, L. Amat, and R. Carbó-Dorca, *J. Comput-Aided Mol. Des.* **13**, 259 (1999).
63. L. Amat, R. Carbó-Dorca, and R. Ponec, *J. Comput. Chem.* **19**, 1575 (1998).
64. R. Ponec, L. Amat, and R. Carbó-Dorca, *J. Phys. Org. Chem.* **12**, 447 (1999).
65. P. G. Mezey, R. Ponec, L. Amat, and R. Carbó-Dorca, *Enantiomer* **4**, 371 (1999).
66. L. Amat, R. Carbó-Dorca, and R. Ponec, *J. Med. Chem.* **42**, 5169 (1999).
67. D. Robert, X. Gironés, and R. Carbó-Dorca, *J. Comput.-Aided Mol. Des.* **13**, 597 (1999).
68. X. Gironés, L. Amat, and R. Carbó-Dorca, R., *SAR QSAR Environ. Res.* **10**, 545 (1999).
69. D. Robert and R. Carbó-Dorca, *SAR QSAR Environ. Res.* **10**, 401 (1999).
70. C. Hansch, B. H. Venger, and A. Panthananickal, *J. Med. Chem.* **23**, 459 (1980).
71. S. E. Sherman and S. J. Lippard, *Chem. Rev.* **87**, 1153 (1987).
72. M. J. Frisch, G. W. Trucks, H. B. Schlegel, P. M. W. Gill, et al. Gaussian-94, (Revision E.2), Gaussian, Inc., Pittsburgh (1995).
73. E. Besalú, R. Carbó, J. Mestres, and M. Solà, *Top. Curr. Chem.* **173**, 31 (1995).
74. R. Carbó and E. Besalú, *J. Math. Chem.* **20**, 247 (1996).
75. R. Carbó, E. Besalú, L. Amat, and X. Fradera, *J. Math. Chem.* **19**, 47 (1996).

76. D. Robert and R. Carbó-Dorca, *J. Chem. Inf. Comput. Sci.* **38**, 469 (1998).
77. P. Constans, L. Amat, and R. Carbó-Dorca, *J. Comput. Chem.* **18**, 826 (1997).
78. R. Carbó-Dorca, L. Amat, E. Besalú, and M. Lobato, in *Advances in Molecular Similarity*, Vol. 2, R. Carbó-Dorca and P. G. Mezey (eds.), JAI Press, Greenwich (1998).
79. L. A. Zadeh, *Inform. Control* **8**, 338 (1965).
80. E. Trillas, C. Alsina, and J. M. Terricabras, *Introducción a la Lógica Difusa*, Ariel Matemática, Barcelona (1995).
81. R. Carbó-Dorca, *J. Math. Chem.* **22**, 143 (1997).
82. R. Carbó-Dorca, E. Besalú, and X. Gironés, Extended Density Functions and Quantum Chemistry, Technical Report: IT-IQC-99-02. *Adv. Quantum Chem.* in press.
83. J. S. Bell, *Speakable and Unspeakable in Quantum Mechanics*, Cambridge University Press, Cambridge (1993).
84. T. M. Vinigradov (ed.) *Encyclopaedia of Mathematics*, Vol. 8, Reidel-Kluwer, Dordrecht (1987).
85. J. von Neumann, *Mathematical Foundations of Quantum Mechanics*, Princeton University Press, Princeton (1955).
86. D. Bohm, *Quantum Theory*, Dover, New York (1989).
87. S. Goldstein, *Physics Today*, March, 42 (1988) and April, 38 (1988).
88. P. O. Löwdin, *Phys. Rev.* **97**, 1474 (1955).
89. R. McWeeny, *Revs. Mod. Phys.* **32**, 335 (1960).
90. R. McWeeny, *Proc. Roy. Soc.* **A 253**, 242 (1959).
91. E. R. Davidson, *Reduced Density Matrices in Quantum Chemistry*, Academic Press, New York (1976).
92. J. Mestres, M. Solà, M. Duran, and R. Carbó, *J. Comput. Chem.* **15**, 1113 (1994).
93. P. Constans and R. Carbó, *J. Chem. Inf. Comput. Sci.* **35**, 1046 (1995).
94. P. Constans, L. Amat, X. Fradera, and R. Carbó-Dorca, in *Advances in Molecular Similarity*, Vol. 1, R. Carbó-Dorca and P. G. Mezey (eds.), JAI Press, Greenwich (1996).
95. L. Amat, X. Fradera, and R. Carbó, *Scientia Gerundensis* **22**, 97 (1996).
96. L. Amat and R. Carbó-Dorca, *J. Comput. Chem.* **18**, 2023 (1997).
97. L. Amat and R. Carbó-Dorca, *J. Comput. Chem.* **20**, 911 (1999).
98. E. V. Ludeña, in *Química Teórica, Nuevas Tendencias*, Vol. 4, S. Fraga (ed.), CSIC, Madrid (1987).
99. I. Shavitt, in *Methods of Electronic Structure Theory, Modern Theoretical Chemistry*, Vol. 3, H. F. Schaefer III (ed.), Plenum Press, New York (1977).
100. A. Lichnerowicz, *Algèbre et Analyse Linéaires*, Masson & Cie. Éditeurs, Paris (1947).
101. C. G. J. Jacobi, *J. Reine Angew. Math.* **30**, 51 (1846).
102. R. Carbó and E. Besalú, in *Strategies and Applications in Quantum Mechanics*, Y. Ellinger and M. Defranceschi (eds.), Kluwer, Dordrecht (1996).
103. R. Carbó and E. Besalú, *J. Math. Chem.* **18**, 37 (1995).
104. R. Carbó and E. Besalú, *Comput. Chem.* **18**, 117 (1994).
105. R. Carbó and E. Besalú, *J. Math. Chem.* **13**, 331 (1993).
106. P. Hohenberg and W. Kohn, *Phys. Rev.* **B136**, 864 (1964).
107. K. D. Sen, E. Besalú, and R. Carbó-Dorca, *J. Math. Chem.* **25**, 253 (1999).
108. P. G. Mezey, *J. Math. Chem.* **23**, 65 (1998).
109. S. K. Berberian, *Introducción al Espacio de Hilbert*, Teide, Barcelona (1970).
110. A. Jeffrey, *Handbook of Mathematical Formulas and Integrals*, Academic Press, New York (1995).
111. R. E. Moss, *Advanced Molecular Quantum Mechanics*, Chapman & Hall, London (1973).

112. W. Greiner, *Relativistic Quantum Mechanics*, Springer-Verlag, Berlin (1997).
113. R. Dedekind, *Essays on the Theory of Numbers*, Dover, New York (1963).
114. R. Carbó, B. Calabuig, E. Besalú and A. Martínez, *Molecular Engineering* **2**, 43 (1992).
115. R. G. Parr, *The Quantum Theory of Molecular Electronic Structure*, W. A. Benjamin Inc., New York (1963).
116. R. Carbó and J. M. Riera, *A general SCF Theory*, Lecture Notes in Chemistry, Vol. 5, Springer-Verlag, Berlin (1978).
117. T. Akutsu, *IEICE Trans. Inf. Sys.* **E-75D**, 95 (1992).
118. T. Akutsu, *IEICE Trans. Fund. Elec. Com. Comput. Sci.* **E-76A**, 9 (1993).
119. T. Asano, *Theo. Comput. Sci.* **38**, 249 (1985).
120. D.J. Klein, *J. Chem. Inf. Comput. Sci.* **37**, 656 (1997).
121. I.T. Jolliffe, *Principal Component Analysis*, Springer, New York (1986).
122. S. Wold, E. Johansson, and M. Cocchi, in *3D QSAR in Drug Design*, H. Kubinyi (ed.), ESCOM, Leiden (1993).
123. S. Wold, in *Chemometric Methods in Molecular Design*, H. van de Waterbeemd (ed.), VCH, Weinheim (1995).
124. I.Borg and P. Groenen, *Modern Multidimensional Scaling*, Springer, New York (1997).
125. C.M. Cuadras and C. Arenas, *Commun. Statist. Theor. Meth.* **19**, 2261 (1990).
126. D.M. Allen, *Technometrics* **16**, 125 (1974).
127. S. Wold and L. Ericksson, in *Chemometric Methods in Molecular Design*, H. van de Waterbeemd (ed.), VCH, New York (1995).
128. L. Amat, P. Constans, E. Besalú, and R. Carbó, MOLSIMIL 97, Institute of Computational Chemistry, University of Girona (1997).
129. L. Amat, D. Robert, and E. Besalú, TQSAR-SIM, Institute of Computational Chemistry, University of Girona (1997).
130. K. Yutani, K. Ogasahara, T. Tsujita, K. Kanemoto, et al., *J. Biol. Chem.* **262**, 13429 (1987).
131. K. Yutani, K. Ogasahara, K. Aoki, T. Kakuno, and Y. Sugino, *J. Biol. Chem.* **259**, 14076 (1984).
132. K. Yutani, K. Ogasahara, and Y. Sugino, *Adv. Biophys.* **20**, 13 (1985).
133. K. Yutani, K. Ogasahara, T. Tsujita, and Y. Sugino, *Proc. Natl. Acad. Sci. USA* **84**, 4441 (1987).
134. J. Damborský, *Prot. Eng.* **11**, 21 (1998).
135. PC Spartan Plus, Wavefunction Inc., Irvine CA, 92612 USA.
136. O. Exner, *Correlation Analysis of Chemical Data*, Plenum Press, New York (1988).
137. Y. Nozaki and C. Tanford, *J. Theor. Biol.* **67**, 567 (1971).
138. M. R. Sweet and D. Eisenberg, *J. Mol. Biol.* **171**, 479 (1983).
139. S. Rackovsky and H. A. Scheraga, *Proc. Natl. Acad. Sci. USA* **74**, 5248 (1977).
140. R. D. Cramer III, D. E. Patterson, and J. D. Bunce, *J. Am. Chem. Soc.* **110**, 5959 (1988).
141. R. F. Rekker, in *Pharmacochemistry Library*, W. T. Nauta and R. F. Rekker (eds.), Elsevier, New York (1977).
142. E. Urrestarazu, W. H. J. Vaes, H. J. M. Verhaar, and J. L. M. Hermens, *J. Chem. Inf. Comput. Sci.* **38**, 845 (1998).
143. S. Karabunarliev, O. G. Mekenyan, W. Karcher, C. L. Russom, and S. P. Bradbury, *Quant. Struct.-Act. Relat.* **15**, 311 (1996).
144. L. H. Hall, L. B. Kier, and G. Phipps, *Environ. Toxicol. Chem.* **3**, 355 (1984).
145. B. D. Gute and S. C. Basak, *SAR QSAR Environ. Res.* **7**, 117 (1997).

146. D. L. Geiger, L. T. Brooke, and D. J. Call (eds.) *Acute toxicities of Organic Chemicals to Fathead Minnows (Pimephales promelas)*, Vol. 5, Center for Lake Environmental Studies, University of Wisconsin-Superior, Superior (1990).
147. M. J. S. Dewar, E. G. Zoebisch, E. F. Healy, and J. J. P. Stewart, *J. Am. Chem. Soc.* **107**, 3902 (1985).
148. AMPAC 6.0, 1994 Semichem, 7128 Summit, Shawnee, KS 66216 D.A.
149. H.J. M. Verhaar, C. J. van Leeuwen, and J. L. M. Hermens, *Chemosphere* **25**, 471 (1992).
150. P.G. Mezey, *Molec. Phys.* **96**, 169 (1999).
151. N. S. E. Sargent, D. G. Upshall, and J. W. Bridges, *Biochem. Pharmac.* **31**, 1309 (1982).
152. R. W. Taft, *J. Am. Chem. Soc.* **86**, 5175 (1968).
153. L. P. Hammett, *J. Am. Chem. Soc.* **59**, 96 (1937).
154. A. Streitweiser, *Molecular Orbital Theory for Organic Chemists*, Wiley, New York (1961).
155. A. K. Saxena, *Quant. Struct.-Act. Relat.* **14**, 142 (1995).
156. M. Randic, *J. Am. Chem. Soc.* **97**, 6609 (1975).
157. L. B. Kier, W. J. Murray, M. Randic, and L. M. Hall, *J. Pharm. Sci.* **65**, 1226 (1976).
158. C.K. Chu, R. F. Schinazi, M. K. Ahn, G. V. Ullas, and Z. P. Gu, *J. Med. Chem.* **32**, 612 (1989).
159. A. T. Balaban, *J. Chem. Inf. Comput. Sci.* **35**, 339 (1995).
160. M. Randic, B. Jerman-Blacic, and N. Trinajstic, *Comput. Chem.* **14**, 237 (1990).
161. Z. Mihalic and N. Trinajstic, *J. Chem. Educ.* **69**, 701 (1992).
162. M. Wagener, J. Sadowski, and J. Gasteiger, *J. Am. Chem. Soc.* **117**, 7769 (1995).
163. http://iqc.udg.es/cat/similarity/QSAR/steroids.htm.
164. H. Hosoya, *Bull. Chem. Soc. Jap.* **45**, 2332 (1971).
165. H. Hosoya, *Bull. Chem. Soc. Jap.* **45**, 3415 (1971).
166. D. C. Montgomery and E. A. Peck, *Introduction to Linear Regression Analysis*, Wiley, New York (1992).
167. B.D. Silverman and D. E. Platt, *J. Med. Chem.* **39**, 2130 (1996).
168. D. A. Pierre, *Optimization Theory with Applications*, Wiley, New York (1969).
169. R. Carbó, L. Molino, and B. Calabuig, *J. Comput. Chem.* **13**, 155 (1992).
170. M. R. Spiegel, *Mathematical Handbook*, McGraw-Hill, New York (1968).
171. J.S. Binkley, J. A. Pople, and W. J. Hehre, *J. Am. Chem. Soc.* **102**, 939 (1980).
172. M. S. Gordon, J. S. Binkley, J. A. Pople, W. J. Pietro, and W. J. Hehre, *J. Am. Chem. Soc.* **104**, 2797 (1982).
173. R. Carbó-Dorca, ATOMIC Program 1995, based on: B. Roos, C. Salez, A. Veillard, and E. Clementi, *A General Program for Calculation of SCF Orbitals by the Expansion Method*, IBM Research/RJ518(#10901) (1968).
174. ASA coefficients and exponents can be downloaded from: http://iqc.udg.es/cat/similarity/ASA/func6-21.html
175. R. Bonaccorsi, E. Scrocco, and T. Tomasi, *J. Chem. Phys.* **52**, 5270 (1970).
176. H. E. Bethe and E. E. Salpeter, *Quantum Mechanics of One- and Two-Electron Systems*, Springer-Verlag, London (1957).
177. R. E. Moss, *Advanced Molecular Quantum Mechanics*, Chapman & Hall, London (1973).
178. R. McWeeny, *Methods of Molecular Quantum Mechanics*, Academic Press, London (1978).
179. J. Almlöf and O. Gropen, *Revs. Comput. Chem.* **8**, 203 (1996).
180. M. D. Miller, R. P. Sheridan, and S. K. Kearsley, *J. Med. Chem.* **42**, 1505 (1999).

181. G. Sello, *J. Chem. Inf. Comput. Sci.* **38**, 691 (1998).
182. D. D. Robinson, P. D. Lyne, and W. G. Richards, *J. Chem. Inf. Comput. Sci.* **39**, 594 (1999).
183. J. Mestres, D. C. Rohrer, and G. M. Maggiora, *J. Comput. Chem.* **18**, 934 (1997).
184. J. A. Grant, M. A. Gallardo, and B. T. Pickup, *J. Comput. Chem.* **17**, 1653 (1996).
185. M. J. Sippl and H. Stegbuchner, *Comput. Chem.* **15**, 73 (1991).
186. P. K. Redington, *Comput. Chem.* **16**, 217 (1992).
187. A. T. Brint and P. Willett, *J. Chem. Inf. Comput. Sci.* **27**, 152 (1987).
188. P. A. Bath, A. R. Poirrette, P. Willett, and F. H. Allen, *J. Chem. Inf. Comput. Sci.* **34**, 141 (1994).
189. D. J. Wild and P. Willett, *J. Chem. Inf. Comput. Sci.* **34**, 224 (1994).
190. D. A. Thorner, D. J. Wild, P. Willett, and P. M. Wright, *J. Chem. Inf. Comput. Sci.* **36**, 900 (1996).
191. S. Handschuh, *J. Chem. Inf. Comput. Sci.* **38**, 220 (1998).
192. G. Bravi, E. Gancia, P. Mascagni, M. Pegna, R. Todeschini, and A. Zalianni, *J. Comput-Aided Mol. Des.* **11**, 79 (1997).
193. G. E. Kellogg, L. B. Kier, P. Gaillard, and L. H. Hall, *J. Comput-Aided Mol. Des.* **10**, 513 (1996).
194. F. H. Allen, S. Bellard, M. D. Brice, B. A. Cartwright, et al., *Acta Cryst.* **B35**, 2331 (1979).
195. P. J. Artymiuk, P. A. Bath, H. M. Grindley, C. A. Pepperrell, et al., *J. Chem. Inf. Comput. Sci.* **32**, 617 (1992).
196. R. Attias, *J. Chem. Inf. Comput. Sci.* **23**, 102 (1983).
197. C. W. Crandell and D. H. Smith, *J. Chem. Inf. Comput. Sci.* **23**, 186 (1983).
198. J. K. Cringean and M. F. Lynch, *J. Inf. Sci.* **15**, 211 (1989).
199. A. J. Gushurst, J. G. Nourse, W. D. Houndshell, B. A. Leland, and D. G. Raich, *J. Chem. Inf. Comput. Sci.* **31**, 447 (1991).
200. T. R. Hagadone, *J. Chem. Inf. Comput. Sci.* **32**, 515 (1992).
201. E. M. Rasmussen, G. M. Downs, and P. Willett, *J. Comput. Chem.* **9**, 378 (1988).
202. P. Willet, *J. Chem. Inf. Comput. Sci.* **25**, 114 (1985).
203. P. Willet, *J. Inf. Sci.* **15**, 223 (1989).
204. P. Willet, V. Winterman, and D. Bawden, *J. Chem. Inf. Comput. Sci.* **26**, 36 (1983).
205. P. Willet, V. Winterman, and D. Bawden, *J. Chem. Inf. Comput. Sci.* **26**, 109 (1983).
206. H. Zhang, M. Minoh, and K. Ikeda, *J. Inf. Proc.* **15**, 108 (1992).
207. E. Garcia and L. M. Reyes, *J. Mol. Struct. (Theochem)* **101**, 101 (1993).
208. P. Kuner and B. Ueberreiter, *Int. J. Pat. Recog. Art. Intel.* **2**, 527 (1988).
209. E. Bienenstock and C. von der Malsburg, *Europhys. Lett.* **4**, 4 (1987).
210. R. Sitaraman and A. Rosenfeld, *Patt. Recog.* **22**, 331 (1989).
211. A. Tomlin (ed.), *The Pesticide Manual: a World Compendium:Incorporating the Agrochemicals Handbook*, Royal Society of Chemistry, London (1995).
212. B. Böhm, J. Stürzbecher, and G. Klebe, *J. Med. Chem.* **42**, 458 (1999).
213. W. E. Lorensen and H. E. Cline, *Comput. Graphics* **21**, 163 (1987).
214. A. Watt and M. Watt, *Advanced Animation and Rendering Techniques*, Addison-Wesley, Reading (1992).
215. Marching cubes algorithm Internet site: http://exaflop.org/docs/marchcubes
216. Marching cubes algorithm Internet site: http://panda.uchc.edu/htbit/indiv/software_docs/marching_cubes.html
217. A Fortran 90 procedure for the MCA can be downloaded from: http://iqc.udg.es/cat/similarity/ASA/mca.html

218. GiD, Geometry and Data, a pre/post processor graphical interface. Internet site: http://gatxan.upc.es. International Center for Numerical Methods in Engineering (CIMNE), Barcelona (1999).
219. X. Gironés, L. Amat, and R. Carbó-Dorca, *J. Mol. Graph. Model.* **16**, 190 (1998).
220. I. M. Vinogradov (ed.) *Encyclopaedia of Mathematics*, Vol. 7, Reidel-Kluwer, Dordrecht (1987).
221. R. Carbó-Dorca and K. D. Sen, *J. Mol. Struct. (Theochem)*, **501**, 173 (2000).
222. C. Roos, T. Terlaky, and J.-P. Vial, *Theory and Algorithms for Linear Optimization*, Wiley, New York (1997).
223. I. M. Vinogradov (ed.) *Encyclopaedia of Mathematics*, Vol. 4, Kluwer, Dordrecht (1989).
224. LF 95 Language Reference, Lahey Computer Systems, Incline Village NV (1998). For more details browse at: http://www.lahey.com.
225. R. Carbó-Dorca, *J. Math. Chem.* **23**, 365 (1998).
226. H. Eyring, J. Walter, and G. E. Kimball, *Quantum Chemistry*, Wiley, New York (1940).
227. L. Pauling and E. B. Wilson Jr., *Introduction to Quantum Mechanics*, Dover, New York (1985).
228. For a recent matrix formulation, see: E. Besalú and R. Carbó-Dorca, *J. Math. Chem.* **22**, 85 (1997).

Chapter 13

Self-Organizing Molecular Field Analysis (SOMFA): A Tool for Structure-Activity Studies

P. J. Winn, D. D. Robinson, and W. G. Richards
Physical and Theoretical Chemistry Laboratory, The University of Oxford, Oxford, UK

1. INTRODUCTION

Quantitative structure activity relations (QSAR), and three-dimensional quantitative structure-activity relations (3D-QSAR) have had a profound impact on medicinal chemistry [1-6]. The ability to produce quantitative correlations between three-dimensional properties of molecules and the biological activity of these compounds is of inestimable value in deciding upon the choice of future synthetic chemistry.

The SOMFA [7] method has obvious parallels with both CoMFA [8] and molecular similarity [9-11] QSAR. It has affinities with the Free-Wilson method [12] and also the work of Doweyko [13-16] but is conceptually simpler and more comprehensive. Like CoMFA, a grid-based approach is used. However no probe interaction energies need to be evaluated. Like the similarity methods, it is the intrinsic molecular properties (such as shape) that are used to develop the QSAR models. Further, because of its inherent simplicity we believe the method has great potential for development, particularly as regards the alignment and conformational problems inherent in 3D-QSAR. Here we explain the perspective from which the SOMFA

Fundamentals of Molecular Similarity, Edited by Carbó-Dorca *et al.*
Kluwer Academic/Plenum Publishers, New York 2001

method is developed. We then contrast it with the CoMFA methodology and assess its potential for extension.

2. METHOD

The aim of any QSAR method is to relate activity to structure, that is

$$\text{Activity} = f(\text{Structure}) \tag{1}$$

For example, if an area of steric bulk is important for activity it would be present in many, if not all, high activity compounds and not present in the low activity compounds. Thus, in the absence of any target crystal structure, the activity represents all we know about the interaction of the ligand with the target. The question is thus: How do we optimally extract that information from the activity data for a given set of molecules?

As with all QSAR techniques, a model is built from a set of molecules of known activity; these molecules constitute the training set. Crucial to SOMFA is the notion of the 'mean centred activity.' By subtracting the mean activity of the molecular training set from each molecule's activity, we obtain a scale where the most active molecules have positive values and the least active molecules have negative values. There is no need to standardize any of the data to unit variance, nor is there any need to mean centre the descriptor data.

Three-dimensional grids are created, as in other QSAR techniques, with values at the grid points representing the shape or electrostatic potential. Shape values are given a value of one inside the van der Waals envelope, zero outside. Electrostatic potential values at grid points are calculated in the normal manner from the partial charges distributed across the atom centres. The most important step is that the value of the shape or electrostatic potential at every grid point for a given molecule is multiplied by the mean centred activity for that molecule. This weights the grid points so that the most active and least active molecules have higher values than the less interesting molecules close to the mean activity. It thus acts a form of descriptor filtering.

Figure 1 shows the structures of four molecules (row 1) together with their affinities (row 2) for a hypothetical binding site. The example is designed such that a hydroxyl group enhances binding and excess steric bulk adjacent to the hydroxyl reduces binding (i.e., the o-methyl groups reduce affinity). The first step in the SOMFA process is to calculate the mean centred affinities/activities for the three molecules with known activity (the compound of unknown activity will be used to highlight the ability of the model to predict activity). In doing this we achieve a scale of activity where all high activity compounds have a positive activity, while all low activity compounds have a negative activity. The molecules are aligned by superimposing the rings in the orientation they are presented in Figure 1.

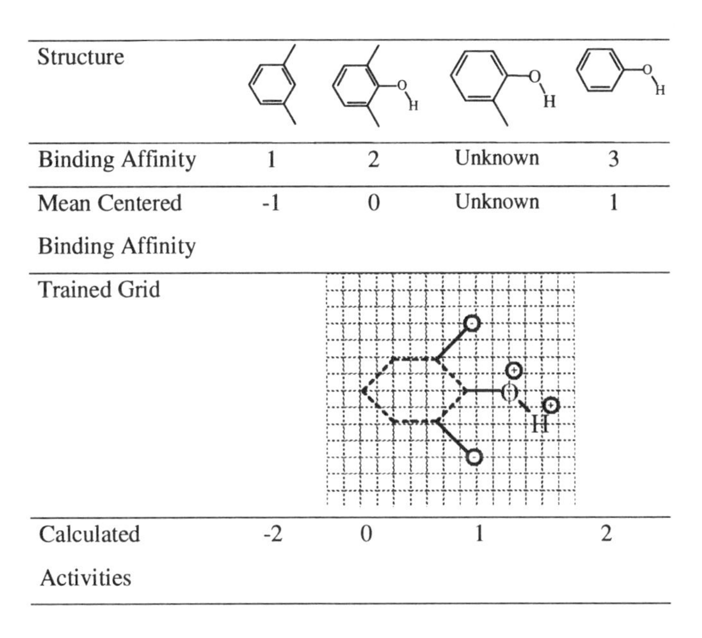

Structure				
Binding Affinity	1	2	Unknown	3
Mean Centered Binding Affinity	-1	0	Unknown	1
Trained Grid				
Calculated Activities	-2	0	1	2

Figure 1. A simple shape-based example of the SOMFA concept. The first row shows four ligands for a hypothetical receptor, the second row shows their hypothetical binding affinity. The third row shows the mean centred binding affinity, and the fourth row shows the SOMFA grid trained using shape and mean centred activity. The final row shows the relative activity of the different compounds as predicted by the trained grid. See the main text for further explanation.

We now create a grid around them. In our simplified example we add the mean centred binding affinity of each molecule to every grid point crossed by the molecule. This gives a grid trained on shape as shown in row three of Figure 1. The common OH group present in the two more active molecules is left with a net positive grid value associated with it, and the methyl groups have a negative grid value associated with them. Thus the mean centred activities have been used as a filtering mechanism to highlight the features which differentiate the high-affinity and low-affinity compounds. An important point is that the benzene ring is not found to be significant. This set of compounds is not diverse enough for any conclusion to be drawn about this structural feature, and SOMFA correctly highlights this and filters it from the analysis. The final result is a grid-based map that can be used to aid molecular design of compounds with enhanced binding affinity for the active site. Because the method is grid-based, there is no need for the compounds under study to be analogues, only that they can be suitably aligned. We would expect any number of diverse compounds with a common binding mode to be suitable for treatment by this method.

Finally, the trained grid can be used to predict the relative activity of a set of compounds, and hence the activity of novel compounds, such as the third compound in Figure 1. When this structure of unknown activity is overlaid on the SOMFA lattice, we can see that the molecule is associated with only one area of poor steric overlap, while simultaneously possessing the hydroxyl group that we have identified as being the most important feature for activity. It is thus ranked according to relative activity as shown in the final row of Figure 1. This is intuitively in agreement with our knowledge of the hypothetical binding site.

In general, a SOMFA grid can be trained on any calculable molecular property. The grids for each molecule in the training set are combined to give master grids for each property. The value of a SOMFA master grid point at a given x,y,z is defined by:

$$SOMFA_{x,y,z} = \sum_{i}^{Training\_Set} Property_i(x,y,z) Mean\_Centred\_Activity_i \quad (2)$$

In this equation, the mean centred activity was represented on a logarithmic scale. The values at each point of a property master grid can be displayed to highlight features favourable or unfavourable to activity. For example, the shape master grid is a template of the areas of steric bulk which enhance or detract from activity.

The properties used in here are shape, electrostatic potential, and hydrophobicity. The shape property is defined for each grid point as 1 inside the van der Waals envelope and 0 outside. The electrostatic potential property for each grid point is calculated from the atomic partial charges according to Coulomb's law. The hydrophobicity is defined using the scheme of Lemmen et al. [17]. In this scheme, an atom is defined as hydrophobic if the partial charge on that atom is below a given threshold, for that element, if the partial charge is above a given threshold the atom is defined as hydrophilic, otherwise the atom is defined as 'ambiguous.' In the SOMFA implementation; the hydrophobic property is defined as 1 for any grid point within the van der Waals radius of a hydrophobic atom, -1 within the van der Waals radius of a hydrophilic atom and 0 within the van der Waals radius of an 'ambiguous' atom.

A QSAR relating a property, such as shape, to molecular activity can be derived from the property master grid in the following way. For every molecule (i) in the training and prediction set an estimate of the activity of the molecule as defined by a certain. property can be obtained by using Equation 3.

$$SOMFA_{property,i} = \sum_x \sum_y \sum_z Property_i(x, y, z) SOMFA_{x,y,z} \tag{3}$$

Linear regressions between the $SOMFA_{property,i}$ values and the logarithms of the experimental activities for the training set are then derived. Calculating the correlation coefficient indicates the potential importance of a given property. The linear equations produced can be used to predict the activity of compounds in the prediction set from their $SOMFA_{property,i}$ values. A better method is to combine the predictive power of the different $SOMFA_{property,i}$. Here we combined the individual property predictions using a weighted average of the shape and electrostatic potential based QSAR, using a mixing coefficient (c_1) as in Equation 4.

$$Activity = c_1 Activity_{Shape} + (1 - c_1) Activity_{ESP} \tag{4}$$

Clearly multiproperty predictions could have been obtained through multiple linear regression. Using Equation 4 instead gives greater insight into the resultant model by allowing the study of the variation in predictive power with different values of c_1.

When developing he SOMFA methodology, we started with a philosophical idea of what was required for a 3D-QSAR. From ideas in close alignment with the Free-Wilson method, the HASL technique and similarity indices, we derived a set of equations which are mathematically similar to PLS, and thus CoMFA. This highlights the similarities between all the main 3D-QSAR techniques.

It is important to contrast the SOMFA method with the CoMFA method to see the potential advantages of it. The first key difference is that in SOMFA the descriptor data is not mean centred, and none of the data is normalized. Not mean centring is potentially important for a property like electrostatic potential where it seems sensible to keep the positive potentials positive and the negative potentials negative (even for a neutral compound, mean centring may alter the zero of electrostatic potential). It also simplifies de novo predictions and cross validation, and gives marginal computational savings during model development. Mean centring of descriptor data is intrinsic to the derivation, of PLS and thus to CoMFA. The mean centring and normalization differences aside, SOMFA is comparable to one latent variable in a CoMFA analysis, in the limit of one column of activity data. However, this should be considered in light of a significant philosophical difference: SOMFA presumes that active compounds have common features and we are performing an analysis to produce a composite molecular picture based on an activity weighted average. In contrast, CoMFA looks for a component of the activity and a component of the descriptor data that correlate. The residuals of the activity and descriptor data are iteratively re-analysed for further components that correlate, thus producing further latent variables in the model. The number of latent variables in a model is usually selected by cross-validation, such that the r^2(CV) is maximised. Such an analysis is not performed with SOMFA and may explain its ability to handle finer grids. It was considered during SOMFA's development that anything more complex might make the process more susceptible to the errors associated with biological assays. Further to this, the SOMFA method allows the descriptors to be represented by Gaussian functions, and thus allows a considerable increase in computational speed and effectively an infinite grid resolution. Initial results suggest that these subtle differences between SOMFA and CoMFA give SOMFA more robust results, and time savings that could be utilised as part of a strategy for determining optimal alignments.

3. TEST SYSTEMS

For the test systems given in this chapter, a grid spacing of 0.5Å was used. Previous work suggests that a SOMFA analysis converges at this resolution.

3.1 Steroids with affinity for corticosteroid-binding globulin

The 'benchmark' steroids were aligned using a PCA alignment of their common core. Also a further twelve compounds were added to the set and used for testing. Further details are described elsewhere [7].

3.2 Cyclic urea derivatives as HIV-1 protease inhibitors

Here we used the 93 training compounds and 25 test compounds of Debnath [18], who kindly gave us coordinates of their structures, as aligned and optimised within the HIV-1 protease active site.

3.3 Inhibitors of Escherichia coli dihydrofolate reductase

There were 35 aligned benzyl pyrimidine derivative structures obtained according to the protocol of Seri-Levy et al. [19].

3.4 Sulfonamide endothelin inhibitors

Here we used the 35 compounds presented by Krystek et al. [20]. In contrast to the work of Krystek et al., we [7] used 17 training compound and 18 test compounds 'Krystek et al. used all 35 compounds for training and judged the quality of the model by the r^2(CV).

4. RESULTS

4.1 Steroids with affinity for Corticosteroid-binding globulin

For the prediction of all 10 steroid test compounds SOMFA outperforms all the other available techniques, as judged by the SDEP (Table 1). This is true for both the hydrophobic model and the combined shape and electrostatics model. Ignoring the steroid 31, which is considered to have anomalous behaviour, we see that SOMFA (shape and ESP) compares very favourably with other techniques. It outperforms the conventional CoMFA analysis (which uses two latent variables) and produces results comparable to that of a fractional factorial design (FFD) CoMFA (which uses three variables). This highlights that whilst similar to CoMFA the SOMFA protocol produces significantly different results.

To provide a feel for the utility of the hydrophobic model used here, we have included a graph of calculated activity against observed activity for the extended test set of steroid compounds (Figure 2). We see that the overall trend in activity is predicted remarkably well for such a crude model.

Table 1. Benchmark steroid results [21]. For the SOMFA analysis 21 molecules were used for training, 10 molecules were used for testing. A ratio of 6:4 shape:ESP was used in the predictions. (a) rms error in prediction of 9 test compounds (b) Excludes steroid 31.

	CoMFA	CoMFA (FFD)	Similarity Matrix Analysis	COMPASS	MS-WHIM	SOMFA Shape and ESP	SOMFA Hydro-phobic
SDEP[(a)]	0.837	0.716	0.640	0.705	0.662	0.584	0.622
	0.486[b]	0.356[b]	0.385[b]	0.339[b]	0.411[b]	0.367[b]	0.553[b]

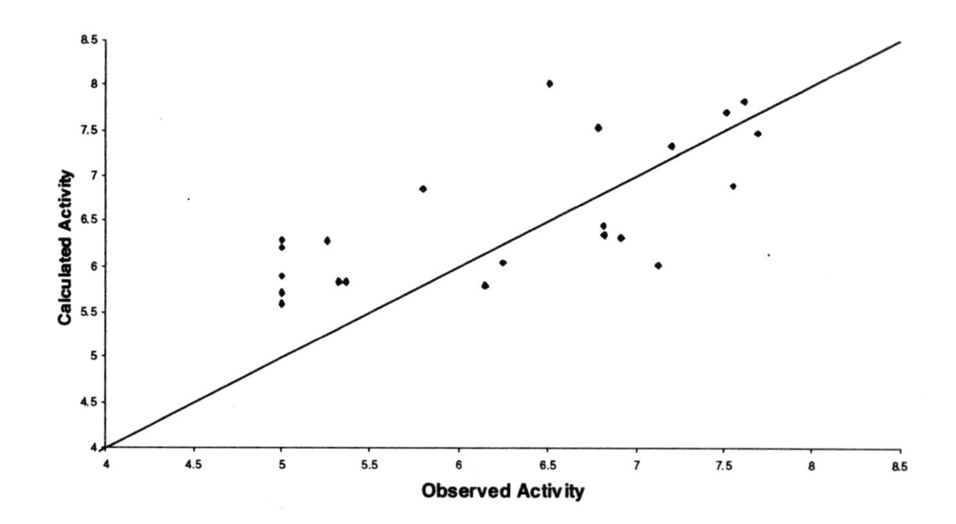

Figure 2. Predictions of the activity of a test set of 22 steroids using a SOMFA hydrophobicity model

4.2 Cyclic urea derivatives as HIV-1 protease inhibitors

The SOMFA predictions for the electrostatic and shape property models are of the same quality as the CoMFA results, as judged by predictions on a test set (Table 2). Cross-validation led Debnath to use a model with 7 latent variables.

Table 2. Comparison of predictions for the cyclic urea HIV-1 protease inhibitiors

	SOMFA (electrostatic)	SOMFA (Hydrophobic)	SOMFA (Shape)	CoMFA
r^2	0.531	0.372	0.514	0.555

4.3 Inhibitors of escherichia coli dihydrofolate reductase

SOMFA shape analysis compares well with conventional Hansch analysis and the shape similarity method for this set of inhibitors (Table 3). Genuine predictions with the SOMFA model show that the leave one out cross-validation method overpredicts the quality of the model.

Table 3. QSAR results for 35, 5-benzyl pyrimidine derivatives that inhibit E-coli dihydrofolate reductase. The SOMFA analysis splits the 35 compounds used by Seri-Levy et al. into 18 training compounds and 17 test compounds. Results for the 17 test compounds are given for SOMFA (shape), and also for the test compounds without an outlier (denoted [a]).

	Shape similarity	Hansch analysis	SOMFA (Shape)	SOMFA (ESP)
R^2(CV)	0.92	0.85	0.86	0.50
R^2			0.44	
			0.67 [a]	

4.4 Sulfonamide endothelin inhibitors

The SOMFA results for the sulfonamide endothelin inhibitors (Table 4) again compare favourably with other QSAR methods, this time a CoMFA using 6 latent variables. An interesting point here is that the optimum ratio of shape to electrostatic potential in both the SOMFA and CoMFA models was found to be 7:3.

Table 4. Comparison of CoMFA and SOMFA models for a set of sulfonamide endothelin inhibitors.

	CoMFA	SOMFA
r^2(CV)	0.70	
r^2		0.71 (training set)
		0.59 (test set)
Optimal shape:ESP	69:31	7:3

5. DISCUSSION

For the test systems presented here SOMFA performs as well as any of the other QSAR methods. It performs as well as the CoMFAs with many latent variables. For the steroid set CoMFA uses 2 latent variables and CoMFA (FFD) uses 3 latent variables, for the HIV-1 protease inhibitor set CoMFA uses 7 latent variables, and the for endothelin inhibitors it uses 6 latent variables. Given the similarity of the CoMFA and SOMFA protocols it is unclear why CoMFA needs so many latent variables to achieve the same quality of result. It could be the ability of SOMFA to utilise fine grid resolution and still show convergent behaviour. The CoMFA analyses used 1Å or 2Å grids because finer grids tend to produce worse results, in contrast the SOMFA analyses here used 0.5Å grid and SOMFA results tend to converge as the grid resolution increases. Indeed, a 2Å grid would appear to be a poor degree of grid resolution for looking at molecular structure, hence the problems of alignment with respect to grid that occur with CoMFA. Another potential explanation is the absence of mean centring on the descriptor data. Finally it could be that cross-validation is for some reason not appropriate as a method for selecting the optimal number of latent variables in a CoMFA analysis. The grid resolution problem would appear to be the most likely of these. These results raise questions about the current CoMFA protocols.

6. FUTURE DIRECTIONS

It is clear that we need more careful comparisons of the CoMFA and SOMFA protocols to establish their relative differences and similarities. However, beyond this, the aim of looking at the 3D-QSAR problem from first principles was to see whether it was possible to develop a method that could encompass some of the outstanding challenges. The most fundamental problem in 3D-QSAR is still one of molecular conformation and relative molecular alignment. Many people in the field seem to toy with the assumption that the true binding conformations and relative molecular alignments of a set of ligands provide the optimum QSAR model. This leads to the question: Can we use the predictivity of a model as a measure of the quality of the alignment?

It is clear that a full systematic search of alignment and conformation is not feasible for most molecules of interest. We thus need a method (or methods) for quickly generating intelligent guesses for likely conformations and relative alignments. Further, we need a fast method to generate QSAR models for each probable relative alignment, from which the most predictive can be selected as the optimum alignment. Novel compound alignments would be trickier to determine though selection of the orientation that predicts highest activity would seem to be the most sensible idea.

Towards this end we have developed a version of SOMFA that uses Gaussian functions as property descriptors. These are computationally easy to manipulate. This implementation typically takes fractions of a second per molecule. As well as achieving the levels of speed we desired, Gaussian descriptors are better suited to describe the shape of an atom since they are soft. They can also approximate the inverse square of Coulomb's law.

Equally impressive they effectively provide an infinite resolution grid. We are also working on different techniques for fast, automated alignment.

7. CONCLUDING REMARKS

SOMFA is a method developed by considering the basic requirements and assumptions of 3D-QSAR. It is closely aligned to the common 3D-QSAR techniques and is mathematically similar to CoMFA. Given this mathematical similarity to one latent variable of a CoMFA, the SOMFA results presented here are surprisingly similar in quality to CoMFA with many latent variables. It is not clear why this should be but one possibility is the ability of SOMFA to handle much finer grids than CoMFA.

Beyond this the SOMFA grid can easily be replaced by a set of Gaussian functions yielding considerable speed advantage and effectively infinite grid resolution. This has potential implications for automated alignment of the active compounds under analysis. It is clear that further work is required to verify and develop the SOMFA method, however it appears to show some promise.

REFERENCES

1. Hansch and T. Fujita, *J. Am. Chem. Soc.* **86**, 1616 (1964).
2. Kubinyi, *3D QSAR in Drug Design: Theory, Methods and Applications*, ESCOM, Leiden, (1993).
3. Kubinyi, *3D QSAR in Drug Design: Ligand-Protein Interactions and Molecular Similarity*; Kluwer Academic Publishers, Dordecht (1998).
4. Kubinyi, *3D QSAR in Drug Design: Recent Advances*, Kluwer Academic Publishers, Dordecht (1998).
5. H. van de Waterbeemd, *Chemometric Methods in Molecular Design*, Vol. 2, VCH, Weinheim (1995).
6. H. van de Waterbeemd, *Advanced Computer Assisted Techniques in Drug Discovery*, Vol. 3, VCH, Weinheim (1995).
7. D. Robinson, P. J. Winn, P. D. Lyne, and W. G. Richards, *J. Med. Chem* **43**, 573 (1999).
8. R. D. Cramer III, D. E. Patterson, and J. D. Bunce, *J. Am. Chem. Soc.* **110**, 5959 (1988).
9. C. Good, S.-S. So, and W. G. Richards, *J. Med. Chem.* **36**, 433 (1993).
10. E. Hodgkin and W. G. Richards, *Int. J. Quantum Chem. Quantum Biol. Symp.* **14**, 105 (1987).
11. G. Burt, P. Huxley, and W. G. Richards, *J. Comput. Chem.* **11**, 1139 (1990).
12. S. M. Free Jr and J. W. Wilson, *J. Med. Chem.* **7**, 395 (1964).
13. M. Doweyko, *J. Med. Chem.* **31**, 1396 (1988).
14. M. Doweyko, and W. B. Mattes, *Biochemistry* **31**, 9388 (1992).
15. M. Doweyko, *J. Med. Chem.* **37**, 1769 (1994).
16. J. Kaminski and A. M. Doweyko, *J. Med. Chem.* **40**, 427 (1997).
17. Lemmen, T. Lengauer, and G. Klebe, *J. Med. Chem.* **41**, 4502 (1998).
18. A. Debnath, *J. Med. Chem.* **42**, 249 (1999).

19. Seri-Levi, R. Salter, S. West, and W. G. Richards, *Eur. J. Med. Chem.* **29**, 687 (1994).
20. S. R. Krystek, J. T. Hunt, P. D. Stein, and T. R. Stouch, *J. Med. Chem.* **38**, 659 (1995).
21. G. Bravi, E. Gancia, P. Mascagni, M. Pegna, R. Todeschini, and A. Zaliani, *J. Comput - Aided Mol. Des.* **11**, 79 (1997).

Chapter 14

Similarity Analysis of Molecular Interaction Potential Distributions. The MIPSIM Software

F. Sanz, M. De Càceres
Grup de Recerca en Informàtica Mèdica, Institut Municipal d'Investigació Mèdica, Universitat Pompeu Fabra, C/ Dr. Aiguader 80, E-08003 Barcelona, Spain

J. Villà
Grup de Recerca en Informàtica Mèdica, Institut Municipal d'Investigació Mèdica, Universitat Pompeu Fabra, C/ Dr. Aiguader 80, E-08003 Barcelona, Spain; Department of Chemistry, University of Southern California, Los Angeles, CA, 90089, USA

1. INTRODUCTION

The Molecular Interaction Potentials (MIP) are standard tools for the comparison of series of compounds in the framework of Structure-Activity Relationship (SAR) studies [1]. These potentials are interaction energies between the considered compounds and relevant probes, which are usually computed in the points of a grid defined around the compounds (see Fig. 1). The historically most used probe is the proton, and the resulting MIP is called Molecular Electrostatic Potential (MEP). In previous works, we developed a computational system (MEPSIM) for the analysis and comparison of MEP distributions computed at the quantum mechanic level [2-5]. The two main modules of the MEPSIM software were MEPMIN and MEPCOMP. MEPMIN allowed the automatic characterisation of the MEP minima around a molecule (values, positions and geometric relationships). MEPCOMP dealt with the computation of a similarity measure (a Spearman

Fundamentals of Molecular Similarity, Edited by Carbó-Dorca *et al.*
Kluwer Academic/Plenum Publishers, New York 2001

rank correlation coefficient) between two MEP distributions. MEPCOMP also carried out a gradient-driven automatic search of the relative positions of the compared molecules that maximised the similarity measure. The present work presents an improved and extended version of the MEPSIM approach that, among other new features, allows the consideration of any MIP definition. This new software is called MIPSIM (Molecular Interaction Potentials SIMilarity analysis) [6].

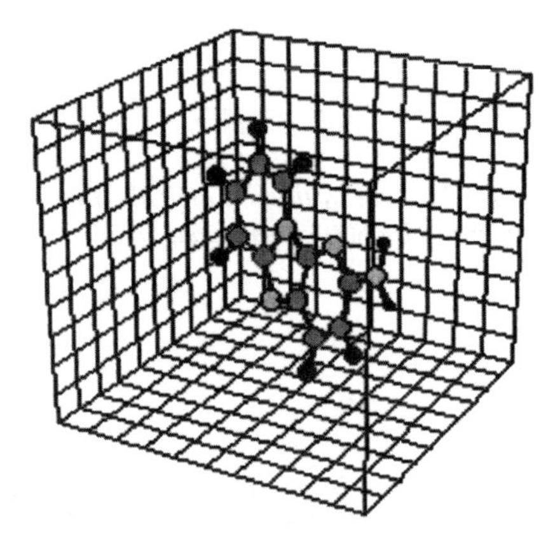

Figure 1. Grid of points around a molecule

2. MIPSIM DESCRIPTION

The MIPSIM software has analogous modules than MEPSIM: MIN to find MIP minima and COMP to compare MIP distributions of pairs of molecules and to search the best molecular alignments. Nevertheless, MIPSIM has interesting new features in comparison to MEPSIM:

- The program is transparently integrated with other programs, like GAMESS [7] or GRID [8], which allow the computation of the potentials to be analysed or compared, or with programs that allow the statistical analysis of the results, as is the case of GOLPE [9].
- In addition to the Spearman rank correlation coefficient used in MEPSIM, MIPSIM incorporates additional similarity measures like the Pearson correlation coefficient, as well as the possibility of averaging several similarity measures (i.e., those obtained using different probes).
- During the molecular alignment optimisation process, one of the molecules is kept fixed while the other is free to move in terms of

translation and rotation. As the free molecule moves, the grids around both molecules fail to coincide in space. In this situation, if using the standard definition of a correlation coefficient, recalculation of the potential on new coincident points would be compulsory. This strategy is computationally expensive, especially if the property to be compared is the quantum mechanic MEP. In MIPSIM we have implemented a modified similarity index, which copes with non-coincident grids [10-12]. In the definition of this index, which is summarised in Fig. 2, all the possible combinations of grid points are considered in the computation of a Pearson-like correlation coefficient. The corresponding product terms p_ip_j are weighted by a coefficient ($e^{-d_{ij}^{**2}}$) that decreases when the distance between the points (d_{ij}) grows. This strategy skips the need of recalculating the MIP values at each relative movement of the molecules. Therefore, the MIPSIM alignment optimisations are much faster than those made with MEPSIM.

$$A_{ab} = \sum_{i=1}^{n_a}\sum_{j=1}^{n_b} p_i p_j e^{-\alpha r_{ij}^2}$$

$$Sim_{ab} = \frac{A_{ab}}{\sqrt{A_{aa}^{self} \cdot A_{bb}^{self}}}$$

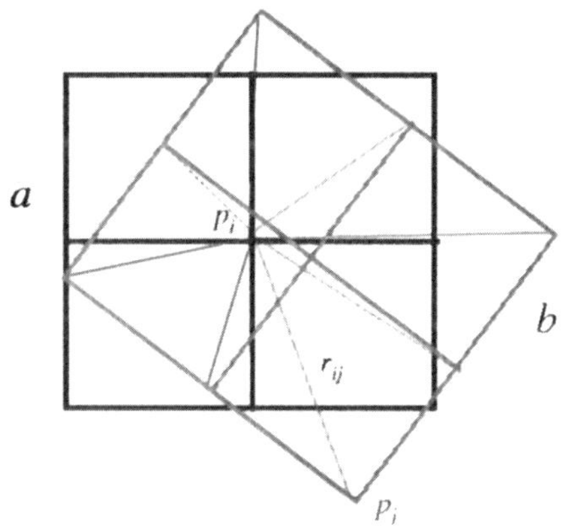

Figure 2. Scheme of the MIP distributions similarity computation

- In order to avoid the problem of local similarity maxima, MIPSIM can perform intensive explorations of the relative positions of two molecules. The program randomly generates initial relative positions of the compared molecules that are then optimised using any of the previously mentioned optimisation procedures. In order to save computational effort, the program eliminates useless solutions (i.e., those placing one of the molecules faraway from the other) and it controls the convergence of optimisation trajectories with previously explored ones.
- MIPSIM can handle a series of molecules in a single run. In the case of the COMP module, it can compare all the possible pairs of molecules, generating the corresponding similarity matrix, or it can compare the first

molecule, which is considered the reference one, with the rest of the compounds of the series.
- MIPSIM incorporates a restart facility useful for long runs.

3. MIPSIM EXAMPLES

Fig. 3 shows plots of the MIP minima of theophylline, obtained using the MIN module of MIPSIM and two different MIP definitions (quantum mechanic MEP computed with GAMESS, and GRID/MIP computed using a alkylic OH as probe).

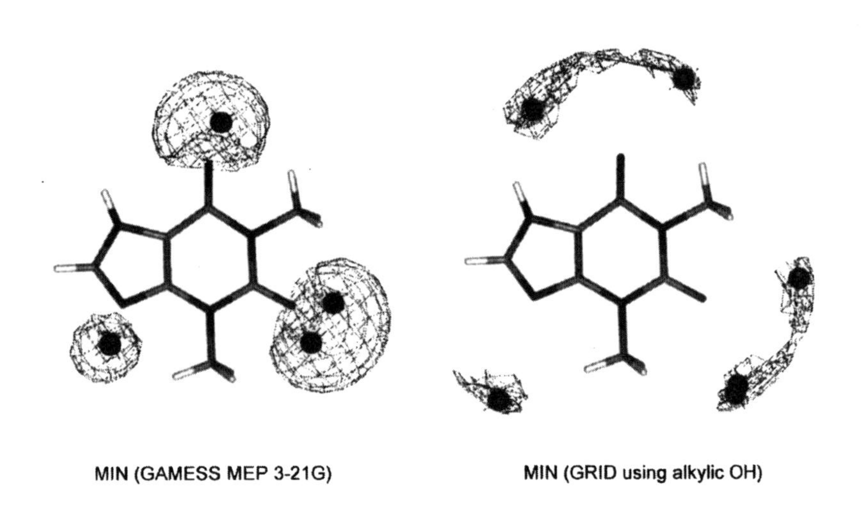

Figure 3. MIP minima of theophilline

Fig. 4 shows two of the maximum similarity relative positions of theophylline, which were obtained with the COMP module of MIPSIM by comparing two molecules of teophylline on the basis of their quantum mechanics MEP. The first solution is a trivial one but the second suggests an interesting possibility of multiple binding positions of this compound.

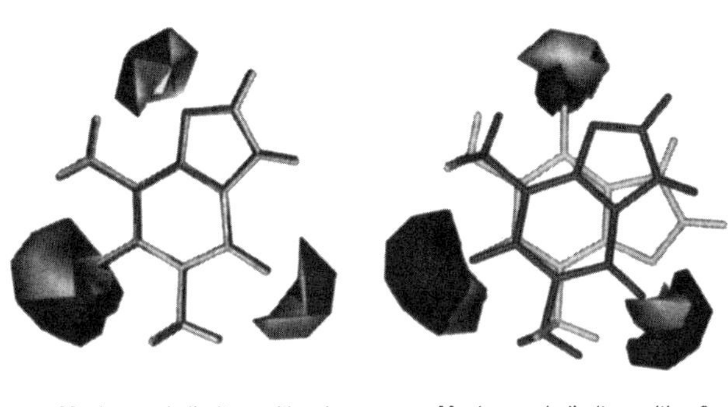

Figure 4. Maximum similarity relative positions of theophilline

ACKNOWLEDGMENTS

The present work has been supported in part by CESCA and CICYT (SAF98-0148-C04-02) grants.

REFERENCES

1. M. Orozco and F. J. Luque, in Molecular Electrostatic Potentials: concepts and applications, Theoretical and Computational Chemistry, Vol. 3, J. S. Murray and K. Sen (eds.), Elsevier (1996), p. 181.
2. F. Manaut, F. Sanz, J. José, and M. Milesi, J. Comput.-Aided Mol. Des. **5**, 371 (1991).
3. F. Sanz, F. Manaut, J. A. Sánchez, and E. Lozoya, J. Mol. Struct. (THEOCHEM) **230**, 437 (1991).
4. F. Sanz, F. Manaut, J. Rodríguez, E. Lozoya, and E. López de Briñas, J. Comput.-Aided Mol. Des. **7**, 337 (1993).
5. J. Rodríguez, F. Manaut, and F. Sanz, J. Comput. Chem. **14**, 922 (1993).
6. M. De Càceres, J. Villà, J. J. Lozano, and F. Sanz, MIPSIM: Similarity Analysis of Molecular Interaction Potentials. Bioinformatics (in press).
7. M. W. Schmidt, K. K. Baldridge, J. A. Boatz, S. T. Elbert, M. S. Gordon, J. H. Jensen, S. Koseki, N. Matsunaga, K. A. Nguyen, S. J. Su, T. L. Windus, M. Dupuis, and J. A. Montgomery, J. Comput. Chem. **14**, 1347 (1993).
8. P. Goodford, J. Med. Chem. **28**, 849 (1985).
9. M. Baroni, G. Costantino, G. Cruciani, D. Riganelli, R. Vagli, and S. Clementi, Quant. Struct.-Act. Relat. **12**, 9 (1993).
10. S. K. Kearsley and G. M. Smith, Tetrahedron Comput. Methodol. 3, 615 (1990).

11. G. Klebe, T. Mietzner, and F. Weber, J. Comput.-Aided Mol. Des. 8, 751 (1994).
12. C. K. Bagdassarian, V. L. Schramm, and D. Schwartz, J. Am. Chem. Soc. **118**, 8825 (1996).

Index